Architectural Structure

Architectural Structure

Editor

Luís Filipe Almeida Bernardo

MDPI • Basel • Beijing • Wuhan • Barcelona • Belgrade • Manchester • Tokyo • Cluj • Tianjin

Editor
Luís Filipe Almeida Bernardo
University of Beira Interior
Portugal

Editorial Office
MDPI
St. Alban-Anlage 66
4052 Basel, Switzerland

This is a reprint of articles from the Special Issue published online in the open access journal *Applied Sciences* (ISSN 2076-3417) (available at: https://www.mdpi.com/journal/applsci/special_issues/Architectural_Structure).

For citation purposes, cite each article independently as indicated on the article page online and as indicated below:

LastName, A.A.; LastName, B.B.; LastName, C.C. Article Title. *Journal Name* **Year**, *Article Number*, Page Range.

ISBN 978-3-03936-994-2 (Hbk)
ISBN 978-3-03936-995-9 (PDF)

Cover image courtesy of Piotr Baszyński.

Contents

About the Editor

Luís Filipe Almeida Bernardo (Prof.) received his Ph.D. in civil engineering (with expertise in mechanics of structures and materials) in 2004 at the University of Coimbra, Portugal. Currently, he is an Assistant Professor with Aggregation and a research member at C-MADE (Centre of Materials and Building Technologies) in the Department of Civil Engineering and Architecture at the University of Beira Interior in Covilhã, Portugal. His research interests include: structural analysis and design, numerical modelling and optimization of concrete structures, new structural materials, and building systems. He has been a researcher in several national research projects, and he collaborates with researchers from Brazil. He has authored/coauthored more than fifty articles in international peer-reviewed journals. He has been working as a reviewer for several international scientific journals. He has supervised several MSc and Ph.D. theses in his research field. He has also developed activities as consultant and civil engineer for the structural design of buildings and special structures in Portugal.

Editorial

Special Issue on "Architectural Structure"

Luís Filipe Almeida Bernardo

Department of Civil Engineering and Architecture, Centre of Materials and Building Technologies (C-MADE), University of Beira Interior, 6201-001 Covilhã, Portugal; lfb@ubi.pt

Received: 23 July 2020; Accepted: 29 July 2020; Published: 31 July 2020

This Special Issue on "Architectural Structure" aims to gather general advances in human-made constructions which simultaneously are driven by aesthetic and structural engineering considerations. Such advances include the analysis of architectural typologies, the study of the mechanical performance of structural materials, structural systems and components, the proposal of techniques to evaluate the mechanical performance in existing structures and also new construction techniques. The aim of this Special Issue is also to inspire researchers and practitioners to share their knowledge and findings in these fields, and also to help explore new trends for the future. This Special Issue brings together twelve contributions covering the previously referred topics.

The global performance of an architectural structure strongly depends on the mechanical performance of its components and also on the used structural materials. Accordingly, the majority of the published articles in this Special Issue focus on the experimental and/or numerical behavior of structures, structural components and structural materials, including innovative ones. In [1], an innovative composite shear wall built with double steel-concrete, able to substitute classic reinforced concrete walls, was studied and a design method was proposed. A refined model, aiming to contribute to the optimum and economical design of thin-walled steel beams, nowadays widely used in architectural steel structures, was proposed in [2]. The incorporation of fly ash in both cementitious and alkali-activated concretes presents environmental advantages and allows to obtain concrete members with glossy and black surfaces, which might be aesthetically appealing for architectural structures with exposed concrete. In [3], the results of a study on the mechanical performance of reinforced beams built with mortar incorporating fly ash was presented, pointing out some important aspects to be further investigated in order to allow for the structural application of such a material. A contribution to the better knowledge of the mechanical performance of a geopolymer obtained by alkali-activation of a new binder was presented in [4], in order to, in the near future, enable the use of this environmental material in innovative architectural structures with finishes of different colors and textures. Although high strength concrete is nowadays used in practice, some particular aspects of the structural behavior of members built with this material still need to be checked for optimum design. This is the case of structural concrete members under primary torsion. In [5] a study on the mechanical performance of prestressed high strength concrete hollow beams under torsion was presented, the results of which can help in the design of box bridges. Half precast solutions have been widely used for structural applications. In [6] the mechanical performance of a recent structural system of half precast concrete slabs with inverted multi-ribs was investigated and guides for a design method were proposed. Block masonry has been used since ancient times as the main component in constructions and is still used nowadays throughout the world. Recently, the manufacture of such components has evolved based on environmental requirements. The mechanical performance and environmental benefits of recycled aggregate concrete hollow blocks were studied and guides for design were also proposed in [7]. Tall buildings are some of the most emblematic architectural structures. One of the main challenges for the designer is to control the lateral displacements. For this, in [8] the efficiency of an innovative outrigger system made of reinforced concrete wall with multiple openings was modeled and studied, and some guide rules for design were proposed.

Many existing structures must be evaluated for maintenance and rehabilitation concerns. In some projects, the structural performance of their structural materials and members must be checked to ensure the structural safety. In the past few years, self-compacting concrete has been widely used due to, for instance, its ease of placement in geometrically complicated formworks and also due to the obtained smooth and well-finished surfaces after concreting. These aspects are important to fulfill many architectural requirements. In [9], the applicability of non-destructive tests to estimate the compressive strength of self-compacting concrete was studied and useful correlations were presented for practice. Additionally, the performance of structural members in existing structures may have to be evaluated in light of current codes of practice. In this sense, in [10] a study was presented to evaluate the real cyclic load bearing of a traditional ceramic-reinforced slab incorporated in an existing building. The testing methodology and the results of the analysis were presented, which could be useful for practitioners.

Construction systems have evolved in the past few years, namely for geometrically complex structures. Two emergent moldless fabrication techniques for complex spatial forms of natural fiber-reinforced polymer structures were presented and validated in [11]. Such techniques could be, in the near future, applied to build larger building elements for more sustainable building systems.

Finally, the analysis of existing architectural typologies may help a future generation of designers to think about new typologies for architectural structures. In [12], the content of the spatial Rudolf Steiner's architecture, using reinforced concrete in architectural structures with complex geometries, and which is considered a unique case in the history of architectural heritage, is determined and discussed.

To end this editorial note, I would like to express my sincere gratitude to all the contributors of the articles submitted to this Special Issue, as well as to the editor-in-chief of *Applied Sciences*, Prof. Dr. Takayoshi Kobayashi, Mr. Melon Zhang as Managing Editor, and the editorial staff for their efforts and support.

Funding: This research received no external funding.

Conflicts of Interest: The author declares no conflicts of interest.

References

1. Zhang, P.; Guo, Q.; Ke, F.; Zhao, W.; Ye, Y. Axial and Bending Bearing Capacity of Double-Steel-Concrete Composite Shear Walls. *Appl. Sci.* **2020**, *10*, 4935. [CrossRef]
2. Szychowski, A.; Brzezińska, K. Local Buckling and Resistance of Continuous Steel Beams with Thin-Walled I-Shaped Cross-Sections. *Appl. Sci.* **2020**, *10*, 4461. [CrossRef]
3. Lopes, A.V.; Lopes, S.M.R.; Pinto, I. Experimental Study on the Flexural Behavior of Alkali Activated Fly Ash Mortar Beams. *Appl. Sci.* **2020**, *10*, 4379. [CrossRef]
4. Lopes, A.V.; Lopes, S.M.; Pinto, I. Influence of the Composition of the Activator on Mechanical Characteristics of a Geopolymer. *Appl. Sci.* **2020**, *10*, 3349. [CrossRef]
5. Bernardo, L.; Lopes, S.; Teixeira, M. Experimental Study on the Torsional Behaviour of Prestressed HSC Hollow Beams. *Appl. Sci.* **2020**, *10*, 642. [CrossRef]
6. Han, S.-J.; Jeong, J.-H.; Joo, H.-E.; Choi, S.-H.; Choi, S.; Kim, K.S. Flexural and Shear Performance of Prestressed Composite Slabs with Inverted Multi-Ribs. *Appl. Sci.* **2019**, *9*, 4946. [CrossRef]
7. Liu, C.; Zhu, C.; Bai, G.; Quan, Z.; Wu, J. Experimental Investigation on Compressive Properties and Carbon Emission Assessment of Concrete Hollow Block Masonry Incorporating Recycled Concrete Aggregates. *Appl. Sci.* **2019**, *9*, 4870. [CrossRef]
8. Kim, H.-S.; Huang, Y.-T.; Jin, H.-J. Influence of Multiple Openings on Reinforced Concrete Outrigger Walls in a Tall Building. *Appl. Sci.* **2019**, *9*, 4913. [CrossRef]
9. Nepomuceno, M.C.S.; Bernardo, L.F.A. Evaluation of Self-Compacting Concrete Strength with Non-Destructive Tests for Concrete Structures. *Appl. Sci.* **2019**, *9*, 5109. [CrossRef]
10. Albareda-Valls, A.; Rivera-Rogel, A.; Costales-Calvo, I.; García-Carrera, D. Real Cyclic Load-Bearing Test of a Ceramic-Reinforced Slab. *Appl. Sci.* **2020**, *10*, 1763. [CrossRef]

11. Costalonga Martins, V.; Cutajar, S.; van der Hoven, C.; Baszyński, P.; Dahy, H. FlexFlax Stool: Validation of Moldless Fabrication of Complex Spatial Forms of Natural Fiber-Reinforced Polymer (NFRP) Structures through an Integrative Approach of Tailored Fiber Placement and Coreless Filament Winding Techniques. *Appl. Sci.* **2020**, *10*, 3278. [CrossRef]
12. Kiuntsli, R.; Stepanyuk, A.; Besaha, I.; Sobczak-Piąstka, J. Metamorphosis of the Architectural Space of Goetheanum. *Appl. Sci.* **2020**, *10*, 4700. [CrossRef]

Article

Axial and Bending Bearing Capacity of Double-Steel-Concrete Composite Shear Walls

Peiyao Zhang [1], Quanquan Guo [1,*], Fei Ke [2], Weiyi Zhao [3] and Yinghua Ye [1]

1 School of Transportation Science and Engineering, Beihang University, Beijing 100191, China; zhang.py@buaa.edu.cn (P.Z.); yhye@buaa.edu.cn (Y.Y.)
2 Department of building structure, China Nuclear Power Engineering Co., LTD, Beijing 100840, China; kefei@buaa.edu.cn
3 School of Civil Engineering, Qingdao University of Technology, Qingdao 266033, China; zhaoweiyi@qut.edu.cn
* Correspondence: qq_guo@buaa.edu.cn

Received: 31 May 2020; Accepted: 15 July 2020; Published: 17 July 2020

Abstract: Double steel-concrete composite shear wall is a novel composite structure. Due to its good mechanical properties, it has been considered as a substitute for reinforced concrete walls in nuclear facilities, marine environmental structures, and high-rise buildings. However, the design method of the double-steel concrete composite shear wall is lacking. The purpose of this paper is to propose the bending capacity formula under large and small eccentric loads. By summarizing the test results of 49 steel-concrete composite double shear walls under cyclic loading from different studies, it was found that the bending failure of double-steel-concrete composite shear walls was featured by the concrete crushing at the bottom. A finite element model was established and it could simulate the axial and bending performance of double steel-concrete composite shear walls reasonably well. According to the experimental results and FE analysis, the primary assumptions for calculating the axial and bending bearing capacity of the double steel-concrete composite shear walls were proposed. Based on these assumptions, the bearing capacity formulas were derived according to the equilibrium theory of the cross section. The calculation results obtained by the bearing capacity formulas were in good agreement with the test results.

Keywords: double-steel-concrete composite shear walls; axial and bending capacity; failure characteristic

---

1. Introduction

Double-steel-concrete composite shear walls (SC walls) mainly consist of two surface steel plates connected by tie bars and filled with concrete. To prevent the buckling of the surface plates and ensure collaborative work between the concrete and steel plates; studs, stiffeners, and other connections are used, except for tie bars. SC walls are developed to reduce wall thickness, to enhance constructability, and to make rapid construction possible by eliminating the use of formwork and reinforcing bars. The filled concrete prevents the early buckling of steel plates, while the steel plates provide the confinement on the concrete, so the SC walls are characterized to have high strength and sufficient ductility for compressive and shear loading. Therefore, the composite effect of steel plates and concrete gives the SC walls significant advantages over the traditional shear walls, such as reinforced concrete shear walls, single steel plate, concrete composite shear walls, profile steel-reinforced concrete composite shear walls, etc.

Previously, the structural characteristics of SC walls have already been studied experimentally and numerically. Depending on the structure, SC walls can be divided into two types. The first type is the SC wall with bending stiffener attached at the edge region, which was subjected to shear failure.

The experiments of the first type of SC walls were mainly finished in Japan (e.g., Usami et al. [1], Takeuchi et al. [2], Niwa et al. [3], Ozaki et al. [4], Kitano et al. [5], Funakoshi et al. [6]). The other type is the SC wall without bending stiffener was subjected to bending failure, and the experiments were mainly finished in China and Korea (e.g., Tae-Sung Eom et al. [7], Wu et al. [8], Nie et al. [9,10], Ji et al. [11], Tian et al. [12]). All the studies and experiments indicated that the SC walls had high strength, good ductility, and high energy-dissipation. In recent years, the construction of CAP1400 nuclear power plant in China has promoted the research progress of SC walls in nuclear engineering, such as Guo et al. [13], Yang et al. [14], Liu et al. [15], Li et al. [16], and Li et al. [17]. Lin et al. [18] tested 12 buckling-restrained shear panel dampers which were equipped with demountable steel-concrete composite restrainers. The influence of key design parameters on seismic behavior was studied and design equations for calculating elastic stiffness and ultimate strength were proposed. Zhao et al. [19] experimentally studied the cyclic behavior of two half-scale concrete stiffened steel plate shear wall specimens including the traditional and the innovation. Both specimens showed highly ductile behavior and stable cyclic yielding performance, and some suggestions for the design were also given based on the experiment results. Behnoosh Rassouli et al. [20] investigated experimentally and numerically the behavior of concrete stiffened steel plate shear wall using precast light weight concrete panels, and three specimens were tested under quasi-static cyclic load. The test results show that the CSPSW can reduce the seismic mass and improve the behavior of steel structures.

To date, some work has been conducted on the calculation method for the axial and bending bearing capacity of SC walls. For example, Varma et al. [21] developed a mechanics model and a detailed nonlinear finite element model, and the two modeling methods were applied to develop a conservative interaction surface in principal forces space that can be used to design or evaluate the adequacy of SC walls subjected to any combinations of in-plane forces. Tae-Sung Eom et al. [7] tested slender isolated walls and coupled walls subjected to cyclic loading. Based on their tested results, they developed the calculation method for the load-carrying capacity. Bo Wu et al. [8] calculated the lateral loading capacity of the SC walls based on the concept of the combined strength of new and old concrete. Ji et al. [11] proposed simplified formulas used to evaluate the flexure strength of the SC walls. However, all the calculation methods were derived using plastic stress distributions at the cross sections, which was the same as [22], and only suitable for calculating the compression members with large eccentricity. When applying to a small eccentric load, the methods would cause larger calculation errors, because the steel plates in tension did not reach the yield strength. Xiaowei Ma et al. [23] developed a model for the elastic and plastic analysis of the axial force-moment capacity of SC walls and analyzed the M-φ curve and the axial force and moment curve. They derived the formula for axial force-moment based on the key factors gained by numerical calculation and parameter analysis. However, the factors were complex and did not take into account the effects of structural measures, so the formula cannot be used widely in practical engineering. Jianguo Nie et al. [10] used the strip method to deduce the calculation formula of normal section bearing capacity, which applied to both small and large eccentric load. However, the strip method was complex and the ultimate compression strain of the concrete was taken as 0.0033, which was inconsistent with the confinement effect of steel plates on the concrete. Papanikolaou et al [24] proposed a confinement-sensitive plasticity constitutive model for concrete in triaxial compression. It incorporates a three-parameter loading surface, uncoupled hardening and softening functions, following the accumulation of plastic volumetric strain and a nonlinear Lode-angle-dependent plastic potential function. Skalomenos et al. [25] created an accurate nonlinear finite element model with the ATENA software, and the influences of different parameters were studied.

Due to the excellent strength and ductility, SC Walls have already been applied to a lot of structures requiring high resistance against severe loads since the 1980s, such as nuclear facilities, or marine environment structures. In China, SC walls are mainly used in high-rise buildings, including Yancheng TV Tower [26] and Guangzhou TV Tower. Due to the lack of detailed design codes, the above structures were mainly carried out based on the overall concept of the structure regarding other composite structures, so a reasonable and correct design method for SC Walls is urgently needed. In this paper, based on the experimental results of 49 SC walls and the results of numerical analysis, the bending failure characteristic of SC walls was studied and summarized, and a primary assumption of SC walls was put forward. Then according to the ultimate equilibrium theory of the cross section, the axial and bending capacity formula was deduced and compared with the existing experimental results.

2. Experimental Results Analysis

SC walls were conceived initially in Japan in the 1980s. Extensive research has been done in many countries to study the behavior and failure modes of SC walls. For example, Eom et al. [7] tested three isolated walls and two coupled walls subjected to cyclic lateral loading. The specimen named DSCW1C in the test failed due to outward buckling of the steel plates and subsequent fracture of the vertical weld joint, followed by the filled concrete crushing and tie bars fracture. Nie et al. [10] tested nine SC wall specimens with different shear span ratios. All the specimens failed due to the buckling of the steel plates and the filled concrete crushing. Ji et al. [11] tested five slender rectangular wall specimens subjected to axial forces and lateral cyclic loading. The specimens failed in a flexural mode, characterized by local buckling of the steel tubes and plates, fracture of the steel tube, and concrete crushing at the wall base. Besides, the tested indicated that the average strain distribution of the surface steel plates agreed with the plane assumption within the range of 300mm above the wall base. Tian [12] tested nine SC wall specimens with the rectangular cross section. The experiment showed all the specimens failed due to the steel plate buckling and concrete crushing.

The following conclusions are obtained by analyzing and summarizing the experimental results of the existing researches.

- The bending failure characteristics of the SC wall are similar. Steel plate at the bottom of the SC wall yields first, followed by the buckling of the steel plate and concrete crushing. All SC wall failures are characterized by concrete crushing at embedded columns. When the studs are closely arranged, the steel plate at the bottom of the wall plate yields before buckling.
- Before the load reaches the peak, the average strain distribution of the surface steel plate at a certain height from the wall base is consistent with the plane assumption.

The ratio of the stud spacing (B) to the steel plate thickness (t) has a significant effect on the steel plate buckling, as shown in Table 1. When the B/t ratio is within a certain range, local buckling occurs when the specimen near reaches the ultimate bearing capacity, and the buckling is accompanied by the pull-out of the concrete, indicating that the slippage between the surface steel plate and the filled concrete is small. When calculating the axial and bending bearing capacity of the SC wall, it is assumed that the steel plate and the concrete can be well unified without relative sliding, and when the SC wall reaches the ultimate bearing capacity, the steel plate does not buckle.

Table 1. The influence of B/t on local buckling of steel plates.

Reference	Specimen	P_{cr} [1]/P_u [2]	Buckling Pattern	B/t [5]	The Limit of B/t
[9]	W0	0.63	buckling1 [3]	58.8	34
[10]	SCW5	0.91	buckling2	40	34
[12]	SCW1-1a	0.95	buckling2 [4]	13.3	33
	SCW1-1b	0.96	buckling2	13.3	33
	SCW1-2a	0.98	buckling2	13.3	33
	SCW1-2b	1.00	buckling2	13.3	33
	SCW1-3	0.91	buckling2	13.3	33
	SCW1-4	0.94	buckling2	20	34
	SCW1-5	0.92	buckling2	10	32
	SCW1-6	0.90	buckling2	26.7	33

[1] P_{cr} is the local buckling load of the column steel plate; [2] P_u is the ultimate load; [3] buckling1 means the concrete at the embedded column is not crushed when the steel plate buckling occurs; [4] buckling2 means the steel plate buckling occurs accompanied with the concrete crushed and pushed out; [5] B/t is calculated by $600\sqrt{f_y}$ referring to [22].

Because the concrete in the SC shear wall is wrapped inside the steel plate, variables such as the ultimate compressive strain of concrete and the average strain of concrete in a certain height range are difficult to measure. To verify whether the strain conforms to the plane assumption in more cases and obtain the strain of the concrete inside the wall under ultimate load, a finite element simulation of the specimen in [12] was conducted. Calculations under various design axial compression ratios were carried out to supplement the test results.

3. Numerical Analysis

3.1. FE Model

In the structure of SC walls, the concrete is covered by the surface steel plates, so it is difficult to measure the ultimate compressive strain of the concrete and the average strain. In this paper, the finite element analysis was done using ABAQUS (Dassault Systèmes, Providence, Rhode Island, USA, 2010) to obtain the strain distribution of the concrete, to verify the plane assumption, and to carry out the parametric analysis. There are few experiments on small eccentric compression failure, so the SC walls having small eccentric compression failure were simulated by ABAQUS.

Nonlinear finite element analysis was performed on all samples in [12] using ABAQUS. The reinforcement diagram of a specimen is shown as an example in Figure 1. The upper and lower parts of the SC wall were embedded in the loading beam and the base beam, respectively, for applying the lateral load and anchoring with the foundation. The parameters of specimens are shown in Table 2. Studs were set on the inner surface of the steel plate and tie bars were arranged between the steel plates. In the test, the specimen was fixed on the ground by two long bolts. The concrete grade was C35. The average cube compressive strength was 42.9 MPa, and the Young's modulus was 33,000 MPa. The average yield strength of the steel plate was 330 MPa, and the Young's modulus was 206 GPa. The quasi-static method was applied in the loading test. First, the vertical load was applied to the specimen until the design axial compression ratio was achieved. Then a lateral load was applied on the top of the wall. Before yielding, the lateral load increased step by step according to 1/10 of the ultimate load, which was estimated, and one cyclic loading was performed at each step.

Table 2. The parameters of specimens.

Specimen	Specimen Size			Steel Plane Thickness	Design Axial Compression Ratio	Stud Spacing
	Height	Width	Thickness			
SCW1-1a	1000	1000	150	3	0.4	40
SCW1-1b	1000	1000	150	3	0.4	40
SCW1-2a	1500	1000	150	3	0.4	40
SCW1-2b	1500	1000	150	3	0.4	40
SCW1-3	2000	1000	150	3	0.4	40
SCW1-4	1000	1000	150	2	0.4	40
SCW1-5	1000	1000	150	4	0.4	40
SCW1-6	1000	1000	150	3	0.4	80

The design axial compression ratio (μ) refers to the ratio of the representative value of the load to the design value of the material strength, which can be expressed as follow:

$$\mu = \frac{N}{f_c A_c + f_y A_s} \tag{1}$$

where N is the design value of the axial pressure of the specimen under the action of the representative value of the gravity load. f_c is the design value of concrete compressive strength. f_y is the design value of the yield strength of the steel plate. A_c and A_s are the concrete cross section area and the steel plate cross section area, respectively.

The ABAQUS model of specimens taken from [12] is shown in Figure 2. The concrete was modeled with the C3D8R solid element; the steel plate was modeled with the S4R shell element; the tie bar was modeled with the T3D2 truss element, which is similar to [27]. The stud was modeled with the SPRING2 element.

In this paper, a three-way zero-length spring was added between the concrete slab node and the steel plate node at the actual position of the stud. The spring stiffness includes shear stiffness and axial stiffness. Aiming at the shear-slip curve, Ollgaard et al. [28] proposed a calculation model, which has been widely recognized. The shear-slip curve is expressed as:

$$V = V_u(1 - e^{-ns})^m \tag{2}$$

where s is the slippage. m and n are parameters. Different researchers used Equation (2) to fit the m and n according to their test results. The value of m is generally between 0.4 and 1.5, and the value of n is generally between 0.5 and 2.0. Gattesco and Giuriani [29] conducted two horizontal push-out shear tests of 19 mm diameter stud. Using the Equation (2) proposed by Ollgaard et al. [28] to fit the experimental value of the shear-slip curve in [29], when $m = 0.425$, $n = 0.5$, the two coincide. Another parameter that needs to be determined is the ultimate shear strength V_u. The test results of [29] are in good agreement with the calculation results of Eurocode-4 [30]. The equation proposed by Eurocode-4 was applied in this paper to calculate V_u, which is expressed as:

$$V_u = \min\left\{0.8 f_u A,\ 0.29 \alpha d^2 \sqrt{f_{ck} E_{cm}}\right\} \tag{3}$$

where f_u is the ultimate tensile strength of the stud, A is the section area of the stud, d is the diameter of the stud, f_{ck} is the compression strength of concrete cylinder, E_{cm} is the young's modulus of concrete, $\alpha = 0.2(h/d + 1)$, and h is the height of the stud. The relationship between shear force and displacement can be obtained by Equations (1) and (2) and can be used as the nonlinear spring stiffness. According to Luis Pallarés et al. 's review [31] of the tensile failure of stud, there are generally three types of failure, namely, stud fracture, concrete cone pull-out failure, and concrete column pull-out failure. According to American Concrete Code 318-08 (ACI 318-08), when the diameter of the stud cap is greater than 1.71 times the diameter of the stud rod, there is a 95% probability that the concrete column will not be pulled out. When the length of the stud is greater than 7.5 times the diameter of the stud, the concrete cone will not be pulled out [31]. Therefore, the brittle failure of the concrete can be completely avoided by construction measures. The damage caused by the stud itself breaking is ductile. The bolt is equivalent to a tension member with one end fixed and one end freely stretched, so the stiffness of the stud can be used to calculate the stiffness of the axial spring, which is expressed as follows:

$$K = \frac{EA}{h} \tag{4}$$

where E is the young's modulus of the stud, A is the section area of the stud and h is the height of the stud.

The steel plates were embedded in the loading beam and the base beam. The studs and tie bars were embedded in the concrete. The contact feature was used between the elements of the concrete wall and the steel plate that was not inserted in the loading beam and the base beam. The contact properties were frictional contact in the tangential direction and hard contact in the normal direction, respectively. The boundary conditions are shown in Figure 2b. The FE model was fixed on the ground by two springs with the spring stiffness of EA/L, where A and L were the bolt section area and length, respectively. The contact between the base beam and the foundation was simulated by an elastic foundation model. The reaction force coefficient of the concrete foundation is generally 7484.6–14,715 kN/mm^3 [32]. In the simulation, the reaction force coefficient was 10,000 kN/mm^3. At both ends of the model foundation beam, the horizontal displacement of the wall in and out of the wall was constrained, and the vertical displacement was not constrained. The concrete damaged plasticity model was applied to simulate the behavior of concrete. The energy equivalence principle of Sidiroff was applied to calculate the damage factor. The ideal elastoplastic stress-strain curve was applied to model the behavior of steel plates and tie bars.

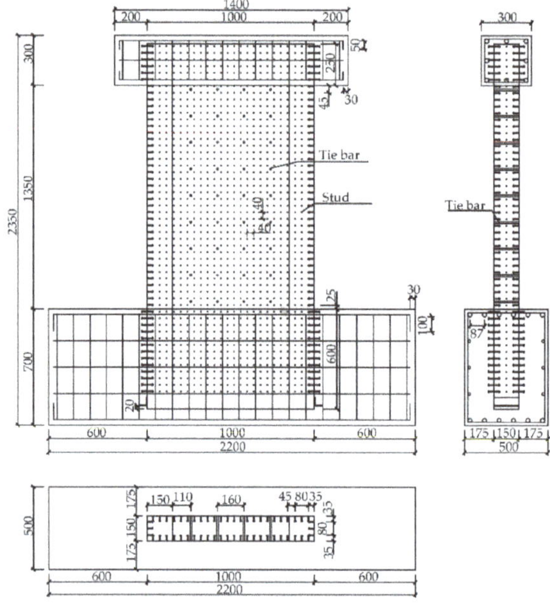

Figure 1. Diagrams of SC walls (SCW1-2) in [33].

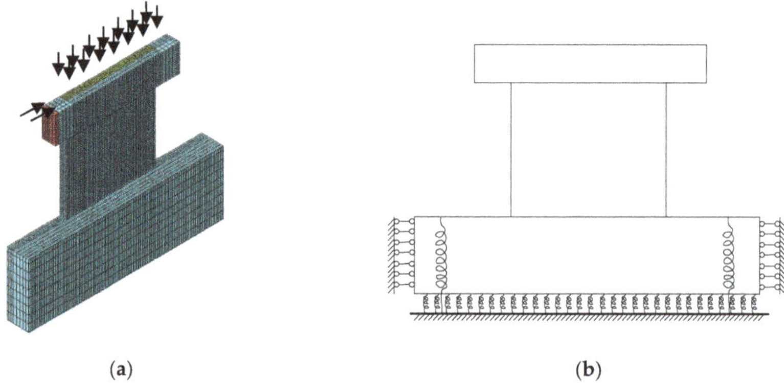

(**a**) (**b**)

Figure 2. FE model in ABAQUS. (**a**) FE models; (**b**) Boundary conditions of FE models.

3.2. FE Calculation Results

The calculated load-displacement skeleton curves were compared with the experimental results. Comparisons between the experimental result and the calculated result are shown in Figure 3, which presents a good agreement. Besides, the average ratio of the calculated value of ultimate bearing capacity to the experimental value is 1.09; the average ratio of the peak displacement between the calculated value and experimental value is 0.9; the average ratio of the steel yield load and displacement are 0.93 and 1.04, respectively. Therefore, it could be concluded that this modeling method could effectively simulate the mechanical property of SC walls under cyclic loads.

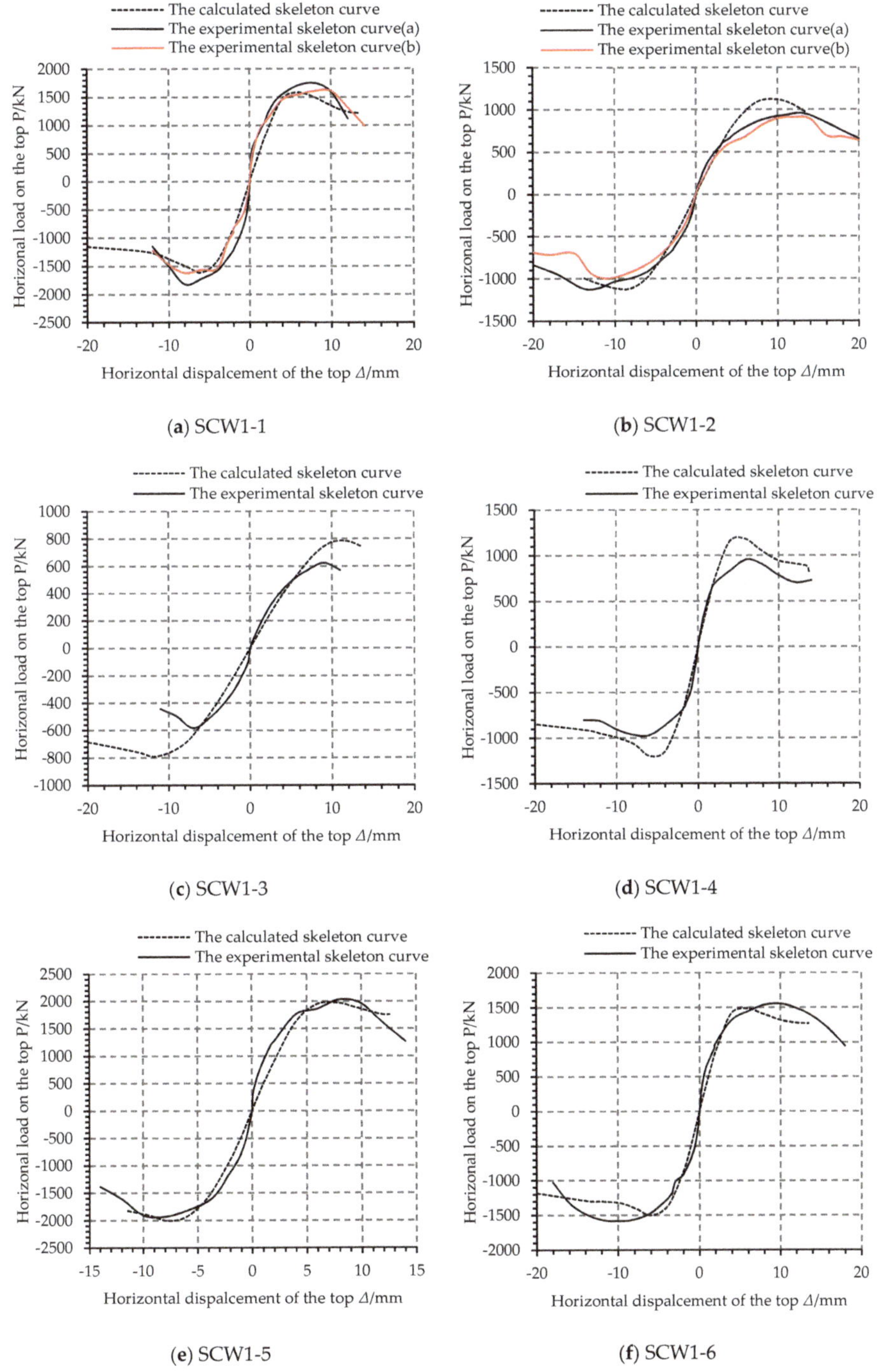

(**a**) SCW1-1 (**b**) SCW1-2

(**c**) SCW1-3 (**d**) SCW1-4

(**e**) SCW1-5 (**f**) SCW1-6

Figure 3. Comparisons between the experimental result and the calculated result.

The specimens in [12] were under the same axial compression ratio. To verify whether the strain meets the plane assumption under different axial compression ratios, three representative specimens with different steel contents and shear span ratios are selected, which are specimen SCW1-1 (steel plate thickness is 3 mm, shear span ratio is 1.0), SCW1-2 (steel plate thickness is 3 mm, shear span ratio is 1.5) and specimen SCW1-4 (steel plate thickness is 4 mm, shear span ratio is 1.0). Nine FE models under different axial compression ratios were simulated using ABAQUS. The results are as follows:

1. Average strain distribution of the cross section

As shown in Figure 4, when the load reaches the ultimate value, under different axial compression ratios, steel ratios and shear span ratios, the average strain at the bottom cross section of the wall is consistent with the plane assumption.

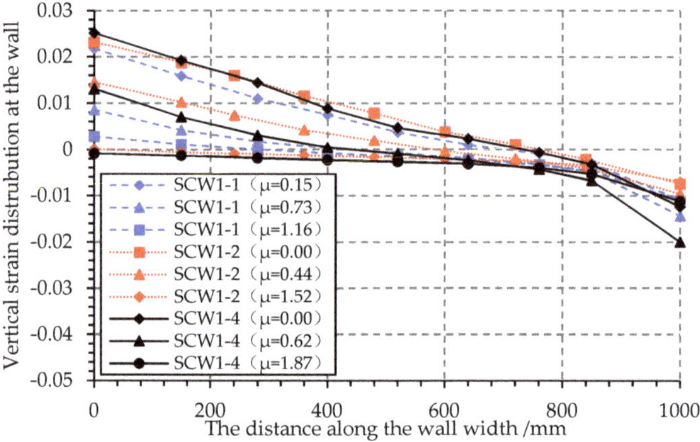

Figure 4. Vertical strain distribution at the wall base for specimens at ultimate load under different axial load ratios (μ is the axial compression ratio).

2. Ultimate compression strain of the concrete

There are few studies on the ultimate compression strain of the confined concrete. In this paper, the ultimate compression strain was obtained by ABAQUS analysis. Figure 5 shows the concrete ultimate compression strain of the specimens with different axial compression ratios. In Figure 5, the theoretical value was calculated referring to [34], in which the ultimate compression strain of concrete was defined as the post-peak strain corresponding to the stress valued $0.5f_c$ in the declining part of the stress-strain curve. For the constitutive relationship of concrete under unidirectional load, refer to the constitutive relationship of rectangular steel tube concrete proposed by Han [35]. As shown in Figure 5, the axial compression ratio, steel ratio, and shear-span ratio influence the concrete ultimate compressive strain; the theoretical value approximated to the lower limit of the results calculated by ABAQUS. Therefore, when calculating the axial and bending bearing capacity for SC walls, the confined concrete compression strain could be calculated conservatively according to the above method.

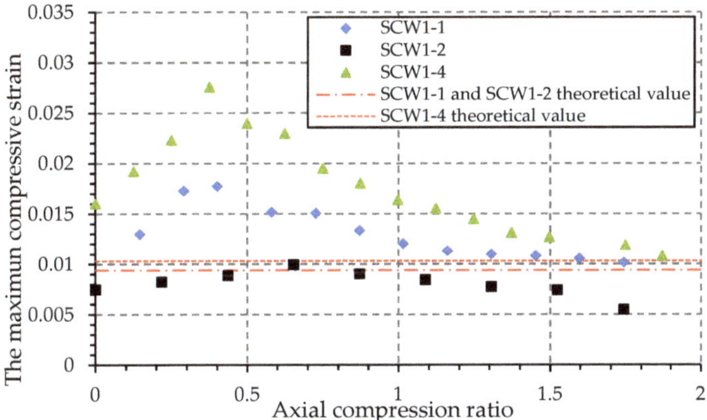

Figure 5. Maximum concrete compression strain versus axial compression ratio of specimens at ultimate load.

4. Derivation of the Axial and Bending Bearing Capacity

4.1. Basic Assumption

Based on the experimental results and finite element analysis, the basic assumptions for calculating the axial and bending bearing capacity of SC walls are proposed.

1. The failure of SC walls is that the outermost concrete in the compression zone reaches the ultimate compressive strain, and the corresponding bearing capacity is taken as the ultimate bearing capacity;
2. The surface steel plate and concrete are well combined without slippage;
3. The average strain distribution of the wall cross section is consistent with the plane assumption;
4. The tensile effect of the concrete is not taken into account;
5. The constitutive of the concrete with the confinement of a rectangular steel tube is introduced to describe the behavior of concrete under a uniaxial load, which was proposed by Han [35]. The ultimate compressive strain of concrete corresponded to a stress value of $0.5f_{cc}$ in the descending part of the stress-strain curve;
6. The ideal elastic-plastic constitutive is introduced to describe the behavior of steel plate, ignoring the strain hardening;
7. The concrete compressive stress distribution is replaced by equivalent rectangular stress distribution. The equivalent stress is $\alpha_1 f_{cc}$ and the equivalent compression zone height is $x = \beta_1 x_n$. α_1 and β_1 are equivalent rectangular stress coefficients, which are related to the concrete grade. The values of α_1 and β_1 are the same as ordinary concrete. f_{cc} is calculated using Equation (5) [34].

$$f_{cc} = \left[1 + \left(-0.0135\xi_1^2 + 0.1\xi\right)\left(\frac{24}{f_c}\right)^{0.45}\right] \tag{5}$$

where f_{cc} is the compressive strength of confined concrete; f_c is the compressive strength of the concrete; and ξ_1 is the confinement effect coefficient.

The confinement effects of the column plate and the steel plate on concrete are different in SC walls. When the connectors between the surface steel plates are enough, ξ_1 can be calculated conservatively using Equation (6).

$$\xi_1 = \frac{A_{s0} f_y}{A_c f_c} \tag{6}$$

where A_{s0} is the total cross-section area of the steel plate in the SC wall (including the vertical partition plates); f_y is the yield strength of the steel plate; and A_c is the total cross-section area of the concrete in the SC wall.

4.2. Section Simplification

Figure 6a shows the configuration of the SC wall. The components used to resist the bending moment include the web plate, column plate, partition plate, and stiffening steel tube (or profile steel). To simplify the calculation, the cross section is simplified, as shown in Figure 6b. The column plate and stiffening steel are simplified to the outermost side of the wall according to the principle of providing the equivalent moment of the cross section centroid axis when the steel yielded. The yield strength of the fringe plate equals the yield strength of the web plate. The web steel plates and partition plates are simplified to web plate according to the principle of equal area.

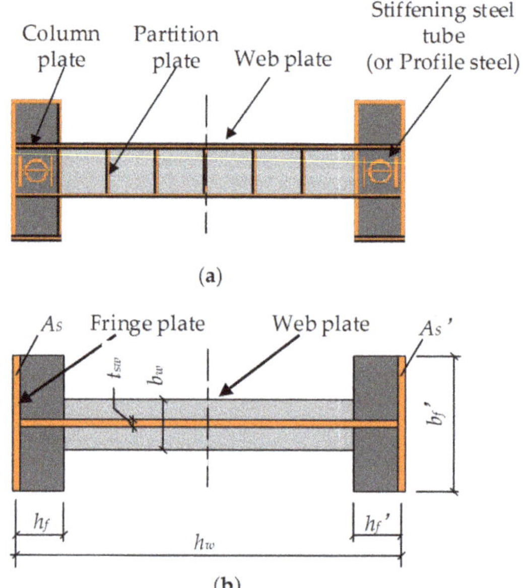

Figure 6. Simplification of the cross section for SC walls. (**a**) The general section structure; (**b**) Cross section simplification.

4.3. Failure Modes

Based on the above assumptions and simplification, the ultimate equilibrium method is applied to derive the axial and bending bearing capacity formula of the SC wall. The balanced failure of the SC wall is defined so that the steel plates in tension reach the yield strength simultaneously when the concrete in the compressive zone reaches the ultimate compressive strain. The relative height of the concrete compressive zone is ξ_b, which can be calculated by Equation (7) referring to [36]. When $\xi \leq \xi_b$, the steel plates in tension could reach the yield strength when the wall breaks, and the failure mode of the SC wall is called large eccentric compression failure. When $\xi > \xi_b$, the steel plates in tension cannot reach the yield strength when the wall breaks; the failure mode of the SC wall is called small eccentric compression failure (as shown in Figure 7).

$$\xi_b = \frac{\beta_1}{1 + \frac{f_y}{\varepsilon_{cu E_s}}} \tag{7}$$

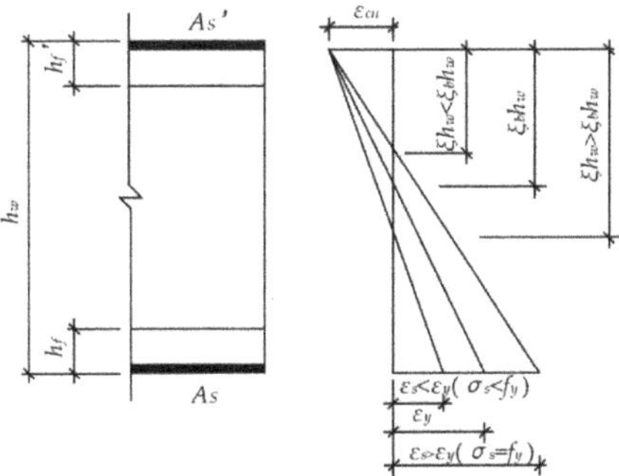

Figure 7. The strain distribution of the SC walls.

4.4. Large Eccentric Compression Failure ($\xi \leq \xi_b$)

Figure 8 shows the distribution of the cross-section stress and strain for the large eccentric SC wall. The equilibrium equation of the cross section is described as Equations (8) and (9).

$$\sum X = 0,\ N = A'_s f_y + C_c + N_{sw} - A_s f_y \tag{8}$$

$$\sum M = 0,\ Ne = A'_s f_y h_w + M_c + M_{sw} \tag{9}$$

where, N is the design value of the axial force; e is the distance from the axial force to the center of the steel plate in tension; $A_s{}'$ is the area of the steel plate in compression; N_{sw} is the resultant force of the web plate (stress in compressive direction is defined positive and stress in tensile direction is negative); M_{sw} is the moment of the web plate normal stress about the center of the steel plate in tension (the clockwise direction is defined as positive, the counterclockwise direction is defined as negative); C_c is the resultant of the concrete compressive stress (stress in compressive direction is defined positive and stress in tensile direction is negative); and M_c is the moment of the concrete normal stress about the center of the steel plate in tension (the clockwise direction is defined as positive, the counterclockwise direction is defined as negative).

If $\beta_1 x_n > h'_f$,

$$C_c = \alpha_1 f_{cc} \beta_1 x_n b_w + \alpha_1 f_{cc}\left(b'_f - b_w\right)h'_f \tag{10}$$

$$M_c = \alpha_1 f_{cc} \beta_1 x_n b_w\left(h_w - \frac{\beta_1 x_n}{2}\right) + \alpha_1 f_{cc}\left(b'_f - b_w\right)h'_f\left(h_w - \frac{h'_f}{2}\right) \tag{11}$$

If $\beta_1 x_n \leq h'_f$,

$$C_c = \alpha_1 f_{cc} \beta_1 x_n b'_f \tag{12}$$

$$M_c = \alpha_1 f_{cc} \beta_1 x_n b'_f\left(h_w - \frac{\beta_1 x_n}{2}\right) \tag{13}$$

where, x_n is the height of the concrete compressive zone; b_w is the thickness of the wall web; $h_f{}'$ is the height of the compression flange; and $b_f{}'$ is the width of the compression flange.

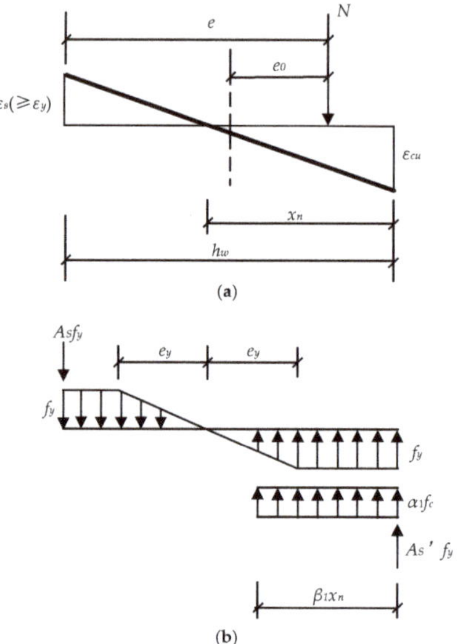

Figure 8. Stress and strain distribution of cross section for the large eccentric SC wall. (**a**) Strain distribution of the cross section; (**b**) Stress distribution of the cross section.

The stress distribution of the web plate is equivalent to the superimposition of one full plastic stress distribution and a triangle stress distribution, as shown in Figure 9. The triangle stress distribution can be seen as a couple. The couple has a negligible effect on the bearing capacity. For example, the horizontal bearing capacity of SCW1-5 in [12] was 1653.57 kN if the couple was considered. The result was 1653.59 kN if the couple was ignored and the difference was only 0.0012%. Therefore, the couple in Figure 9b can be ignored in the process of calculating the bearing capacity. The formula is simplified as Equations (14) and (15).

$$N_{sw} = f_y t_{sw}(2x_n - h_w) \tag{14}$$

$$M_{sw} = f_y t_{sw}\left[\frac{1}{2}h_w^2 - (h_w - x_n)^2\right] \tag{15}$$

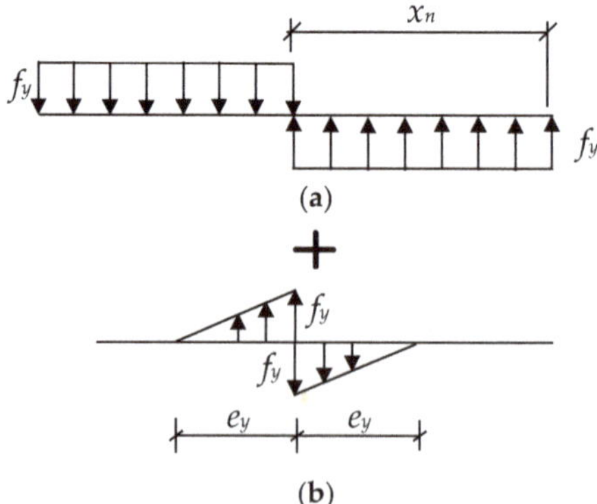

Figure 9. Equivalence of web steel plate for the large eccentric SC wall. (**a**) Full plastic stress distribution; (**b**)Triangle stress distribution.

4.5. Small Eccentric Compression Failure ($\xi > \xi_b$)

Figure 10 shows the distribution of the cross-section stress and strain for the small eccentric SC wall. The equilibrium equations of the section are described as Equations (16) and (17).

$$\sum X = 0,\ N = A_s' f_y h_0 + C_c + N_{sw} - T_s \tag{16}$$

$$\sum M = 0,\ Ne = A_s' f_y h_0 + M_c + M_{sw} \tag{17}$$

where, T_s is the resultant force of the steel plate in tension. T_s is calculated by Equation (15), referring to [36].

$$-f_y A_s < T_s = \frac{x_n - h_w}{x_{nb} - h_w} f_y A_s < f_y A_s \tag{18}$$

where, x_{nb} is the height of the concrete compression zone in the balanced failure mode.

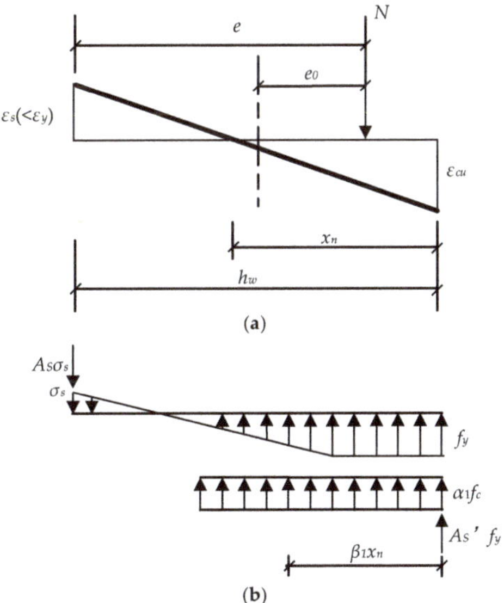

Figure 10. Stress and strain distribution of cross section for the small eccentric SC wall. (**a**) Strain distribution of cross section; (**b**) Stress distribution of cross section.

To simplify the calculation, the web plate stress distribution is simplified, as shown in Figure 11. η is used to revise the compression field height, to reduce the simplification error. In this paper, η was calculated using Equation (19). Therefore, N_{sw} and M_{sw} can be calculated using Equations (20) and (21).

$$\eta = 1 - \frac{2\varepsilon_y}{3\varepsilon_{cu}} \tag{19}$$

$$N_{sw} = f_y t_{sw} \eta x_n \tag{20}$$

$$M_{sw} = f_y t_{sw} \eta x_n (h_w - 0.5\eta x_n) \tag{21}$$

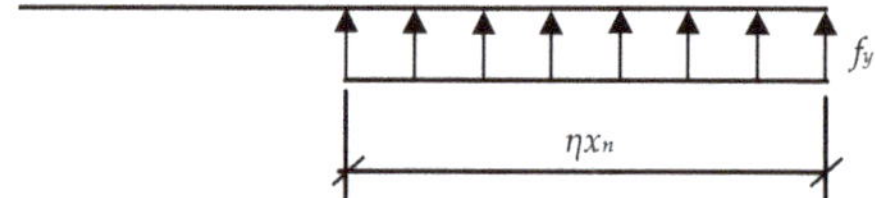

Figure 11. Equivalence of web plate for the small eccentric SC wall.

4.6. Verification

The ultimate bearing capacities of 31 specimens in different researches were calculated by Equations (9)–(21). The calculated results are compared with the experimental results, as shown in Figure 12. Since large eccentric compression failure occurred in all the specimens in the references, therefore, the finite element analysis for three small eccentric specimens was carried out using ABAQUS to verify the effectiveness of the calculation formula of small eccentricity, and the results were compared with the theoretical results from the calculation formula of small eccentricity. As shown in Figure 12, the calculated results agree well with the experimental and finite element analysis results. The average ratio of the calculation results to the experimental results of 31 specimens and the results of 3 FE models is 1.054, and the standard deviation is 0.086. Consequently, the above calculation method can predict the axial and bending bearing capacity of SC walls effectively.

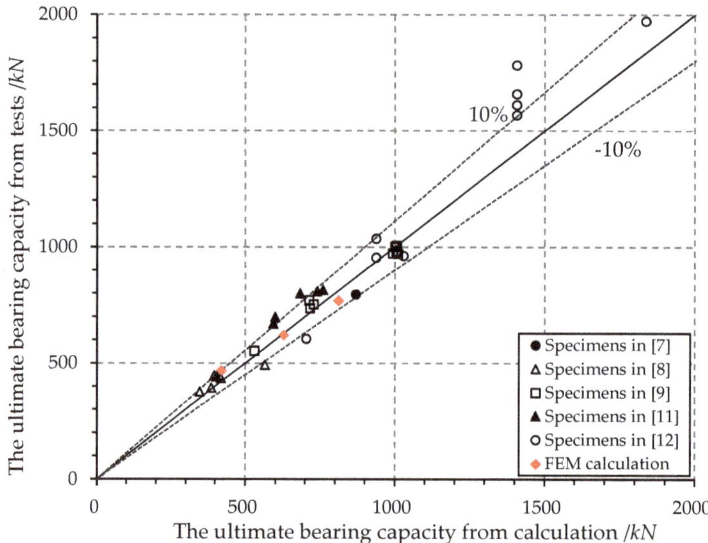

Figure 12. Comparison between test results and calculated results of specimens in the existing literature.

Also, for some specimens in [12], it is clear that the calculation results are much lower than the experimental results. This is mainly because the hardening of the steel plate in tension is not considered in the formulas, but the steel plates of some specimens in [12] had reached the strain-hardening range, and even tension fracture appeared.

5. Conclusions

This paper summarized the experimental results of 49 SC walls subjected to cyclic loading. The parameter analysis using ABAQUS was carried out to supplement the experiments. Based on the analysis of the experimental results and finite element analysis results, the calculation method for the axial and bending bearing capacity of SC walls are proposed. The conclusions of this paper are summarized as follows.

- The bending failure of SC walls featured the concrete crushing at the column and the average strain distribution of the wall cross section agreed with the plane assumption.
- When studs meet certain construction requirements, the surface steel plates and filled concrete are unified well, and the bond-slip between them is relatively small.
- The ultimate compressive strain of the concrete can be defined as the post-peak strain corresponding to the stress valued $0.5f_{cc}$ in the declining part of the stress-strain curve; the constitutive of the concrete with the confinement of a rectangular steel tube is applied to descript the behavior of concrete.
- The basic assumption for calculating the axial and bending bearing capacity of SC walls is put forward and the calculation formula applying to both small and large eccentric load is deduced. The calculated results are compared with the experimental results, which present good agreements.

Author Contributions: Conceptualization, Q.G.; Data curation, P.Z.; Formal analysis, P.Z. and F.K.; Funding acquisition, Q.G.; Investigation, W.Z. and F.K.; Methodology, W.Z. and Y.Y.; Project administration, Q.G. and Y.Y.; Writing—original draft, F.K.; Writing—review & editing, P.Z. All authors have read and agreed to the published version of the manuscript.

Funding: This work was funded by National Natural Science Foundation of China (Grant No. 51778032).

Acknowledgments: This work was supported National Natural Science Foundation of China (Grant No. 51778032). The authors greatly appreciate the financial support.

Conflicts of Interest: The authors declare no conflict of interest. The funders had no role in the design of the study; in the collection, analyses, or interpretation of data; in the writing of the manuscript, or in the decision to publish the results.

References

1. Usami, S.; Akiyama, H.; Suzuki, T.; Sekimoto, H.; Kobayashi, M. Experimental study on concrete filled steel shear wall: Part 2 bensing shear test. In *Summaries of Technical Papers of Annual Meeting*; Structures. II; Architectural Institute of Japan: Tokyo, Japan, 1991; pp. 1661–1662.
2. Takeuchi, M.; Narikawa, M.; Matsuo, I.; Hara, K.; Usami, S. Study on a concrete filled structure for nuclear power plants. *Nucl. Eng. Des.* **1998**, *179*, 209–223. [CrossRef]
3. Niwa, N.; Sasaki, N.; Akiyama, H.; Narikawa, M.; Hara, K.; Takeuchi, M.; Usami, S. Study on concrete filled steel structures for nuclear power plants: Part 3 outline of bending shear tests. In *Summaries of Technical Papers of Annual Meeting*; B-2; Structures II; Structural Dynamics Nuclear Power Plants; Architectural Institute of Japan: Tokyo, Japan, 1995; pp. 987–988.
4. Ozaki, M.; Akita, S.; Osuga, H.; Nakayama, T.; Adachi, N. Study on steel plate reinforced concrete panels subjected to cyclic in-plane shear. *Nucl. Eng. Des.* **2004**, *228*, 225–244. [CrossRef]
5. Kitano, T.; Akita, S.; Nakazawa, M.; Fujino, Y.; Ohta, H.; Yamaguchi, T.; Nakayama, T. Experimental study on a concrete filled steel structure: Part 4 shear tests (outline of the experimental program nad the results). In *Summaries of Technical Papers of Annual Meeting*; B-2; Structures II, Structural Dynamics Nuclear Power Plants; Architectural Institute of Japan: Tokyo, Japan, 1997; pp. 1057–1058.
6. Funakoshi, A.; Akita, S.; Matsumoto, N.; Hara, K.; Hara, K.; Hayashi, N. Experimental study on a concrete filled steel structure: Part 7 bending shear tests(outline of the experimental program nad the results). In *Summaries of Technical Papers of Annual Meeting*; B-2; Structures II; Structural Dynamics Nuclear Power Plants; Architectural Institute of Japan: Tokyo, Japan, 1997; pp. 1063–1064.
7. Eom, T.-S.; Park, H.-G.; Lee, C.-H.; Kim, J.-H.; Chang, I.-H. Behavior of double skin composite wall subjected to in-plane cyclic loading. *J. Struct. Eng. ASCE* **2009**, *135*, 1239–1249. [CrossRef]
8. Wu, B.; Liu, C.; Zhao, X.; Liu, Q. Experimental study on seismic behavior of double thin skin hybrid walls filled with demolished concrete lumps. *J. Build. Struct.* **2011**, *32*, 116–125.
9. Nie, J.; Tao, M.; Fan, J.; Bu, F.; Hu, H.; Ma, X.; Li, S.; Liu, F. Research advances of composite shear walls with double steel plates and filled concrete. *Build. Struct.* **2011**, *41*, 52–60.

10. Nie, J.; Bu, F.; Fan, J. Experimental research on seismic behavior of low shear-span ratio composite shear wall with double steel plates and infill concrete. *J. Build. Struct.* **2011**, *32*, 74–81.
11. Ji, X.; Jiang, F.; Qian, J. Seismic behavior of steel tube–double steel plate–concrete composite walls: Experimental tests. *J. Constr. Steel Res.* **2013**, *86*, 17–30. [CrossRef]
12. Tian, C.; Cheng, W. *Report of the Experimental Research of Steel Plate and Concrete Composite Shear Wall in Htr*; Beihang University: Beijing, China, 2013.
13. Guo, X.; Qiu, L.; Luo, Y.; Li, J.; Dong, N.; Zhang, J. Seismic performance of concrete-filled double skin composite shear wall with opening. *J. Harbin Inst. Technol.* **2015**, *47*, 69–76.
14. Yang, Y.; Liu, J.; Nie, X.; Fan, J. Experimental research on out-of-plane cyclic behavior of steel-plate composite walls. *J. Earthq. Tsunami* **2016**, *10*, 1650001. [CrossRef]
15. Liu, Y.; Yang, Q.; Liu, J.; Liao, Y. Experimental research on axial compressive behavior of shear wall with double steel plates and filled concrete. *J. Sichuan Univ.* **2016**, *48*, 83–90.
16. Li, X.; Li, X.; Shen, L.; Liu, J. Experimental study of composite shear walls with double steel plates and filled concrete for a nuclear island structure under low cyclic loading. *J. Beijing Univ. Technol.* **2015**, *42*, 1498–1508.
17. Li, X.; Li, X. Study on in-plane flexural behavior of double steel plates and concrete infill composite shear walls for nuclear engineering. *Eng. Mech.* **2017**, *34*, 43–53.
18. Lin, X.; Wu, K.; Skalomenos, K.A.; Lu, L.; Zhao, S. Development of a buckling-restrained shear panel damper with demountable steel-concrete composite restrainers. *Soil Dyn. Earthq. Eng.* **2019**, *118*, 221–230. [CrossRef]
19. Zhao, Q.; Astaneh-Asl, A. Cyclic behavior of traditional and innovative composite shear walls. *J. Struct. Eng. ASCE* **2004**, *130*, 271–284. [CrossRef]
20. Rassouli, B.; Shafaei, S.; Ayazi, A.; Farahbod, F. Experimental and numerical study on steel-concrete composite shear wall using light-weight concrete. *J. Constr. Steel Res.* **2016**, *126*, 117–128. [CrossRef]
21. Varma, A.H.; Malushte, S.R.; Sener, K.C.; Lai, Z. Steel-plate composite (sc) walls for safety related nuclear facilities: Design for in-plane forces and out-of-plane moments. *Nucl. Eng. Des.* **2014**, *269*, 240–249. [CrossRef]
22. Japanese Electric Association. *Techinical Guidelines for Aseismic Design of Steel Plate Reinforced Concrete Structures-Buildings and Structures*; Japanese Electric Association: Tokyo, Japan, 2005; Volume JEAG4618-2005.
23. Ma, X.; Nie, J.; Tao, M.; Bu, F. Numerical model and simplified formula of axial force-moment capacity of composite shear wall with double steel plates and infill concrete. *J. Build. Struct.* **2013**, *34*, 99–106.
24. Papanikolaou, V.K.; Kappos, A.J. Confinement-sensitive plasticity constitutive model for concrete in triaxial compression. *Int. J. Solids Struct.* **2007**, *44*, 7021–7048. [CrossRef]
25. Skalomenos, K.A.; Hatzigeorgiou, G.D.; Beskos, D.E. Parameter identification of three hysteretic models for the simulation of the response of cft columns to cyclic loading. *Eng. Struct.* **2014**, *61*, 44–60. [CrossRef]
26. Ding, Z.; Jiang, H.; Zeng, J.; Zhang, H.; Du, G. An innovative application of scs composite wall: Structural design of yancheng tv tower. *Build. Struct.* **2011**, *41*, 87–91.
27. Shafaei, S.; Farahbod, F.; Ayazi, A. Concrete stiffened steel plate shear walls with an unstiffened opening. *Structures* **2017**, *12*, 40–53. [CrossRef]
28. Ollgaard, J.G.; Slutter, R.G.; Fisher, J.W. Shear Strength of Stud Connectors in Lightweight and Normal-Weight Concrete, AISC Eng'g Jr., April 1971 (71-10). 1971, pp. 55–64. Available online: https://preserve.lehigh.edu/cgi/viewcontent.cgi?article=3009&context=engr-civil-environmental-fritz-lab-reports (accessed on 15 July 2020).
29. Gattesco, N.; Giuriani, E. Experimental study on stud shear connectors subjected to cyclic loading. *J. Constr. Steel Res.* **1996**, *38*, 1–21. [CrossRef]
30. The European Committee for Standardization (CEN). *Design of Composite Steel and Concrete Structures, Part 2: Composite Bridges*; CEN: Brussels, Belgium, 1997.
31. Pallarés, L.; Hajjar, J.F. Headed steel stud anchors in composite structures, part II: Tension and interaction. *J. Constr. Steel Res.* **2010**, *66*, 213–228. [CrossRef]
32. CSSC. *Calculation of Beam and Rectangular Plate on Elastic Foundation*; National Defense Industry Press: Beijing, China, 1983.
33. Guo, Q. *Report of the Theoretical Research of the Seismic Performance of High-Temperature Gas-Cooled Reactor (HTR-PM) Steel-Concrete Composite Shear Wall*; Beihang University: Beijing, China, 2013.

34. MOHURD. *Code for Design of Concrete Structures*; China Architecture & Building Press: Beijing, China, 2010; Volume GB50010-2010.
35. Han, L. *Concrete Filled Steel Tube Structure-Theory and Practice*; Science Press: Beijing, China, 2007.
36. Teng, Z.; Zhu, Z. *Concrete Structure and Masonry Structure*; China Architecture & Building Press: Beijing, China, 2003.

Article

Local Buckling and Resistance of Continuous Steel Beams with Thin-Walled I-Shaped Cross-Sections

Andrzej Szychowski * and Karolina Brzezińska

Faculty of Civil Engineering and Architecture, Kielce University of Technology, al. Tysiąclecia Państwa Polskiego 7, 25-314 Kielce, Poland; k.brzezinska@tu.kielce.pl
* Correspondence: aszychow@tu.kielce.pl

Received: 8 April 2020; Accepted: 23 June 2020; Published: 28 June 2020

Featured Application: The research results included in the paper can be widely used in designing thin-walled steel elements for use in construction. The use of the presented calculation model for the calculation of continuous beams allows for more optimal and economical design of sections of this element class under resistance conditions. In this model, the real behavior of thin-walled beams under load was taken into account much more precisely in relation to the Eurocode. This made it possible to use the resistance reserves resulting from the effect of longitudinal stress variation and the elastic restraint of the thin-walled cross-section walls.

Abstract: In modern steel construction, thin-walled elements with Class 4 cross-sections are commonly used. For the sake of the computation of such elements according to European Eurocode 3 (EC3), simplified computational models are applied. These models do not account for important parameters that affect the behavior of a structure susceptible to local stability loss. This study discussed the effect of local buckling on the design ultimate resistance of a continuous beam with a thin-walled Class 4 I-shaped cross-section. In the investigations, a more accurate computational model was employed. A new calculation model was proposed, based on the analysis of local buckling separately for the span segment and the support segment of the first span, which are characterized by different distributions of bending moments. Critical stress was determined using the critical plate method (CPM), taking into account the effect of the mutual elastic restraint of the cross-section walls. The stability analysis also accounted for the effect of longitudinal stress variation resulting from the varied distribution of bending moments along the continuous beam length. The results of the calculations were compared with the numerical simulations using the finite element method. The obtained results showed very good congruence. The phenomena mentioned above are not taken into consideration in the computational model provided in EC3. Based on the critical stress calculated as above, "local" critical moments were determined. These constitute a limit on the validity of the Vlasov theory of thin-walled bars. Design ultimate resistance of the I-shaped cross-section was determined from the plastic yield condition of the most compressed edge under the assumptions specified in the study. Detailed calculations were performed for I-sections welded from thin metal sheets, and for sections made from two cold-formed channels (2C). The impact of the following factors on the critical resistance and design ultimate resistance of the midspan and support cross-sections was analyzed: (1) longitudinal stress variation, (2) relative plate slenderness of the flange, and (3) span length of the continuous beam. The results were compared with the outcomes obtained for box sections with the same contour dimensions, and also with those produced acc. EC3. It was shown that compared with calculations acc. EC3, those performed in accordance with the CPM described much more accurately the behavior of the uniformly loaded continuous beam with a thin-walled section. This could lead to a more effective design of structures of this class.

Keywords: thin-walled I-section; continuous beam; local buckling; longitudinal stress variation; design ultimate resistance of the cross-section

1. Introduction

In modern metal construction, thin-walled lightweight components that are sensitive to local stability loss are increasingly being used. Local buckling of a thin-walled member causes lateral (in relation to the direction of compressive stress) displacement of the component walls of the cross-section. This reduces the load capacity of the critical cross-section, which reduces the resistance of the entire structural member. In the European Eurocode 3 (EC3) [1–3] standards, cross-sections that are subjected to local buckling (in the elastic range) are included in Class 4 cross-sections. The classification of a steel cross-section into Class 4 is based on a comparison of slenderness (b/t) of separated component walls of the section with the limit values for Class 3. In the event the limits of at least one of the component walls being exceeded, the cross-section is qualified as Class 4, which implies the need to include the effects of local buckling in the calculation model. However, an interest in thin-walled members is currently growing due to a significant reduction in the weight of structural steel cross-sections. For example, hot-formed bisymmetric sections (e.g., IPE, HEB), much lighter cold-formed open sections (monosymmetric or compound cross-sections, e.g., two-branch cross-sections), or bisymmetrical box sections can be used for supporting elements of light halls, floor beams, trusses, columns, or purlins. This approach reduces the weight of the structure. The use of steels with ever higher strength as well as modern automatic welding technologies also affects the development of thin-walled welded I-sections. Such Class 4 bisymmetrical and monosymmetrical welded cross-sections are characterized by low weight, with an analogous degree of reliability of the structural member in relation to hot-formed members.

From the point of view of the occurring instability phenomena, I-section beams may be subject to local buckling and lateral torsional buckling, including flexural–torsional loss of stability with the so-called forced rotation axis (e.g., wind suction purlin or from gravity loads in the supporting segment of continuous beams). This is due to the fact that thin-walled open cross-sections are characterized by low wall stiffness to local buckling and low rigidity of the whole cross-section to torsion [4,5]. However, a forced rotation axis at lateral–torsional buckling, e.g., with a suitably rigid cover, results in a significant increase in critical moment and, in many cases, the adverse effects of lateral–torsional buckling can be neglected.

Therefore, in the Class 4 thin-walled I-section beams, protected against lateral buckling, the influence of local stability loss remains, especially because thin-walled I-cross-sections are much less resistant to local buckling than, for example, box cross-sections [6]. This results from the fact that in such a cross-section there are cantilever walls, which are much less resistant to buckling in relation to internal walls [7,8] or cantilever walls with a stiffening bend [9]. However, the use of I-sections (in relation to box sections) allows for simpler shaping of connections and nodes due to free and double-sided access, e.g., to bolt joints. Moreover, continuous beam systems, due to the favourable longitudinal distribution of bending moments, enable the effect of longitudinal stress variation to be used in local stability analysis [10].

The resistance of a thin-walled cross-section is determined using the effective width method [3]. This method consists of determining the critical stresses of local buckling ($\sigma_{cr} = k\sigma_E$, where k is the plate buckling coefficient) for individual cross-section walls, assuming their hinged support and constant distribution of stresses along the member length. On this basis, the relative plate slenderness $(\overline{\lambda}_p = \sqrt{f_y/\sigma_{cr}}$, where f_y is the yield strength of steel) and effective widths of individual walls (plates) are determined $b_{eff} = \rho(\overline{\lambda}_p)b$, where ρ is the reduction factor.

Reference [6] presents an analysis of the so-called "local" critical resistance (i.e., determined from the condition of the local buckling) and the design ultimate resistance of a continuous beam with a

Class 4 thin-walled box section. It was shown that the resistance of a five-span beam can be determined from the supporting segment of the first span. Calculations were made using the critical plate method (CPM) [10]. The following were analyzed with respect to their effect on the "local" critical and design ultimate resistance of box cross-sections: (1) the relative plate slenderness of the flange, and (2) the span length of the continuous beam.

In the case of transversely bent beams made of thin-walled I-sections, which are protected against lateral–torsional buckling, the possibility of local buckling of the compression flange must be taken into account along with, for adequately high cross-sections, the possibility of local buckling of the bending and shear web [11].

This study dealt with the determination of the resistance of a continuous beam (e.g., purlins) with a thin-walled I-section (sheet-welded or composed of two cold-formed channel sections). The calculations accounted for (1) the effect of the elastic restraint of the weakest cross-section wall in the stiffening wall, and (2) the effect of longitudinal stress variation caused by the variability of bending moments.

The analysis was based on the fixed I-section along the whole length of the beam (Figure 1). In this case, the resistance of the continuous beam is determined by the end span. The case where the first (extreme) span is additionally reinforced, and the resistance is determined by the intermediate spans, will be the subject of a separate work.

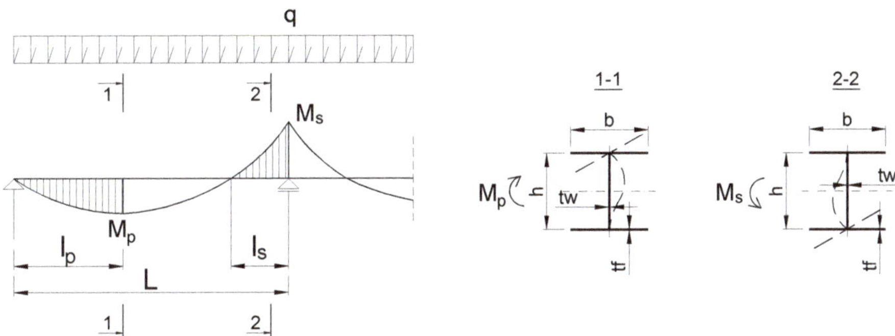

Figure 1. Static model of a continuous beam end span with thin-walled I-section.

2. Computational Model for Local Cross-Section Buckling

The Class 4 cross-section calculation model according to European standards EC3 [1–3] assumes that all component plates of athe cross-section are hinge-supported at their bonding edges. This means that the plates do not interact with each other, and the local buckling of the whole cross-section is determined by the weakest separated plate (cross-section wall).

However, in many technically important cases this model is too conservative, which has been confirmed by numerous experimental studies, e.g., References [5,12–14]; theoretical studies, e.g., References [4,7,15,16]; and numerical simulations, e.g., References [17–23]. This is due to the fact that in real thin-wall cross-sections, there is an effect of mutual, elastic interaction of component plates (walls).

Further, the standard computational model assumes that after determining the relative slenderness of the individual simply supported plates, their respective effective widths are determined, which determine the effective cross-section of the thin-walled member. In the calculation according to EC3, the effect of longitudinal stress variation, which often occurs in practice (e.g., in continuous beams), is also ignored.

In References [19,21], the effect of the mutual elastic restraint of thin-walled section walls was considered. Various load cases were taken into account, ranging from axial compression to pure bending relative to both axes of the section gravity, including interactive loads. Very large sets of simulation results were analyzed via the finite strip method using CUFSM software [18]. Calculations were made for many load cases and approximate analytical formulas were proposed for determining the plate buckling coefficient k [19] and the local buckling half-wavelengths [21] for various types of hot-rolled and welded sections. In Reference [22], in turn, approximation formulas of the k coefficient for hot-rolled sections under simple loads (axial compression, bending relative to the stronger and weaker axis of the cross-section) were given. However, the solutions obtained in References [19,21,22] relate only to constant stress distributions over the length of the structural member.

In References [7,24,25], the impact of longitudinal stress variation in the compressed cantilever plate (which is, for example, a component of an I-cross-section) was taken into account, but only for extreme support conditions (hinge or full restraint).

Meanwhile, in Reference [26], the approximation formulas of the plate buckling coefficient k for an axially compressed cantilever plate with any degree of elastic restraint of the supported edge and longitudinal stress distribution according to the linear function and 2° parabola were derived. At the same time, graphs of k coefficients were determined in References [27,28] for elastically restrained and eccentrically compressed cantilever plates with longitudinal stress variation. Many technically important load cases have been considered.

Reference [10] presents a more accurate computational model of the effect of local buckling on the resistance of thin-walled cross-sections. The critical plate method (CPM) takes into account the effect of both the elastic restraint of the cross-section plates and that of longitudinal stress variation.

3. "Local" Critical Resistance and Design Ultimate Resistance of the Cross-Section

In Reference [10], the bending moment inducing local buckling of a thin-walled cross-section was called the "local" critical moment (M_{cr}^L), which can be determined from the formula

$$M_{cr}^L = \sigma_{cr}^L W_{el,y} / \gamma_{M0}, \tag{1}$$

where σ_{cr}^L—critical buckling stress determined according to a more accurate computational model (CPM) and related to the extreme edge of the cross-section,

$W_{el,y} = I_y / z_c$ – elastic section modulus,

γ_{M0}—partial factor for cross-section resistance.

This moment is a limitation of the pre-buckling range of cross-section behavior and determines the limit of validity of Vlasov's thin-walled bar theory. On its basis, the critical resistance (critical load) of the structure can be determined from the condition of local stability loss of the weakest cross-section.

As regards the reliability of metal building structures, the designer's interest lies in the so-called design resistance [1], which can be determined from the formula

$$M_{eff} = W_{eff} f_y / \gamma_{M0}, \tag{2}$$

where $W_{eff} = I_{eff} / z_{eff}$—effective section modulus.

The "local" critical moment according to (1) can in practice be treated as the resistance of a thin-walled section in the pre-buckling range, assuming unlimited material elasticity. Therefore, where $W_{el}\sigma_{cr}^L > W_{eff} f_y$, the design resistance of a Class 4 cross-section must necessarily be determined using Equation (2).

The so-called design ultimate cross-section resistance defined in References [6,10] (taking into account a more accurate model of local buckling) is a lower (conservative) estimate of the failure load determined for the mechanism of plastic hinge [29]. The resistance of a thin-walled cross-section, which is obtained during the failure phase, cannot be applied in the design of building structures. However, it can be applied, for example, in designing so-called mechanical energy absorbers [30].

4. The Idea of the Critical Plate Method (CPM)

The idea of the critical plate method [10] consists of determining the buckling stress of the weakest component plate of the cross-section, taking into account its elastic restraint in adjacent plates (walls), for which the buckling stresses are higher. In addition, the method allows the effect of longitudinal stress variation to be taken into account. In this computational model, the so-called critical plate (CP) must be identified, which determines the local buckling of the section in a given loading condition.

The elastic restraint against longitudinal edge rotation of the CP results from the action of the adjacent restraining plate (RP). For example, for certain proportions of the bending I-section [10], the critical plate may be a compressed cantilever wall of the width b_s as a outstanding part of the flange width $b_f = 2b_s$, and the restraining plate will be a web of height h_w. The buckling stress for the CP so determined is higher than that determined under the assumption of its simple support. This is especially true for cantilever plates, for which the elastic restraint against rotation of the longitudinal supported edge significantly affects the buckling stress values. The difference in plate buckling coefficients for an uniform compressed cantilever plate between a hinged support ($k_\sigma = 0.43$) [3] and a fully restrained support ($k_\sigma = 1.25$) [7,19] is almost threefold.

In References [19,21], attention was paid to the additional phenomenon of the so-called mutual constraint of the component walls of the cross-section (affecting the increase of buckling stress), despite the fact that the theoretical critical stresses of individual walls (determined under the assumption of their hinged support) are equal. This was found to be due to the different buckling lengths of the internal and cantilever plates. Therefore, in Reference [21], approximation formulas for the local buckling half-wavelengths of the full cross-section were derived. This length is between the buckling length for a separate, simply supported, and fully restrained plate, on which the critical resistance of the full cross-section depends. Knowledge of local buckling half-wavelengths is useful for the direct definition of geometric imperfections in analytical and numerical models.

On the other hand, in Reference [10], it was assumed that in a thin-walled section where thought-separated hinge-supported walls theoretically have the same buckling stresses, the effect of mutual constraint is insignificant (i.e., increases the buckling stress from 5 to 20%) and can be conservatively ignored. On this basis, so-called "zero" cross-sections were defined, in which the thought-separated hinge-supported plates have the same buckling stresses, and the resistance of such cross-sections can be calculated according to the standard procedure [1,3].

The distinction between "zero" and "non-zero" cross-sections significantly simplifies the identification of the critical plate in the latter and allows for a slightly conservative (safe) estimation of the buckling stress. Therefore, the elastic restraint coefficient of the CP in RP can be estimated using Equation (3) [10]

$$\kappa = 1/(1 + 2D_s/b_s C_\theta), \tag{3}$$

where $C_\theta$rotational spring stiffness equal to the bending moment created during rotation by unit angle ($C_\theta = M/\theta$), b_s—CP width, D_s—plate flexural rigidity according to the formula

$$D_s = \frac{Et_s^3}{12(1-\nu^2)} \tag{4}$$

where for $E = 210$ GPa and $\nu = 0.3$, it can be approximately assumed that $D_s = 19,200t_s^3$, t_s—CP thickness.

The restraint coefficient according to Equation (3) varies from $\kappa = 0$ for simple support to $\kappa = 1$ for full restraint.

When the above-mentioned effects are taken into consideration, it is possible to determine the critical stress causing the cross-section local buckling in a more accurate way. Based on the critical stress, the following can be determined: (1) the "local" critical resistance of the cross-section M_{cr}^L, (2) the effective width of the critical plate, and (3) the design ultimate resistance of the cross-section M_{eff}^{CP} according to Reference [10].

5. Algorithm of the CPM Method for Thin-Walled I-Cross-Sections under Bending

For the transverse bending of I-beams of a Class 4 cross-section (Figure 2), their resistance is usually determined by local buckling of the compression flange (of width $b_f \cong 2b_s$ and thickness t_f) or local buckling of the web (of height h_w and thickness t_w) under bending and shear interaction. Because of their geometric topology, those cross-sections were categorised in Reference [10] as part of the group of so-called "simple" cross-sections.

The ultimate height h_0 of the cross-sections at which the compression flange buckling and web in-plane bending with a slight impact of shear stresses ($\tau \leq 0.25\tau_{cr}$) occur almost simultaneously can be determined using Equation (5) [10]

$$h_0 = \sqrt{\frac{k_w^0}{k_f^0} \cdot \left(\frac{t_w b_s}{t_f}\right)} \tag{5}$$

where k_i^0—basic buckling coefficient for the hinged supported i-th plate at given load distribution ($k_f^0 = 0.43$ for axial cantilever plate compression, $k_w^0 = 23.9$ for bending in the plane of the web plate).

For $h < h_0$, it is the compression flange that determines local buckling of the cross-section, whereas for $h > h_0$, the weakest wall is the web under in-plane bending.

For the mean proportion of shear stresses ($0.25\tau_{cr} \leq \tau \leq 0.5\tau_{cr}$) Equaiton (5) can be modified to the form:

$$h_0 = \sqrt{\frac{k_w^0}{k_f^0} \cdot \left(\frac{t_w b_s}{t_f}\right)} \left[1 - \left(\frac{\tau}{\tau_{cr}}\right)^2\right]^{0.25} \tag{6}$$

where: τ_{cr}—critical stress for the web plate in shear.

However, if there is a large proportion of transverse forces in the slender webs (plates supported at four edges) that are part of beam cross-sections, shear stresses τ may need to be considered more closely, as their contribution to local loss of web stability may be significant. Appropriate formulas for elastic strain energy and work done by external forces allowing critical stresses to be determined (using the energy method) for internal plates loaded (within their plane) with bending with high shear were derived in Reference [11].

For thin-walled I-sections, for which $h < h_0$ according to Equation (6), the CPM assumptions are as follows: (1) the compression flange of the cross-section consists of two critical plates (CPs) which width of $b_s \leq b_f/2$ each, which are supported on a web plate of the height h_w, (note: dimensions b_s and h_w can be determined based on the rules given in Reference [10]); (2) a single CP acts as a cantilever plate, with one side elastically restrained against rotation; (3) the CP to RP connection (i.e., the web) is rigid (i.e., on the longitudinal edge of their connection, the conditions of continuity of displacements (rotation angles) and forces (bending moments), are met); (4) the transverse edges of the plates (CP and RP) are simply supported on the segment ends; and (5) the thin-walled bar segment (with the length l_s), as in Reference [6], is defined as follows: (a) for constant longitudinal stress distribution, as the distance between the so-called buckling nodal lines, (b) for longitudinal stress variation, as the distance between transverse stiffeners (diaphragms, ribs, or supports) that maintain a rigid cross-section contour, but not longer than the range of the compression zone in the critical plate [26]. The conditions under which Assumption 3 can be adopted were discussed in Reference [10].

The case of I-sections with high and slender webs (at $h > h_0$ acc. to Equation (6)), where the "local" critical resistance is determined by the buckling of the bent and shear web, will be discussed in a separate study.

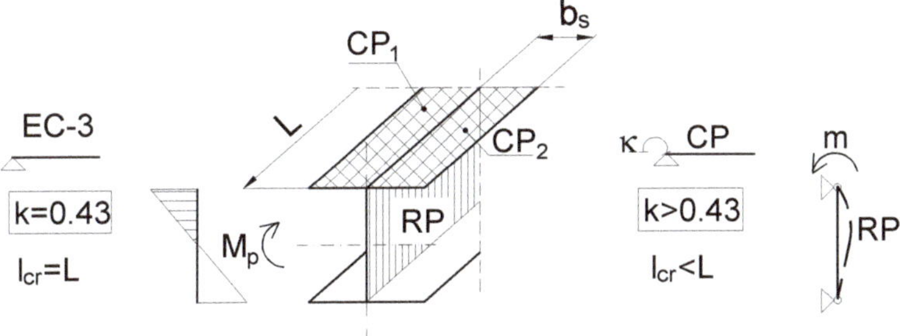

Figure 2. Division of I-cross-section into critical plate (CP) and restraining plate (RP) for $h < h_0$.

The calculation algorithm is as follows:

(1) Division of the cross-section into component plates according to Figure 2;

(2) Identifying CP for a "simple" cross-section based on the condition

$$\sigma_{cr}^{0} = min\{\sigma_{cr,i}^{0}\} \tag{7}$$

$$\sigma_{cr,i}^{0} = k_i^0 \sigma_{E,i} \tag{8}$$

where $\sigma_{E,i}$—Euler stress for the i-th plate according to the formula: $\sigma_{E,i} = 190,000(t_i/b_i)^2$. Note: for $h < h_0$ acc. to Equation (6), the critical plate (CP) is the cantilever wall of the compression flange and the rigid plate (RP) is the web;

(3) Adoption of the initial CP edge restraint coefficient value (for the so-called zero step), e.g., $\kappa_0 = 0.4$;

(4) Estimation of the critical length (l_{cr}) for a single half-wave of CP buckling [26] according to the formula

$$l_{cr} = b_s\left(\frac{2.02 - 0.37\kappa}{\kappa^{0.25}}\right) \tag{9}$$

(5) Determination of the coefficient η [31], depending on the static scheme and the way of forcing (loading) RP by buckling CP (Figure 2) for the critical length according to Equation (9) [10]

$$\eta = \sqrt{33.4 + 50.7(b_r/l_{cr})^2} - 2.78 \tag{10}$$

where b_r—RP width;

(6) Determination of σ_{cr}^{L} according to the formula

$$\sigma_{cr}^{L} = k^* \sigma_{E,s} \tag{11}$$

where k^*—buckling coefficient according to a more accurate calculation model. The k^* coefficient can be determined on the basis of Reference [26] from the following formulas:

- for linear stress distribution (Figure 3a),

$$\begin{gathered} k^*(\kappa, m, \gamma) = k_\infty(\kappa) + \big[0.765m - 0.31m^2 + 0.227m^3 + \\ \left(3.201m - 0.307m^2 - 3.724m^3 + 2.842m^4\right)\kappa^2 + \\ \left(-3.887m - 9.205m^2 + 21.528m^3 - 12.306m^4\right)\kappa^3 + \\ \left(-1.132m + 22.933m^2 - 37.767m^3 + 19.091m^4\right)\kappa^4 + \\ \left(2.559m - 14.341m^2 + 20.975m^3 - 10.023m^4\right)\kappa^5\big] / \gamma_s^{(0.59+0.16m)} \end{gathered} \tag{12}$$

- for non-linear stress distribution (according to 2° parabola, Figure 3b),

$$\begin{gathered} k^{*}(\kappa,m,\gamma)=k_{\infty}(\kappa)+\left[1.096m-0.808m^{2}+0.363m^{3}+\right.\\ \left(20.064m-63.78m^{2}+74.842m^{3}-30.565m^{4}\right)\kappa^{2}+\\ \left(-58.826m+195.88m^{2}-232.857m^{3}+95.684m^{4}\right)\kappa^{3}+\\ \left(61.802m-211.851m^{2}+254.031m^{3}-104.82m^{4}\right)\kappa^{4}+\\ \left.\left(-22.086m+77.761m^{2}-93.977m^{3}+38.925m^{4}\right)\kappa^{5}\right]/\gamma_{s}^{(1.06+0.04m)} \end{gathered} \tag{13}$$

where $m = 1 - \sigma_1/\sigma_0$—longitudinal stress distribution coefficient (Figure 3), $\gamma_s = l_s/b_s$. The plate buckling coefficient $k_\infty(\kappa)$ of an elastically restrained and infinitely long cantilever plate can be determined according to Reference [26], using the formula

$$\begin{gathered} k_{\infty}(\kappa)=0.425+2.893\kappa-19.433\kappa^{2}+83.849\kappa^{3}-195.943\kappa^{4}+250.971\kappa^{5}\\ -165.321\kappa^{6}+43.833\kappa^{7} \end{gathered} \tag{14}$$

On the other hand, Reference [28] presents a simplified formula for $k_\infty(\kappa)$ within the range $0.05 \le \kappa \le 1$ in the form of

$$k_{\infty}(\kappa)=0.49+0.974\kappa-0.822\kappa^{2}+0.632\kappa^{3} \tag{15}$$

(7) Estimation of the critical stress $\sigma_{cr,r}$ for a RP bent in its plane (width b_r) [32] for one half-wave of CP buckling length l_{cr} according to Point 4.

$$\sigma_{cr,r}=\frac{Et_{r}^{2}}{l_{cr}^{2}b_{r}^{4}}\left(11.32l_{cr}^{4}+1.97b_{r}^{4}+12.06l_{cr}^{2}b_{r}^{2}\right) \tag{16}$$

(8) Determination of rotational spring stiffness C_θ according to Equation (17) and the restraint coefficient κ_{i+1} according to Equation (3) for the first (i = 1) and subsequent iteration steps.

$$C_{\theta}=\frac{c_{j}\eta_{j}D_{r}}{b_{r}}\left(1-\frac{\sigma_{cr}^{L}}{\sigma_{cr,r}}\right) \tag{17}$$

where c_j—parameter of geometrical configuration of plates in contact with the j-th edge (for a welded I-section $c_j = 1/2$ [10]; note: for an I-section made up of two channel sections $c_j = 1$, i.e., one web with thickness t_w stabilizes one CP of the flange on one edge), $D_r = 19,200t_r^3$RP flexural rigidity, t_r—RP thickness;

(9) Repetition of Steps (4) to (8) up to the moment when $\kappa_i \approx \kappa_{i+1}$;

(10) $\sigma_{cr}^L(\kappa_{i+1})$ according to Equation (11) is the sought buckling stress for CP;

(11) Determination of the "local" critical resistance of the cross-section according to Equation (1);

(12) Determination of the design ultimate resistance of the cross-section according to Equation (2) for the W_{eff} coefficient calculated with the following assumptions [10]:

    (a) the slenderness of the critical plate (CP) shall be determined from the buckling stress determined by Step 10 (i.e., taking into account the elastic restraint effect of the cantilever plate and the effect of longitudinal stress variation);
    (b) for the web (rigid plate RP), a simple support shall be provided at the same edge;
    (c) the boundary conditions at the other edge of the RP have a slight impact on the result of the calculations (conservatively, simple support can also be assumed here),
    (d) the effect of possible longitudinal stress variation in RP is insignificant and can be ignored,

(e) the effective widths so determined "shall be folded" into an effective cross-section, for which W_{eff} is determined.

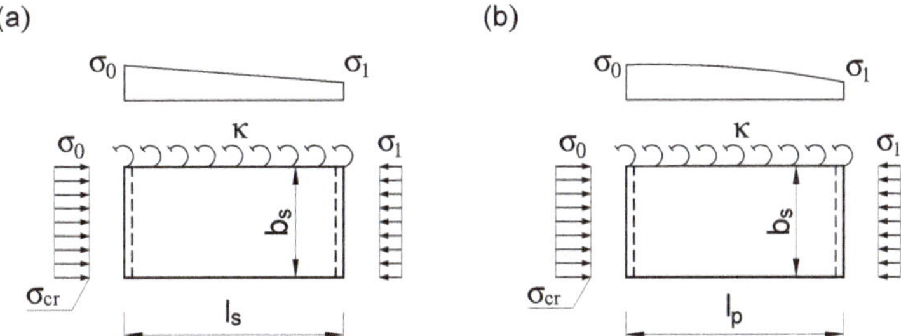

Figure 3. Longitudinal stress distribution in a cantilever plate: (**a**) linear, (**b**) non-linear according to 2° parabola.

The algorithm proposed above, compared to the classical version of the effective width method, allows for a more accurate consideration of the behavior of a thin-walled section, the resistance of which is determined by the local buckling phenomenon. A more detailed comparative analysis on this issue is presented in Reference [10].

6. Calculation Method for a Continuous I-Beam

In the case of a multi-span continuous uniformly loaded beam (Figure 4), a non-linear bending moment distribution M_y occurs in the first (end) span with maximum values (M_p and M_s). For the range ($l_p + c$), the graph M_y is convex, while for the range l_s the graph is concave. Such longitudinal distributions of M_y cause non-linear (along the length of the beam) normal stress distributions σ_x, which may cause local loss of stability [6].

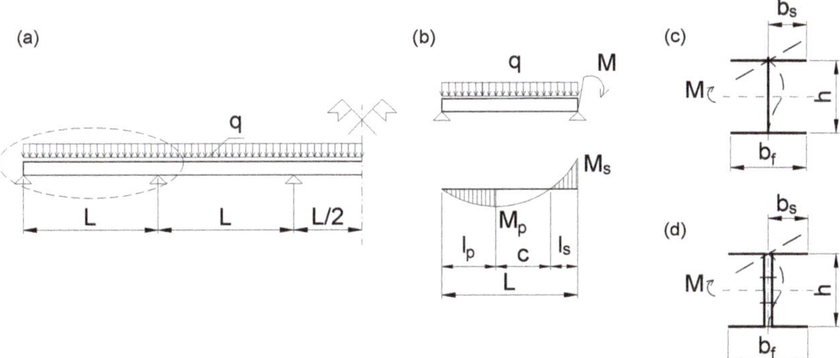

Figure 4. Static scheme of a beam: (**a**) load distribution, (**b**) moment diagram for the first (extreme) span, (**c**) welded I-section, (**d**) 2C compound I-section.

Reference [33] showed that in the case of non-linear stress distribution for the full span range $l_k = l_p + c$ (Figure 5), the k coefficient takes basically the same values as those determined for the reduced range l_p (e.g., for $l_k/b_s \geq 4$ differences not exceeding 2%). However, in Reference [26], it was shown that if the stress mark changes, the design segment length can be limited to the range of the compression zone. This is applicable to the support zone l_s of a continuous beam. Therefore, similarly to Reference [6], in order to take into account the effect of longitudinal stress variation when determining the span cross-section resistance, it is sufficient to assume the segment length to be l_p, and for the support cross-section to be a segment with the length of l_s.

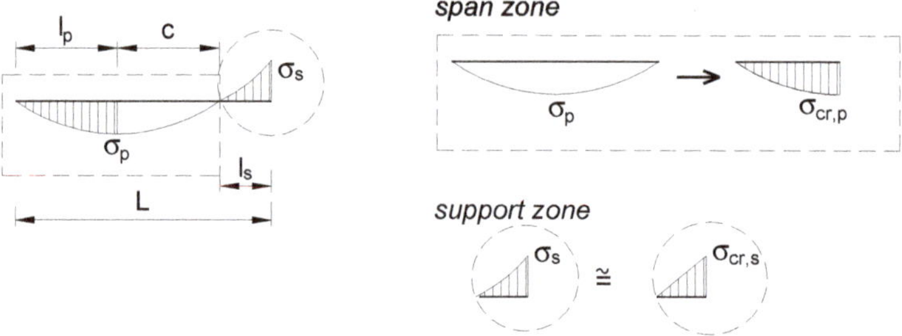

Figure 5. Calculation model of the first (end) span of a uniformly loaded continuous beam.

For the span segment, there is a non-linear (according to 2° parabola) stress distribution. For the support segment, however, the concave stress distribution can be conservatively approximated by the linear distribution.

7. Example of Using CPM to Determine the Resistance of a Continuous I-Beam

7.1. Statics and Cross-Sectional Forces

The calculations were carried out, among others, for five-span continuous beams (e.g., purlins) (cf. Figure 4) with a constant span length L and thin-walled I-sections either (a) welded from thin sheets (I; Figure 4b,c) or (b) composed of two cold-formed channel sections (type 2C; Figure 4d). It was assumed that the above-mentioned sections after welding (I) or cold-forming (2C) were subjected to appropriate heat treatment to reduce the impact of residual stresses. In References [23,34], the influence of remaining welding stress on the behavior and resistance of welded I-sections was investigated. It was shown that this effect is greatest in the case of axial compression, but in some cases, bending elements should also be considered. Therefore, the effect of residual stresses on the local buckling of continuous beams with welded I-sections will be described in a separate study.

The beams were under continuous, uniformly distributed load. It was assumed that the elements were structurally protected against lateral–torsional buckling. In this case, the resistance of the beam depends on the resistance of the most stressed cross-section.

The relationship between M_s and M_p (Figure 4) in the first span was $u = M_s/M_p = 1.351$, irrespective of the type of cross-section and the span length. In the support cross-sections of the beam, two-sided stiffening ribs were used to transfer the support reactions without causing distortion of the thin-walled cross-section.

7.2. Welded Section

7.2.1. Calculation of Critical Resistance According to CPM

Detailed calculations were carried out for a five-span continuous beam with a span length $L = 4\ m$ and a thin-walled I-section welded from sheets of dimensions I-300×t_w×250×t_f (I-h×t_w×b×t_f) for t = t_f = t_w = 5, 6, 7, and 8 mm made of S355-grade steel. It should be noted that in this case, adopting a smaller web thickness, i.e., $t_w < t_f$, led to a significant reduction in the elastic restraint of the compressed flange in the plate of the web. This resulted in a much greater reduction in the cross-sectional resistance than that resulting from a reduction in the elastic section modulus caused solely by a reduction in the thickness of t_w.

Table 1 shows the results of the calculation of "local" critical moments for the ruling cross-sections: span $M^L_{cr,p}$, support $M^L_{cr,s}$, and the critical loads q_{cr} determined from them. The symbol $q^{CP}_{cr,s}$ indicates the critical load value determined from the condition of reaching $M^L_{cr,s}$ in the support cross-section, and the symbol $q^{CP}_{cr,p}$ indicates that determined from the condition of reaching $M^L_{cr,p}$ in the span cross-section. (Note: in Table 1, the values of the moments $M^L_{cr,s}$ and $M^L_{cr,p}$ are given as absolute values). The critical resistance of the beam is determined by the minimum load $q^{CP}_{cr,min} = min\{q^{CP}_{cr,p}, q^{CP}_{cr,s}\}$.

Table 1. Summary of critical results according to critical plate method (CPM) and European Eurocode 3 (EC3) for the I-300×t×250×t section and four wall thickness variants.

Wall Thickness (mm)	t=	8	7	6	5
Slenderness	λ=b/t=	15.63	17.86	20.83	25.00
Euler stress (N/mm^2)	Flange $\sigma_{E,i}$=	778.24	595.84	437.76	304.00
	Web $\sigma_{E,i}$=	135.11	103.44	76.00	52.78
Buckling coefficient	for l_p: k*=	0.800	0.800	0.800	0.800
	for l_s: k*=	0.950	0.950	0.950	0.950
Critical stress acc. CPM (N/mm^2)	for l_p: $\sigma_{cr,p}$=	622.60	476.67	350.21	243.20
	for l_s: $\sigma_{cr,s}$=	739.08	565.86	415.73	288.70
Critical stress acc. EC3 (N/mm^2)	$\sigma_{cr,0}$=	334.64	256.21	188.24	130.72
Local critical resistance (kNm)	for l_p: $M^L_{cr,p}$=	448.36	300.35	189.13	109.45
	for l_s: $M^L_{cr,s}$=	532.24	356.54	224.52	129.93
	M^{EC3}_{cr} =	240.99	161.44	101.66	58.83
Critical load acc. CPM (kN/m)	for l_p: $q^{CP}_{cr,p}$=	359.72	240.97	151.74	87.81
	for l_s: $q^{CP}_{cr,s}$=	316.02	211.70	133.31	77.14
Critical load acc. EC3 (kN/m)	q^{EC3}_{cr} =	143.09	95.85	60.36	34.93
Percentage increase of resistance (%)	$q^{CP}_{cr} / q^{EC3}_{cr}$ =	120.86	120.86	120.86	120.86

Furthermore, in Table 1, the symbol M^{EC3}_{cr} indicates the critical cross-sectional moment determined according to σ_{cr} calculated according to EC3 [3], and the symbol $q^{EC3}_{cr,min}$ stands for the minimum value of critical load determined on the basis of M^{EC3}_{cr}. Due to the simplified calculation model used in the standard, M^{EC3}_{cr} does not depend on the longitudinal stress distribution or on the degree of mutual elastic restraint of the cross-section component walls. It therefore determines the critical resistance of both the span cross-section (p) and the support cross-section (s).

Note: in the example under consideration, determined on the basis of the $q^{CP}_{cr,min}$ parameter h_0 according to Equation (6), $h_0 = 880$ mm and was almost three times greater than the cross-section height $h = 300$ mm. Of course, the determination of the parameter h_0 according to Equation (6) was iterative because the shear stresses τ and τ_{cr} depend on the cross-section height, among other things. This check calculation allows the correct identification of the critical plate in a given cross-section loading condition to be confirmed.

A comparison of the results in Table 1 shows that $M^L_{cr,s}$ was approx. 18.7% higher than $M^L_{cr,p}$. This was due to different stress distributions and different reliable lengths of the span (l_p) and support (l_s) segments. Despite the fact that $M^L_{cr,p} < M^L_{cr,s}$, the critical resistance of the beam was determined by the support zone, because: $u = 1.351 > M^L_{cr,s}/M^L_{cr,p} = 1.187$. The percentage increase of $M^L_{cr,s}$ in relation to M^{EC3}_{cr} was about 121%. The same relationship occurred for the corresponding critical loads q^{min}_{cr}.

7.2.2. Finite Element and Finite Strip Analysis

The critical stresses listed in Table 1 ($\sigma_{cr,s}$, $\sigma_{cr,p}$) were verified via the finite element method (FEM) using ABAQUS software [35], and the finite strip method (FSM) using CUFSM software [18].

The FEM enables the analysis of local buckling of the extreme span of a continuous beam by taking into account both the effect of mutual elastic restraint of cross-section's components plates, as well as the effect of longitudinal stress variation. However, the commonly used finite strip method (FSM) allows the effect of mutual elastic restraint of cross-section's component plates to be considered, but with the assumption of a constant stress distribution over the length of the member.

For the numerical simulations in ABAQUS v. 6.12, the first (extreme) span of a continuous beam was modeled. Due to the small wall thickness (compared to other dimensions), the cross-section's geometry was simplified to so-called centerline. S4R shell elements (four nodes with six degrees of freedom in the node) were used, and the dimensions of the finite element were assumed to be 12.5 × 12.5 mm. Such division made it possible to obtain technically sufficient accuracy while reducing calculation time. The transverse load was applied in the form of uniform pressure ($p = 0.1$ N/mm^2) on the upper flange with a width of $b = 250$ mm, which corresponded to a uniform load of $q = 25$ kN/m. Conversely, the loading with the "corresponding" support moment of the value of $M_s = 0.105 \times qL^2 =$ 42 kNm (where 0.105 is the coefficient from Winkler's tables) was performed by means of loading the support cross-section with a stress block. The boundary conditions in the support cross-sections were assumed to be continuous constraints perpendicular to each of the component plate. An additional tie securing the beam against displacement along the longitudinal axis was applied in the gravity axis of one of the support sections. In Figure 6b, a general view of FEM model of the I-300×5×250×5 cross-section in ABAQUS software is presented. Figure 6 shows the support ties used, and Figure 6c shows the method of loading with the support moment M_s. The critical load calculations were performed using the "buckling" procedure.

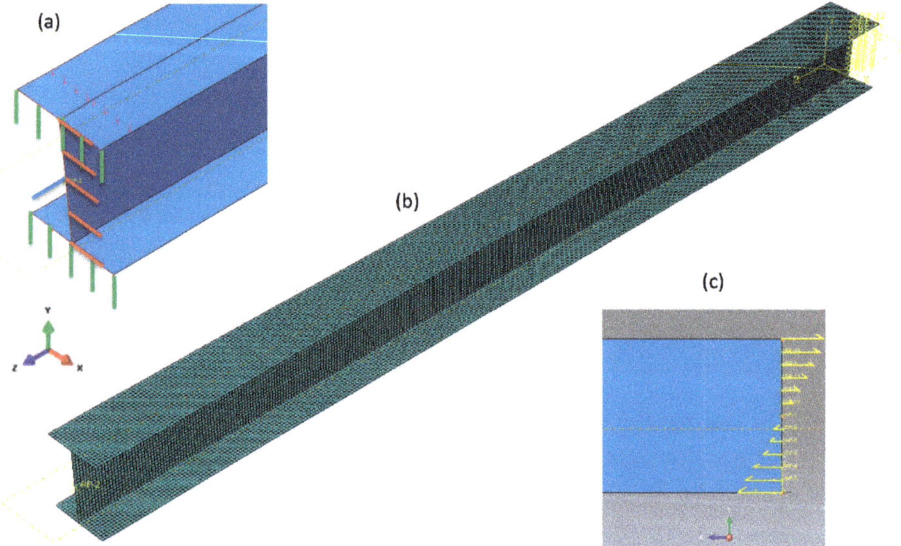

Figure 6. Finite element method (FEM) model of the I-300×5×250×5 beam: (**a**) boundary conditions, (**b**) division into finite elements, (**c**) method of loading with the support moment.

Figure 7 shows the local buckling mode (axonometric view and side view) corresponding to the first (smallest) eigenvalue of the critical load multiplier ($q_{cr,s} = 3.121 \times q = 78.02$ kN/m). Maximum local deflections of the cross-section occurred in the support zone.

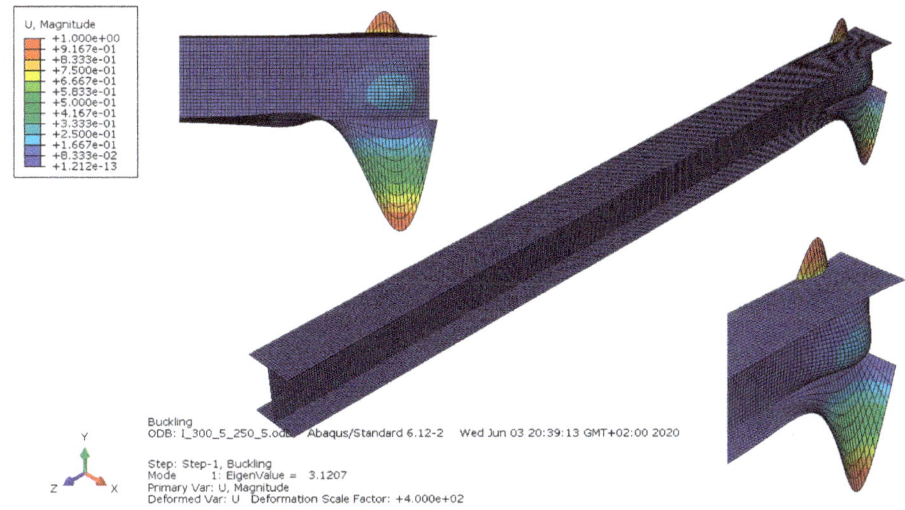

Figure 7. Local buckling of I-300×5×250×5 beams in the support zone.

Figure 8 shows the local buckling mode (axonometric view and side view) corresponding to the third (second positive) eigenvalue of the critical load multiplier ($q_{cr,p} = 3.568 \times q = 89.21$ kN/m). In this case, the maximum local deflections of the cross-section occurred in the span zone.

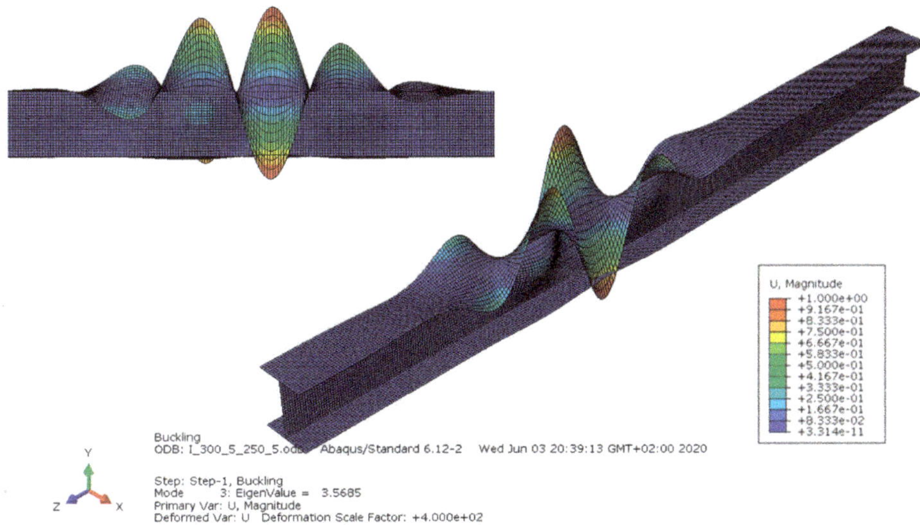

Figure 8. Local buckling of I-300×5×250×5 beams in the span zone.

For numerical simulations in CUFSM v.5.04, the I-section was also simplified to centerline geometry, and then divided into 48 finite strips (16 per flange and 16 in the web). The load was applied in the form of a stress block, causing bending about the stronger axis of the cross-section.

7.2.3. Comparison of Critical Stress Results

Table 2 compares the values of critical stresses for supporting $\sigma_{cr,s}$ and span $\sigma_{cr,p}$ sections, determined according to: (1) ABAQUS, (2) CPM, (3) CUFSM, (4) Reference [19], (5) Reference [22], and (6) EC3. It should be emphasised that comparison of stresses determined according to Reference [22] was for reference only, because the formulas derived in Reference [22] apply only to hot-rolled sections, and not to the welded ones.

Table 2. Comparison of critical stresses determined by several methods for the I-300×t×250×t section and four wall thickness variants.

t	s/p	ABAQUS	CPM	CUFSM	[19]	[22]	EC3
1		2	3	4	5	6	7
5	s	291.98	288.70	235.53	223.98	191.64	130.72
	p	247.16	243.20				
6	s	418.32	415.73	339.16	322.53	275.96	188.24
	p	355.00	350.21				
7	s	566.31	565.86	461.62	439.00	375.59	256.21
	p	481.86	476.67				
8	s	735.48	739.08	602.92	573.38	490.58	334.64
	p	627.53	622.60				

From the comparison of the values given in Table 2, it follows that the new calculation model adopted in this work using the critical plate method (CPM) gave very good congruence with the results obtained from the ABAQUS software, which were adopted as a reference. This applied to both the support section (first eigenvalue, compliance approx. 0.99) and the span section (second eigenvalue, congruence approx. 0.985). The calculations obtained from CUFSM also gave very good results in the span section (congruence approx. 0.96), while in the support section, where a large stress gradient occurred, and on which the critical resistance of the beam depends, congruence was around 0.81. A similar situation occurred when comparing the results obtained from the formulas according to Reference [19] (span section about 0.91, support section about 0.77). However, a worse congruence of the results obtained according to Reference [22] (span section about 0.78, support section about 0.66) resulted from the different, in relation to hot-rolled sections, geometrical proportions of the sections considered in this study. The largest differences (span section about 0.53, support section about 0.45) were obtained from calculations made according to EC3. The local buckling half-wavelengths determined in this case were: (1) according to CPM, Equation (9) l_{cr} = 307 mm; (2) according to CUFSM, l_{cr}= 340 mm; and (3) according to Reference [21] l_{cr} = 353 mm. It should be noted that in this case, the following rule applies: for higher critical stresses, there is a shorter local buckling half-wavelength.

The ompared computational methods showed that very good congruence of results in relation to finite element analysis (ABAQUS) was obtained by the critical plate method (CPM), which takes into account both effects. Methods that consider the effect of mutual elastic restraint of cross-section component plates give a good approximation of critical stresses in those beam zones where there is no significant variation of stresses in the longitudinal direction (e.g., span cross-section). In contrast, calculations according to EC3 are the most conservative, because they omit both effects.

7.2.4. Design Ultimate Resistance

Table 3 shows the design values of the ultimate resistance for the reliable cross-sections: span $M^{CP}_{eff,p}$, support $M^{CP}_{eff,s}$, and the design ultimate loads (q_{eff}) determined from them according to CPM. The symbol $q^{CP}_{eff,s}$ indicates the ultimate load value determined from the condition of reaching $M^{CP}_{eff,s}$ in the support cross-section, and the symbol $q^{CP}_{eff,p}$ from the conditions needed to achieve $M^{CP}_{eff,p}$ in the span cross-section. (Note: in Table 3, the values of the moments $M^{CP}_{eff,s}$ and $M^{CP}_{eff,p}$ are given as absolute values). The ultimate resistance of the beam was determined by the minimum load $q^{CP}_{eff,min} = min\left\{q^{CP}_{eff,p}, q^{CP}_{eff,s}\right\}$.

Table 3. Summary of ultimate resistance results according to CPM and EC3 for the I-300×t×250×t section and four wall thickness variants.

Wall Thickness (mm)	t=	8	7	6	5
Slenderness	λ=b/t=	15.63	17.86	20.83	25.00
Design ultimate resistance (kNm)	for l_p: M^{CP}_{eff}=	244.81	201.32	159.99	121.57
	for l_s: M^{CP}_{eff}=	245.78	210.24	167.37	127.27
	$M^{EC3}_{c,Rd}$=	208.72	170.89	135.56	103.16
Ultimate load acc. CPM (kN/m)	for l_p: $q^{CP}_{eff,p}$=	196.41	161.52	128.36	97.54
	for l_s: $q^{CP}_{eff,s}$=	145.93	124.83	99.38	75.57
Ultimate load acc. EC3 (kN/m)	for l_s: q^{EC3}_{eff} =	123.93	101.47	80.49	61.25
Percentage increase of resistance (%)	$q^{CP}_{eff}/q^{EC3}_{eff}$ =	17.76	23.02	23.46	23.37

Furthermore, in Table 3, the symbol $M^{EC3}_{c,Rd}$ indicates the design cross-section resistance determined according to EC3 [1,3], and the symbol $q^{EC3}_{eff,min}$ indicates the minimum value of ultimate load determined on the basis of $M^{EC3}_{c,Rd}$.

Comparison of the results displayed in Table 3 showed that for a wall thickness of t = 5, 6, and 7 mm, the resistance $M^{CP}_{eff,s}$ was approx. 4.5% higher than $M^{CP}_{eff,p}$. This was due to different stress distributions and different lengths of the span (l_p) and support (l_s) segment. Despite the fact that $M^{CP}_{eff,p} < M^{CP}_{eff,s}$, the design ultimate resistance of the beam is determined by the support zone, because $u = 1.351 > M^{CP}_{eff,s}/M^{CP}_{eff,p} = 1.045$. However, for wall thickness t = 8 mm, the difference between $M^{CP}_{eff,s}$ and $M^{CP}_{eff,p}$ was only 0.4%. This resulted from the fact that the cross-section loaded in this way, calculated according to CPM, can be classified as a Class 3 cross-section, for which the value of the reduction factor ρ is close to one. In contrast, the same cross-section calculated according to EC3 (due to low $\sigma_{cr,0}$) is still treated as a Class 4cross-section (in this case, $\rho = 0.79$). The percentage increase of $M^{CP}_{eff,s}$ in relation to $M^{EC3}_{c,Rd}$ was about 23% for t = 5, 6, and 7 mm, and about 18% for t = 8 mm. The same relationship occurred for the corresponding ultimate loads q^{min}_{eff}.

Figure 9 shows the effective cross-sections of the I-300×5×250×5 section determined in the span and support cross-sections according to CPM (Figure 9a,b) and EC3 (Figure 9c,d). For example, the sum of the effective widths $2b_{eff}$ of the compressed bottom flange of the support cross-section (which determines the resistance of the beam) determined according to CPM was about 39% larger than that determined according to EC3 (see Figure 9b,d). The shift of the neutral axis of the effective cross-section according to CPM in relation to the gross cross-section was e = 12.2 mm and was 51% smaller than e = 24.9 mm calculated according to EC3. This effect additionally affected the differentiation of the effective modulus W_{eff} of the reliable cross-sections and, consequently, the differences in the design ultimate resistance of the beam determined according to CPM and EC3.

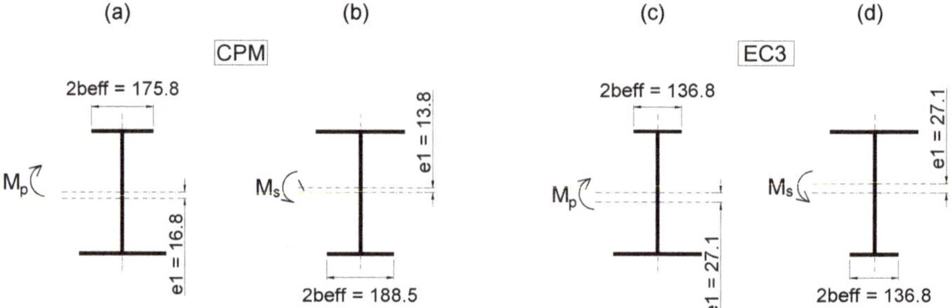

Figure 9. Effective cross-sections of the I-300×5×250×5 section according to CPM and EC3 methods. (**a**) span cross-section acc. CPM, (**b**) support cross-section acc. CPM, (**c**) span cross-section acc. EC3, (**d**) support cross-section acc. EC3.

7.3. I-Section Consisting of Two Cold-Formed Channel Sections (2C)

For compound sections (type 2C), detailed calculations were made for five-span beams with a constant span length for variants of L = 4, 5, 6, and 7 m made of S355-grade steel. For comparison purposes, the analyzed dimensions of 2C sections (Figure 10) with the contour of Sk250×250×t box sections (for t = 2, 3, 4, and 5 mm) presented in Reference [6] were used. In this case, the elastic gross section bending moduli W_{el} for Sk and 2C sections were equal.

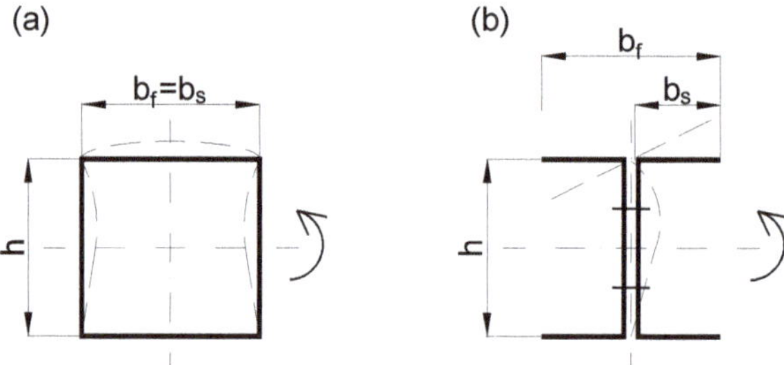

Figure 10. Compared cross-sections of continuous beams: (**a**) box cross-section according to Reference [6], (**b**) compound cross-section type 2C.

Table 4 presents the results of the analysis of the effect of the span length on the "local" critical moment value M^L_{cr} and the resistance M_{eff} calculated according to CPM and EC3 [1,3] for the 2C-250×250×4 (2C-h×b×t) section. For the support segment decisive of the beam resistance, the difference between the critical load value q_{cr} obtained according to CPM method in relation to EC3 ranged from + 158% for $L = 4\ m$ up to + 141% for $L = 7\ m$. The difference between the load ultimate value q_{eff} ranged from + 22% for $L = 4\ m$ up to +20% for $L = 7\ m$.

Table 4. Summary of results according to CPM and EC3 for the 2C-250×250×4 section and four span-length range variants.

Span Length (m)	$L=$	4	5	6	7
Slenderness	$\lambda=b/t=$		30.25		
Euler stress (N/mm^2)	flange $\sigma_{E,i}=$		207.64		
	web $\sigma_{E,i}=$		50.23		
Length of ranges (m)	$l_p=$	1600	2000	2400	2800
	$l_s=$	843	1050	1263	1474
Buckling coefficient	for l_p: k*=	0.94	0.93	0.92	0.92
	for l_s: k*=	1.11	1.08	1.05	1.04
Critical stress acc. to CPM (N/mm^2)	for l_p: $\sigma_{cr,s}=$	195.42	193.34	191.98	191.03
	for l_s: $\sigma_{cr,s}=$	230.47	223.70	218.80	215.20
Critical stress acc. to EC3 (N/mm^2)	$\sigma_{cr,0}=$		89.28		
"Local" critical resistance (kNm)	for l_p: $M^L_{cr,p}=$	64.13	63.45	63.00	62.69
	for l_s: $M^L_{cr,s}=$	75.63	73.41	71.80	70.62
	$M^{EC3}_{cr}=$		29.30		
Critical load acc. to CPM (kN/m)	for l_p: $q^{CP}_{cr,p}=$	51.45	32.58	22.47	16.42
	for l_s: $q^{CP}_{cr,s}=$	44.91	27.90	18.95	13.69
Critical load acc. to EC3 (kN/m)	$q^{EC3}_{cr}=$	17.40	11.13	7.73	5.68
Percentage increment of resistance (%)	$q^{CP}_{cr}/q^{EC3}_{cr}=$	158.13	150.54	145.06	141.03
Design ultimate resistance (kNm)	for l_p: $M^{CP}_{eff}=$	89.15	88.94	88.80	88.71
	for l_s: $M^{CP}_{eff}=$	92.43	91.82	91.38	91.04
	$M^{EC3}_{c,Rd}=$		75.77		
Ultimate load acc. to CPM (kN/m)	for l_p: $q^{CP}_{eff,p}=$	71.52	45.67	31.67	23.24
	for l_s: $q^{CP}_{eff,s}=$	54.88	34.89	24.11	17.65
Ultimate load acc. to EC3 (kN/m)	for l_s: $q^{EC3}_{eff}=$	44.99	28.79	19.99	14.69
Percentage increase of resistance (%)	$q^{CP}_{eff}/q^{EC3}_{eff}=$	21.99	21.19	20.60	20.16

Note: for comparison, Table 4 also includes the values of q_{cr} and q_{eff} determined according to CPM for the case where the resistance of the beam is decided by the span section. This would be the case, for example, in protection against local buckling of the support segment.

Table 5 compares the "local" critical resistances (M^L_{cr}, q_{cr}) of beams with box Sk250×250×4 sections according to Reference [6] and 2C-250×250×4 compound I-sections depending on the span length of the continuous beam (L = 4, 5, 6, and 7 m).

Comparison of the values presented in Table 5 shows that for 2C sections, where critical resistance is determined by cantilever walls, "local" critical moments M^L_{cr} determined according to CPM were lower on average by about 25–26% in relation to corresponding box sections [6], and the reduction M^{EC3}_{cr} was as much as 54%. The same relationship occurred when comparing the respective critical loads q_{cr}. Meanwhile, the relation q^{CP}_{cr}/q^{EC3}_{cr} for 2C sections in relation to Sk sections increased by as much as 62%. This effect resulted from a much larger reserve of the critical resistance of the elastically restrained cantilever plate (which determines the resistance of the 2C section) in relation to the internal plate of the compression flange, which in turn determines the resistance of the box section. In fact, this is the difference between the critical stresses for the plates (cantilever and internal) that are simply supported in relation to the same plates that are elastically restrained at one or two edges, respectively. For comparison, the relation between the buckling coefficient for the compressed cantilever plate when fully restrained ($k_u = 1.25$ [7,19]) and simply supported ($k_p = 0.43$ [3]) was $k_u/k_p = 1.25/0.43 = 2.91$. For an internal plate, on the other hand, the same relation was much smaller and was $k_u/k_p = 6.97/4 =$

1.74. Of course, this does not change the fact that the critical and ultimate resistance of an internal plate is much higher than that of a cantilever plate with the same geometrical and material parameters.

Table 5. Comparison of "local" critical resistances of box sections (Sk) and compound I-sections (2C) depending on the span length.

L (m)		$M^L_{cr,p}$	$M^L_{cr,s}$	M^{EC3}_{cr}	$q^{CP}_{cr,s}$	$q^{EC3}_{cr,s}$	$q^{CP}_{cr,s}/q^{EC3}_{cr}$
4	SK	87.09	101.27	63.85	60.13	37.91	1.59
	2C	64.13	75.63	29.30	44.91	17.40	2.58
	%	−26.36	−25.32	−54.11	−25.32	−54.11	62.74
5	SK	86.32	98.61	63.85	37.47	24.26	1.54
	2C	63.45	73.41	29.30	27.90	11.13	2.51
	%	−26.49	−25.55	−54.11	−25.55	−54.11	62.23
6	SK	85.81	96.74	63.85	25.53	16.85	1.52
	2C	63.00	71.80	29.30	18.95	7.73	2.45
	%	−26.58	−25.78	−54.11	−25.78	−54.11	61.75
7	SK	85.45	95.33	63.85	18.48	12.38	1.49
	2C	62.69	70.62	29.30	13.69	5.68	2.41
	%	−26.63	−25.92	−54.11	−25.92	−54.11	61.43

In turn, Table 6 compares the design ultimate resistances (M^{CP}_{eff}, q^{CP}_{eff}) of the same box section beams [6] and 2C compound I-sections, also depending on the span length. For the 2C sections, the design ultimate resistances M^{CP}_{eff} determined according to CPM were on average 4–5% lower than the corresponding box sections. The reduction, however, of M^{EC3}_{eff} was nearly 14%. The same relationship occurred when comparing the respective ultimate loads q_{eff}. In this case, the relation $q^{CP}_{eff}/q^{EC3}_{eff}$ for the 2C sections in relation to Sk sections increased by 11%.

Table 6. Comparison of design ultimate resistances of box sections (Sk) and I-sections (2C) depending on the span length.

L (m)		$M^{CP}_{eff,p}$	$M^{CP}_{eff,s}$	$M^{EC3}_{c,Rd}$	$q^{CP}_{eff,s}$	q^{EC3}_{eff}	$q^{CP}_{eff,s}/q^{EC3}_{eff}$
4	SK	93.62	96.59	87.79	57.35	52.13	1.10
	2C	89.15	92.43	75.77	54.88	44.99	1.22
	%	−4.78	−4.31	−13.69	−4.31	−13.70	10.88
5	SK	93.45	96.06	87.79	36.50	33.35	1.09
	2C	88.94	91.82	75.77	34.89	28.79	1.21
	%	−4.83	−4.41	−13.69	−4.41	−13.68	10.73
6	SK	93.33	95.68	87.79	25.25	23.17	1.09
	2C	88.80	91.38	75.77	24.11	19.99	1.21
	%	−4.86	−4.50	−13.69	−4.50	−13.70	10.66
7	SK	93.25	95.39	87.79	18.49	17.02	1.09
	2C	88.71	91.04	75.77	17.65	14.69	1.20
	%	−4.87	−4.56	−13.69	−4.56	−13.69	10.58

Table 7 presents the results of the analysis of the slenderness effect (b_s/t) of the compressed cantilever wall of 2C section on the value of "local" critical moment M^L_{cr} and the design ultimate resistance M_{eff} determined according to CPM and EC3. The calculations were made for a continuous beam with a span length of $L = 4$ m. With the reduction of section wall thickness, the critical load q_{cr} decreased, whereby the difference between the results obtained according to CPM and EC3 was about +158% regardless of the slenderness of the critical plate. The reduction in thickness also reduced the design ultimate resistance M_{eff} and the ultimate load q_{eff}, whereby the difference between the results according to CPM and EC3 was about +23% for t = 5 mm and about +16% for t = 2 mm.

Table 7. Summary of results according to CPM and EC3 for the 2C-250×250×t section and four wall thickness variants.

Wall Thickness (mm)	$t =$	**5**	**4**	**3**	**2**
Slenderness	λ=b/t=	24.00	30.25	40.67	61.50
Euler stress (N/mm^2)	plate $\sigma_{E,i}$=	329.86	207.64	114.89	50.23
	web $\sigma_{E,i}$=	79.13	50.23	28.03	12.36
Buckling coefficient	for l_p: k*=	0.94	0.94	0.94	0.94
	for l_s: k*=	1.11	1.11	1.11	1.11
Critical stress acc. to CPM (N/mm^2)	for l_p: $\sigma_{cr,s}$=	310.46	195.42	108.13	47.28
	for l_s: $\sigma_{cr,s}$=	366.13	230.47	127.52	55.76
Critical stress acc. to EC3 (N/mm^2)	$\sigma_{cr,0}$=	141.84	89.28	49.40	21.60
"Local" critical resistance (kNm)	for l_p: $M^L_{cr,p}$=	126.49	64.13	26.80	7.86
	for l_s: $M^L_{cr,s}$=	149.17	75.63	31.60	9.28
	M^{EC3}_{cr} =	57.79	29.30	12.24	3.59
Critical load acc. to CPM (kN/m)	for l_p: $q^{CP}_{cr,p}$=	101.49	51.45	21.50	6.31
	for l_s: $q^{CP}_{cr,s}$=	88.57	44.91	18.76	5.51
Critical load acc. to EC3 (kN/m)	q^{EC3}_{cr} =	34.31	17.40	7.27	2.13
Percentage increment of resistance (%)	q^{CP}_{cr}/q^{EC3}_{cr} =		158.13		
Design ultimate resistance (kNm)	for l_p: M^{CP}_{eff}=	122.56	89.15	59.47	34.28
	for l_s: M^{CP}_{eff}=	127.15	92.43	61.51	35.26
	$M^{EC3}_{c,Rd}$ =	103.28	75.77	51.38	30.43
Ultimate load acc. to CPM (kN/m)	for l_p: $q^{CP}_{eff,p}$=	98.33	71.52	47.72	27.50
	for l_s: $q^{CP}_{eff,s}$=	75.50	54.88	36.52	20.94
Ultimate load acc. to EC3 (kN/m)	for l_s: q^{EC3}_{eff} =	61.32	44.99	30.51	18.07
Percentage increment of resistance (%)	$q^{CP}_{eff}/q^{EC3}_{eff}$ =	23.12	21.99	19.71	15.87

Note: for comparison, Table 7 also includes the values of q_{cr} and q_{eff} determined according to CPM for the case where the resistance of the beam is decided by the span section. This would be the case, for example, in protection against local buckling of the support segment.

Table 8 compares the "local" critical resistances (M^L_{cr}, q_{cr}) of beams with box sections [6] and compound I-sections (2C) depending on the wall thickness (t = 5, 4, 3, and 2 mm).

Table 8. Comparison of "local" critical resistances of box sections (Sk) and compound I-sections (2C) depending on the wall thickness.

t (mm)		$M^L_{cr,p}$	$M^L_{cr,s}$	M^{EC3}_{cr}	$q^{CP}_{cr,s}$	$q^{EC3}_{cr,s}$	$q^{CP}_{cr,s}/q^{EC3}_{cr}$
5	**SK**	169.45	197.05	124.23	117.00	73.76	1.59
	2C	126.49	149.17	57.79	88.57	34.31	2.58
	%	−25.35	−24.29	−53.48	−24.29	−53.48	62.74
4	**SK**	87.09	101.27	63.85	60.13	37.91	1.59
	2C	64.13	75.63	29.30	44.91	17.40	2.58
	%	−26.36	−25.32	−54.11	−25.32	−54.11	62.74
3	**SK**	36.88	42.89	27.04	25.46	16.05	1.59
	2C	26.80	31.60	12.24	18.76	7.27	2.58
	%	−27.34	−26.32	−54.72	−26.32	−54.72	62.74
2	**SK**	10.97	12.76	8.04	7.57	4.78	1.59
	2C	7.86	9.28	3.59	5.51	2.13	2.58
	%	−28.31	−27.29	−55.32	−27.29	−55.32	62.74

Comparison of the values presented in Table 8 shows that for 2C sections, where critical resistance is determined by cantilever walls, "local" critical moments $M^L_{cr,s}$ determined according to CPM were lower in relation to corresponding box sections by approx. 24.3% for t = 5 mm, up to 27.3% for t = 2 mm. For $M^L_{cr,p}$, the differences ranged from 25.4% for t = 5 mm to 28.3% for t = 2 mm. For M^{EC3}_{cr}, the reduction was 53.5% for t = 5 mm to 55.3% for t = 2 mm. The same relationship occurred when comparing the respective critical loads q_{cr}. In this case, the relation q_{cr}^{CP}/q_{cr}^{EC3} for 2C sections in relation to Sk box sections increased by 62.7%.

In turn, Table 9 compares the design ultimate resistances (M_{eff}^{CP}, q_{eff}^{CP}) of the same beams with box sections Sk according to Reference [6] and compound I-sections (2C), also depending on the wall thickness. For 2C sections, the design ultimate resistances $M_{eff,s}^{CP}$ determined according to CPM, in relation to the corresponding box sections Sk, were reduced by 3.1% for t = 5 mm to 5% for t = 2 mm. For $M_{eff,p}^{CP}$ the differences ranged from 3.8% for t = 5 mm to 5.1% for t = 2 mm. In contrast, the reduction in M_{eff}^{EC3} ranged from 13.6% for t = 5 mm to 11% for t = 2 mm. The same relationship occurred when comparing the respective ultimate loads q_{eff}. In this case, the relation q_{cr}^{CP}/q_{cr}^{EC3} for 2C sections in relation to Sk box sections increased from 6.9% for t = 2 mm to 12.2% for t = 5 mm

Table 9. Comparison of design ultimate resistances of box sections (Sk) and I-sections (2C) depending on wall thickness.

t (mm)		$M_{eff,p}^{CP}$	$M_{eff,s}^{CP}$	$M_{c,Rd}^{EC3}$	$q_{eff,s}^{CP}$	q_{eff}^{EC3}	$q_{eff,s}^{CP}/q_{eff}^{EC3}$
5	SK	127.41	131.27	119.58	77.94	71.00	1.10
	2C	122.56	127.15	103.28	75.50	61.32	1.23
	%	−3.80	−3.14	−13.64	−3.14	−13.64	12.16
4	SK	93.62	96.59	87.79	57.35	52.12	1.10
	2C	89.15	92.43	75.77	54.88	44.99	1.22
	%	−4.78	−4.31	−13.69	−4.31	−13.69	10.87
3	SK	62.80	64.75	59.05	38.45	35.06	1.10
	2C	59.47	61.51	51.38	36.52	30.51	1.20
	%	−5.29	−5.01	−12.98	−5.01	−12.98	9.16
2	SK	36.12	37.11	34.24	22.04	20.33	1.08
	2C	34.28	35.26	30.43	20.94	18.07	1.16
	%	−5.10	−4.99	−11.12	−4.99	−11.13	6.91

Figure 11 shows the relation between the "local" critical moment $M_{cr,s}$ (Figure 11a) and the design ultimate resistance M_{eff} (Figure 11b) as a function of slenderness $\lambda_{CP} = b_s/t$ of the critical cantilever plate of a compound I-section (2C).

Comparison of the graphs shown in Figure 11a showed that the "local" critical resistance and the design ultimate resistance decreased non-linearly with increasing slenderness of the cross-section wall. The difference between the value $M_{cr,s}$ determined by the CPM method and the standard approach according to EC3 was constant in each case of wall thickness and was +158.1%. In contrast, the decrease of M_{eff} (Figure 11b) was mildly non-linear. The percentage increase of the ultimate resistance according to CPM in relation to EC3 was basically directly proportional to the increase of the cross-section wall thickness from about 16% for t = 2 mm ($\lambda_{CP} = 61.5$) to about 23% for t = 5 mm ($\lambda_{CP} = 24$).

The relation between the critical load q_{cr} (Figure 12a) and the ultimate load q_{eff} (Figure 12b) as a function of the span length is shown in Figure 12.

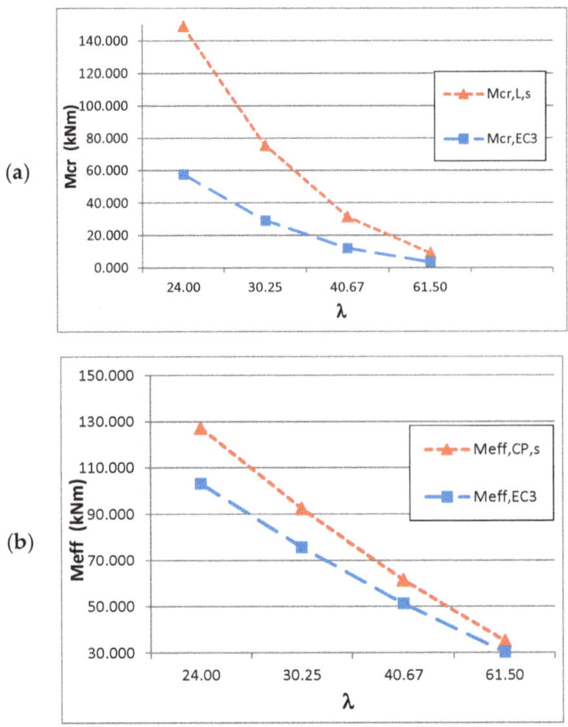

Figure 11. Graphs of (**a**) $M_{cr,s}$ and (**b**) M_{eff} as a function of slenderness λ_{CP}.

Comparison of the graphs shown in Figure 12 showed that an increase in the span length caused a non-linear decrease both in $q_{cr,s}$ and q_{eff}. Moreover, it can be stated that the shorter the span, the greater the percentage increase of the critical resistance determined according to CPM in relation to EC3 (from 158% for short beams $L = 4$ m, to 141% for long beams $L = 7$ m). For ultimate resistance (Figure 12b), the differences in results ranged from 22% for $L = 4$ m to about 20% for $L = 7$ m, respectively).

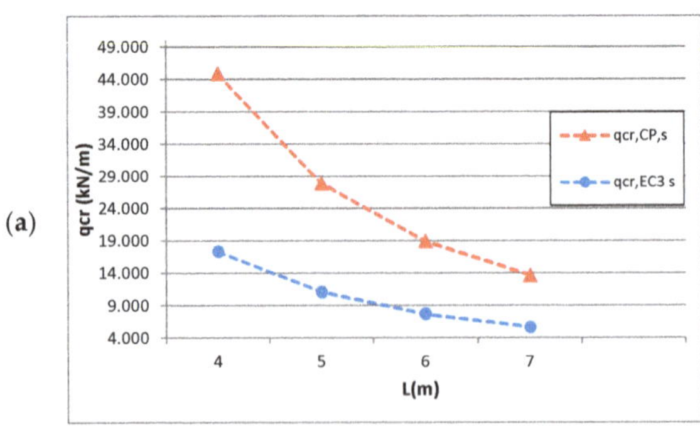

Figure 12. *Cont.*

(b) 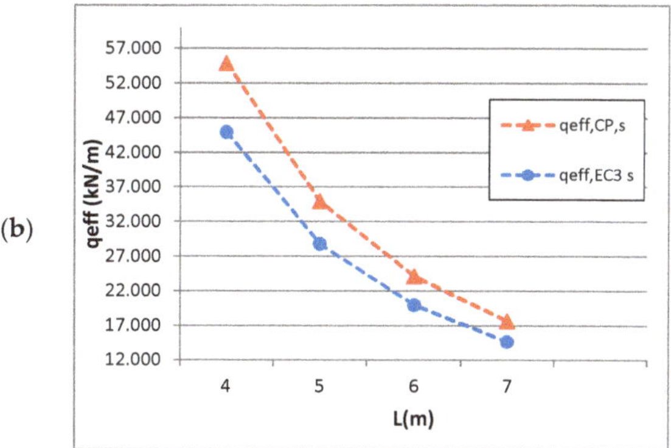

Figure 12. Graphs of (**a**) $q_{cr,s}$ and (**b**) q_{eff} as a function of length L.

8. Summary and Conclusions

The new calculation model presented in this paper for the local stability loss of the first continuous beam span, consisting of separate local buckling analysis for the span segment (loaded with a non-linear "convex" bending moment distribution) and the support segment (in which the "concave" moment distribution was approximated by a linear distribution) was successfully verified by numerical simulations using the finite element method.

To improve the accuracy of the calculation of the critical resistance of a continuous beam with a Class 4 cross-section (in which large stress gradients in the longitudinal direction are found), it was necessary to take into account both the effect of the mutual elastic restraint of the cross-section component plates, and the longitudinal stress variation.

The use of the critical plate method (CPM) according to Reference [10] to determine the "local" critical resistance and the design ultimate resistance of a thin-walled I-section led to a more optimal design of such elements. In this method, a more accurate calculation model is used, which takes into account both the effect of mutual elastic restraint of the cross-section component plates and the effect of longitudinal stress variation. The increase in the resistance that occurs in this case, compared to the EC3 calculation, results from a more faithful representation of the behavior of the thin-walled structural member in the engineering calculation model.

At the same time, it should be noted that taking into account the above effects has only a minor impact on the complexity of calculations, which can be easily algorithmized and presented as relatively simple spreadsheets. This is due, among other things, to the simplified identification of the so-called critical plate and the definition of the so-called "zero" cross-sections [10]. Such assumptions allow for a slightly conservative assessment of the cross-section resistance. In this sense, the CPM can be used both for preliminary design or verification of FEM calculations and, in many technically important cases, for basic design.

The "local" critical resistance defines the range of pre-critical behavior of a section and constitutes the limit of validity of the thin-walled bar theory with a rigid cross-section contour (the Vlasov theory), assuming unlimited elasticity of the material.

To determine the design ultimate resistance of a thin-walled cross-section, the effective width method applied to the individual plates can be used along with the additional assumptions formulated in Section 5, Step 12. The relative slenderness of the component plates is determined from the respective critical stresses. For CP, these are stresses determined by taking into account the restraint coefficient and longitudinal stress variation. For the RP, simple support and a constant stress distribution along the length can be assumed on the same edge. Such assumptions allow for a technically precise and sufficient calculation of the resistance of a thin-walled cross-section.

In the case of a continuous beam under uniform load, the segment that determines the resistance of the structure is the support segment (M_s) of the first span, where the "local" critical resistance is first achieved, followed by the design ultimate resistance.

For the analyzed continuous beam made of a welded section I-300×5×250×5 and calculated according to CPM (Section 7.2), a 121% increase in the critical resistance q_{cr} was determined, which translated into a 23% increase in the design ultimate resistance q_{eff} in relation to the calculations done according to EC3.

In the case of the same continuous beam but made of a compound section 2C-250×250×4 and calculated according to CPM, there was a 158% increase in the critical resistance q_{cr}, which translated into a 22% increase in the design ultimate resistance q_{eff} in relation to the calculations done according to EC3.

The comparison of the resistance parameters of continuous beams with box sections Sk according to Reference [6] and the compound sections 2C compatible with them showed that in the calculation according to CPM, the percentage differences (to the disadvantage of open sections 2C) were as follows: (1) for critical resistance q_{cr} from about 24% to 27% (more than 53% according to EC3), (2) for design ultimate resistance q_{eff} only from 3 to 5%, but according to EC3, more than 13%. In this case, the use of CPM gave greater possibility to use the resistance reserve of the so-called "non-zero" cross-sections, in which the effect of mutual elastic restraint of cross-section component plates can be applied.

The advantage of 2C sections over box sections (assuming that the beam is protected against lateral buckling) is the greater ease of constructing joints and assembly nodes due to free and two-sided access to mechanical connectors (e.g., bolts). In the case of beams sensitive to lateral buckling, on the other hand, the use of closed cross-sections is more optimal due to their much higher torsional stiffness compared to open cross-sections.

If there are significant residual stresses (compressive) in a thin-walled cross-section, the critical stress of local buckling is reduced. This applies especially to axially compressed welded sections [34,35]. Therefore, further testing of welded components with a high proportion of residual stresses must take into account their adverse effects on local buckling of the cross-sections.

In open cold-formed sections (produced without welding), however, the adverse effect of residual stresses caused by bending is usually insignificant [36] and can be omitted in most cases (this effect is further reduced by the beneficial effect of increasing the strength of steel in cold-formed corners).

Beams with thin-walled I-sections, for which the ultimate resistance is determined by local buckling of the bent and shear web, and continuous beams with reinforced end spans will be subject to further study by the authors.

Author Contributions: Introduction was prepared by A.S. Mathematical description were written by A.S., K.B. Results were obtained by K.B., A.S. The analysis of the results and conclusions was written by A.S. All authors have read and agreed to the published version of the manuscript.

Funding: This research received no external funding.

Conflicts of Interest: The authors declare no conflict of interest.

References

1. *EN 1993-1-1, Eurocode 3: Design of Steel Structures—Part 1-1: General Rules and Rules for Buildings*; European Committee for Standardization: Brussels, Belgium, 2005.
2. *EN 1993-1-3, Eurocode 3: Design of Steel Structures—Part 1-3: General rules—Supplementary Rules for Cold-Formed Members and Sheeting*; European Committee for Standardization: Brussels, Belgium, 2006.
3. *EN 1993-1-5, Eurocode 3: Design of Steel Structures—Part 1-5: Plated Structural Elements*; European Committee for Standardization: Brussels, Belgium, 2006.
4. Szychowski, A. A theoretical analysis of the local buckling in thin-walled bars with open cross-section subjected to warping torsion. *Thin-Walled Struct.* **2014**, *76*, 42–55. [CrossRef]
5. Kowal, Z.; Szychowski, A. Experimental determination of critical loads in thin-walled bars with Z-section subjected to warping torsion. *Thin-Walled Struct.* **2014**, *75*, 87–102. [CrossRef]

6. Brzezińska, K.; Szychowski, A. Stability and resistance of steel continuous beams with thin-walled box sections. *Arch. Civ. Eng.* **2018**, *64*, 123–143. [CrossRef]
7. Bulson, P.S. *The Stability of Flat Plates*; Chatto and Windus: London, UK, 1970.
8. Yu, C.; Schafer, B.W. Effect of longitudinal stress gradient on the ultimate strength of thin plates. *Thin-Walled Struct.* **2006**, *44*, 787–799. [CrossRef]
9. Hancock, G.J. Cold-formed steel structures. *J. Constr. Steel Res.* **2003**, *58*, 473–487. [CrossRef]
10. Szychowski, A. Computation of thin-walled cross-section resistance to local buckling with the use of the Critical Plate Method. *Arch. Civ. Eng.* **2016**, *62*, 229–264. [CrossRef]
11. Jakubowski, S. The matrix analysis of stability and free vibrations of walls of thin-walled girders. *Arch. Budowy Masz.* **1986**, *XXXIII*, 357–376. (In Polish)
12. Rusch, A.; Lindner, J. Load carrying capacity of thin-walled I-sections subjected to bending about the z-axis (in German). *Stahlbau* **1999**, *68*, 457–467. [CrossRef]
13. Beale, R.G.; Godley, M.H.R.; Enjily, V. A theoretical and experimental investigation into cold-formed channel sections in bending with the unstiffened flanges in compression. *Comput. Struct.* **2001**, *79*, 2403–2411. [CrossRef]
14. Kotełko, M.; Lim, T.H.; Rhodes, J. Post—Failure behaviour of box section beams under pure bending (an experimental study). *Thin-Walled Struct.* **2000**, *38*, 179–194.
15. Jakubowski, S. Buckling of thin-walled girders under compound load. *Thin-Walled Struct.* **1988**, *6*, 129–150. [CrossRef]
16. Nassar, G. Das Ausbeulen dünnwandiger Querschnitte unter einachsig aussermittiger Druckbeanspruchung. *Der Stahlbau H* **1965**, *10*, 311–316.
17. Papangelis, J.P.; Hancock, G.J. Computer analysis of thin-walled structural members. *Comput. Struct.* **1995**, *56*, 157–176. [CrossRef]
18. Schafer, B.W.; Ádány, S. Buckling analysis of cold-formed steel members using CUFSM: Conventional and constrained finite strip methods. In Proceedings of the Eighteenth International Specialty Conference on Cold-Formed Steel Structures, Orlando, FL, USA, 26–27 October 2006.
19. Gardner, L.; Fieber, A.; Macorini, L. Formulae for calculating elastic local buckling stresses of full structural cross-sections. *Structures* **2019**, *17*, 2–20. [CrossRef]
20. Ciesielczyk, K.; Rzeszut, K. Locla and distortional buckling of axially loaded cold rolled sigma profiles. *Acta Mech. et Autom.* **2016**, *10*, 218–222. [CrossRef]
21. Fieber, A.; Gardner, L.; Macorini, L. Formulae for determining elastic local buckling half-wavelengths of structural steel cross-sections. *J. Constr. Steel Res.* **2019**, *159*, 493–506. [CrossRef]
22. Seif, M.; Schafer, B.W. Local buckling of structural steel shapes. *J. Constr. Steel Res.* **2010**, *66*, 1232–1247. [CrossRef]
23. Couto, C.; Vila Real, P. Numerical investigation on the influence of imperfections in the local buckling of thin-walled I-shaped sections. *Thin-Walled Struct.* **2019**, *135*, 89–108. [CrossRef]
24. Yu, C.; Schafer, B.W. Effect of longitudinal stress gradients on elastic buckling of thin plates. *J. Eng. Mech. ASCE* **2007**, *133*, 452–463. [CrossRef]
25. Szychowski, A. The stability of eccentrically compressed thin plates with a longitudinal free edge and with stress variation in the longitudinal direction. *Thin-Walled Struct.* **2008**, *46*, 494–505. [CrossRef]
26. Szychowski, A. Stability of cantilever walls of steel thin-walled bars with open cross-section. *Thin-Walled Struct.* **2015**, *94*, 348–358. [CrossRef]
27. Szychowski, A. Local buckling of cantilever wall of thin-walled member with longitudinal and transverse stress variation. *Bud. i Archit.* **2015**, *14*, 113–121. (In Polish) [CrossRef]
28. Szychowski, A. Stability of eccentrically compressed cantilever wall of a thin-walled member. *JCEEA* **2015**, *XXXII*, 439–457. (In Polish) [CrossRef]
29. Kotełko, M. *Thin-Walled Structures Resistance and Failure Mechanisms*; Wydawnictwa Naukowo –Techniczne: Warsaw, Poland, 2011. (In Polish)
30. Kotełko, M. Thin-walled profiles with edge stiffeners as energy absorbers. *Thin-Walled Struct. Elsevier* **2007**, *45*, 872–876.
31. Kalyanaraman, V. Local buckling of cold-formed steel members. *J. Struct. Div.* **1979**, *105*, 813–828.
32. Li, L.-Y.; Chen, J.-K. An analytical model for analyzing distortional buckling of cold-formed steel sections. *Thin-Walled Struct.* **2008**, *46*, 1430–1436. [CrossRef]

33. Kowal, Z. *The Stability of Compressed Flange of Plate Girder with a Box Cross-Section*; Zeszyty Naukowe Politechniki Wrocławskiej: Budownictwo, Poland, 1965; pp. 73–85. (In Polish)
34. Rykaluk, K. *Residual Welding Stresses in Chosen Ultimate Bearing Capacity*; Scientific Papers of the Institute of Building Engineering of Wroclaw Technical University, No 29, Series: Monographs No 11; Wroclaw Technical University: Wroclaw, Poland, 1981. (In Polish)
35. Hibbitt, Karlsson & Sorensen Inc. *2000 ABAQUS/CAE User's Manual*; Hibbitt, Karlsson & Sorensen: Pawtucket, RI, USA, 2000.
36. Ignatowicz, R.L. The Theoretical and Experimental Analysis of Resistance of the Cold-Formed Steel Columns with Box cross-Section. Ph.D. Thesis, Instytut Budownictwa Politechniki Wrocławskiej, Series PRE 52/99. Wrocław University of Science and Technology, Wroclaw, Poland, 1999. (In Polish).

Article

Experimental Study on the Flexural Behavior of Alkali Activated Fly Ash Mortar Beams

Adelino V. Lopes [1], Sergio M. R. Lopes [2,*] and Isabel Pinto [2]

[1] Institute for Systems and Computers Engineering, Department of Civil Engineering, University of Coimbra, 3030-788 Coimbra, Portugal; avlopes@dec.uc.pt

[2] Centre for Mechanical Engineering, Materials and Processes, Department of Civil Engineering, University of Coimbra, 3030-788 Coimbra, Portugal; isabelmp@dec.uc.pt

* Correspondence: sergio@dec.uc.pt

Received: 1 June 2020; Accepted: 23 June 2020; Published: 25 June 2020

Featured Application: Fly ash is a byproduct from burning pulverized coal in the electric power industry and can be incorporated in cementitious concretes or in alkali activated concretes (AAC) with obvious environmental advantages. When part of AAC, it will be able to lead to products with glossy, black surfaces, which might be aesthetically appealing for some modern architectural solutions. A reinforced concrete structure is one type of such products. This document presents an experimental study on reinforced mortar beams, half of them made with fly ash and the other half made with Portland cement. Different aspects of the structural behavior are compared for possible applications of fly ash geopolymers in civil engineering structures.

Abstract: This work aims to study the possibility of using alkaline activated fly ash in structural members. The work, of an experimental nature, focuses on the evaluation of the behavior of simply supported beams under two symmetrical loads (four-point tests). For such study, 10 beams were built, of which, five using fly ash and the remaining five using traditional Portland cement. The test results are compared. Conclusions on the practical application of fly ash in structures were explained and, as mention later in this document, there is room for improvement. This is one of very few works on fly ash alkali activated structures and further studies are necessary in the future. Some aspects, such as shrinkage and deformability are presented as some of the negative points concerning the potential use of fly ash. These are two aspects that need more attention in future investigations.

Keywords: experimental study; analytical model; reinforced concrete; beams; fly ash alkali activated; bending

---

1. Introduction

The cement industry is under increasing pressure from public opinion due to its contribution to CO_2 emissions. To bring the cement sector in line with the 2015 Paris Agreement on climate change, its annual emissions will need to fall by at least 16 per cent by 2030 [1]. The awareness of the high level of CO_2 emissions is not new, but only recent international agreements and measures really forced the cement industry to intensify the search for new alternative production technologies and materials. A vision of the producers on this issue can be found in a report by Cembureau [2].

Obviously, non-cementitious materials must be considered as an alternative route of the entire road map to low emission construction production. Various research works on alternative materials to Ordinary Portland Concrete (OPC) have been published in the last three decades, but OPC continues to dominate the market at present. However, the target limits for CO_2 emissions might alter the competitive advantage of OPC and surely increases the demand for research on alternative materials.

This is the case of alkali activated concretes (AAC), which show some positive points concerning the strength values, durability, or environmental impacts [3].

In fact, many works have been published on AAC in the last 2–3 decades. Almost all of them have focused on the development of the material itself, including the mechanical properties. Very few studies were concerned with the structures so far. Since the size of specimens for structural studies is bigger when compared to that of specimens for mechanical properties of the material, some difficulties arise when passing from the material to structural research. Even a recent study [4] has pointed out that there are some studies in mortars, but the need for studies in concretes still persists. This means that the step from mortar to concrete is still not entirely covered and the next step from material specimens to concrete structures is also of great necessity.

From different types of AAC, alkali activated fly ash concrete has its own particularities that need to be known. One of the most important is the workability, which is directly linked with the practical application of the material in structures. Not going deeply into the topic, it should be known that the rheological behavior observed in fly ash is different from that of OPC. The commercial admixtures for improving the workability of OPC do not have similar effects in fly ash. Further explanations on this topic are available in bibliography [5].

Shrinkage is also a characteristic of fly ash concretes, which has deserved some attention. It is known that curing conditions and the liquid-to-ash ratio are very important for shrinkage, with some researchers recommending curing temperatures of around 60–70 °C (Degrees Celsius) for minimizing shrinkage [6–8]. Indeed, the few experimental studies on reinforced beams made with fly ash activated concrete found in literature followed this recommendation [9–12]. Curing conditions at this temperature level are not practical for on-site applications. Some recent studies have concentrated on curing in ambient temperatures [13–15].

Curing at specific temperatures has other effects beyond shrinkage, for instance, on strength. Therefore, it should be considered as a strong option when possible, for instance, in precast plants. However, in the opinion of the authors, the on-site practical construction at ambient temperature must not be ruled out since it is a potential application of this material. The studied reported here was trying to evaluate the possibility of application on reinforced concrete members cast on-site at ambient temperature.

As explained before, the general possibility of replacing the Portland cement with alkaline activated binders, known as geopolymers, should be considered as a strong option given the increased concerns on environmental issues. Fly ash is one of the materials that can be used in alkaline activated products. The authors have tested other materials [16], but the current document reports a study on the behavior of fly ash in structures, particularly in beams under bending. It is interesting, in particular, to evaluate the behavior of these beams under bending action, for different levels of the loading, and to compare it with that of beams built with traditional Portland cement. It should be kept in mind that the behavior in service conditions is also gaining great importance in current regulations, when compared to typical ultimate strength.

To achieve this goal, an experimental procedure was carried out, in which two groups of five beams each were built. From group to group, the beams vary in material, with five built with Portland cement binder and five with fly ash based binder. Within each group of beams, only the ratio of the reinforcement was varied in order to cover an intentional range to cover specific situations explained later. The dimensions are the same for all beams, which are about half the dimensions usually used in the laboratory and in construction.

In addition to this objective, it would be also important to theoretically estimate the behavior of beams built with new materials. A good fitting of the numerical methodologies to the behavior of such new materials needs to be verified [17].

In this work, beams with reduced dimensions were used compared to those usually used in construction. Nevertheless, they are significantly bigger than the geopolymeric specimens generally presented in the bibliography for the study of the material. It is, therefore, important not to ignore the possible scale effects. Some studies indicate that up to a scale of 1 to 3, the differences are negligible [18]. However other studies indicated that some differences in behavior could be important, especially for large size structures [19]. Kim et al. [20] presented a study with reinforced concrete specimens, using a scale of 1:5, where it was concluded that the models present equivalent results when compared to the normal size members. However, other studies indicate some differences. Since both types of beams of this investigation, fly ash and cement, were built with equal dimensions, it would be expected that the size effects, if any, would have similar consequences. Therefore, the comparison of the results can be accepted for beams of these dimensions.

2. Experimental Procedure

2.1. Design

In the construction of the beams, the following dimensions were chosen: length L = 1500 mm, width b = 100 mm, and height h = 150 mm. For the longitudinal reinforcing bars, ϕ6 (6 mm) and ϕ8 sizes were assumed with a 10 mm cover. Transversal ϕ4 stirrups were used (a reinforcement factory accepted to supply this special size). Taking into account the reduced covering, it was decided to use sand (maximum particle size of Dmax ≤ 2 mm) without coarse aggregates. It was intended that the reinforcement to be used, as well as the covering, would also be reduced in the same proportionality as the external dimensions of the beams.

Taking into account the objectives of the test to be carried out, it was intended that the combination of the materials allowed to obtain approximately half of the beams reaching failures by the reinforcement, and the other half by crushing of the mortar. Keeping this in mind, for the five beams to be constructed with each material, the only parameter to be varied, as uniformly as possible, would be the percentage of the reinforcement, which should cover a range of usual values in construction. That is, in addition to the dimensions, the compressed longitudinal reinforcement, the transversal reinforcement and the material in each group of beams remained constant. Obviously, the failure of the middle beam of each group could occur either by the reinforced bars or by crushing of mortar. Table 1 shows the designations adopted for the beams. In this table, ϕ indicates the number and diameter of the tensioned reinforcement of the beam, As the corresponding area, and ρ the percentage of tensioned reinforcement.

Table 1. Designation of the beams.

Binder	Longitudinal Reinforcing Bars				Label
	ϕ [mm]	As [mm^2]	ρ [%]		
Cement	2 ϕ 6	56.5	0.38	Insufficiently reinforced	Cem-2F6
	3 ϕ 6	84.8	0.57	Lightly reinforced	Cem-3F6
	4 ϕ 6	113	0.75	Bellow normally reinforced	Cem-4F6
	3 ϕ 8	151	1.01	Normally reinforced	Cem-3F8
	4 ϕ 8	201	1.34	Above normally reinforced	Cem-4F8
Fly Ash	2 ϕ 6	56.5	0.38	Insufficiently reinforced	FA-2F6
	3 ϕ 6	84.8	0.57	Lightly reinforced	FA-3F6
	4 ϕ 6	113	0.75	Bellow normally reinforced	FA-4F6
	3 ϕ 8	151	1.01	Normally reinforced	FA-3F8
	4 ϕ 8	201	1.34	Above normally reinforced	FA-4F8

One of the initial concerns was to obtain failure by bending and not by shear. Since the critical beam would be the one under the greatest loads, it was decided to fix an amount of stirrups, which would prevent shear failure for such beam, and to use the same amount of transverse reinforcement in all the beams. Following this option, identical conditions of material confinement in all beams would also be guaranteed. This parameter is very important in bending beams. This led to 4 mm diameter transverse reinforcement bars (ϕ4) made by two branches stirrups spaced 70 mm. In the top of the beam, 2 ϕ 6 longitudinal bars were used. Figure 1 shows the reinforcement before casting.

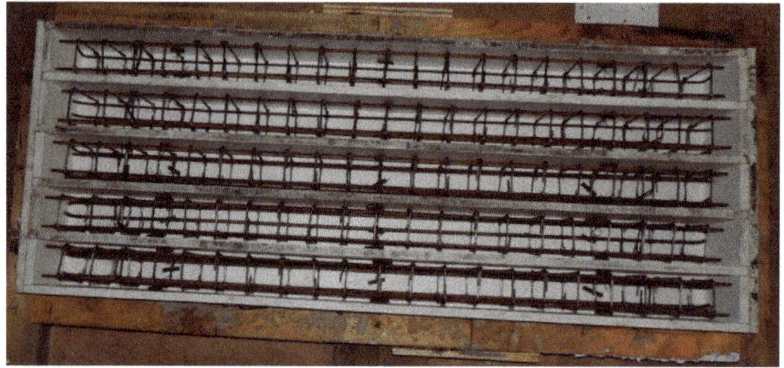

Figure 1. Beams before casting.

2.2. Materials

Apart from the especial size 4 mm bars, which were courtesy of a factory nearby [21], the rest of the reinforcing bars were bought in the market. All of the bars were tested according to standard NP EN 10002-1 [22]. In these tests, four samples of each type of steel, about 40 cm long, were used. Table 2 shows the averages values of yield strength of reinforcement (fsy), ultimate strength of reinforcement (fsu), and strain of reinforcement steel at maximum load (εsu).

Table 2. Mechanical characteristics of steel bars.

Diameter [mm]	fsy [MPa]	fsu [MPa]	fsu/fsy	εsu [%]
4	524	665	1.27	5.0
6	572	823	1.44	5.0
8	614	706	1.15	11.1

The granulometry of the coarse aggregates to be used in the mixtures was conditioned by the size of the specimens to be cast, especially the cover. It was found that the size of the aggregates would not be appropriate and they were not used. A sand was used.

For the mortar production, Normal Portland Type I CEM I 42.5 R cement (Compressive Strength = 42.5 MPa) was used, from CIMPOR. According to the producing company, this material consists essentially of CaO (61.5%) and SiO_2 (21%).

The fly ash came from the Sines thermal power plant and resulted from the burning of coal. The ashes are certified by the National Laboratory of Civil Engineering (LNEC); thus, fulfilling the compliance and performance requirements defined in Annex ZA of standard NP EN450-1 [23]. Its composition essentially contains calcium oxide, CaO (71%) and potassium oxide, K_2O (16%), according to tests carried out at the Department of Earth Sciences of the University of Coimbra. Moreover, for these ashes, the high specific surface leads to great reactivity. An activator needs to be used; it must be dosed and concentrated by taking into account the binder, namely its chemical composition, and the degree of fineness that influence the activation reaction. In this work, it was decided to use an activator composed of sodium hydroxide (NaOH) and sodium silicate (NaSiO3), in the proportion of 1:2, respectively. The NaOH was obtained by mixing caustic soda with water in the appropriate proportions so that the 12.5 M molal concentration of NaOH could be achieved following the indications given by Pinto [24]. This decision also took into account the experience of other works carried out at the host laboratory regarding workability, and also the limits of the molar concentration of NaOH so that the alkaline reaction occurs in full: 12 M according to Granizo [25] and 20 M according to Davidovits [26]. The second component of the activator, sodium silicate ("D40") was purchased from the company "Quimitecnica".

2.3. Production of Beams

The cement mortar was produced at Department of Civil Engineering of the University of Coimbra, as well as the tests carried out on small specimens (cubes and prisms) and on the beams. To follow the initial objectives, an average strength of 35 MPa was targeted for the beams. For this purpose, a mixture was carried out. The mix proportions were 270 kg of sand, 60 kg of cement, 30 kg of water, and 0.54 kg of superplasticizer Sika ViscoCrete 20HE. This mixture was sufficient for casting five beams, twelve 150 mm cubes, and six 40 × 40 × 160 mm prisms. After three days of curing, the beams, cubes and prisms were demolded.

The fly ash based mixture was done with: 200 kg of sand, 80 kg of fly ash, and 40 kg of compound activator. These proportions were those adopted from a base recipe suggested by Pinto [27]. With this mixture, five beams, twelve 40 × 40 × 160 mm prisms, and five 150 mm cubes were cast and demolded five days later.

Table 3 shows the weight and real dimensions, measured after the construction of the beams. The height (h) and the width (b) result from an average of three measurements made in the central area of the beam (the most important area of the tests). L represents the length of the beam. It appears that the greatest deviations occurred for the height (h).

Table 3. Effective dimensions of the beams.

Beam	Weight [kg]	b [mm]	h [mm]	L [mm]
Cem-2F6	52.1	98.1	152	149.5
Cem-3F6	53.9	100	153	149.5
Cem-4F6	54.9	100	154	149.5
Cem-3F8	54.6	101	153	149.5
Cem-4F8	53.4	97.4	152	149.5
FA-2F6	51.9	104	152	149.6
FA-3F6	51.7	99.3	152	149.6
FA-4F6	52.2	99.2	153	149.5
FA-3F8	52.7	102	153	149.5
FA-4F8	53.2	103	152	149.5

Control tests were carried out on cubes and prisms, on different days, in order to evaluate the evolution of the strength.

Regarding the cement mortar tests, the results on cubes gave very close results to the expected average compressive strength of the concrete, an exponential curve proposed by Neville [28] (see Figure 2). In this figure, fcm indicates the average value of the compressive strength of concrete for the time T in days.

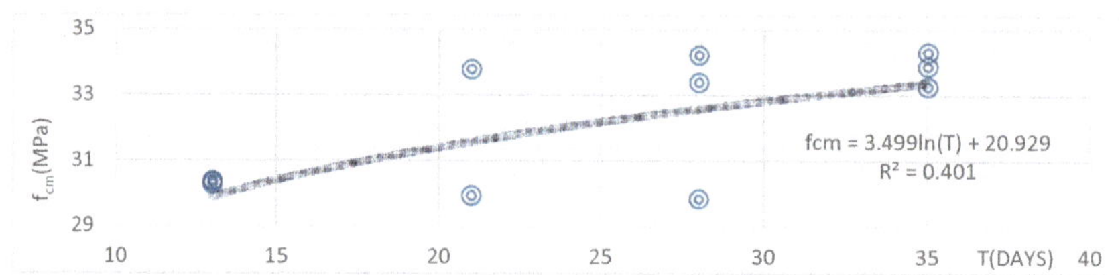

Figure 2. Average strength of the cement specimens.

Regarding the fly ash specimens, compression and/or tension strength tests were performed on twelve 40 × 40 × 160 mm test specimens, following NP EN 196-1 and NP EN12390-5 [29,30], to determine the mechanical characteristics of this material. The value for the modulus of elasticity of this material was about 20 GPa in non-destructive cyclic tests and about 18 GPa in failure tests. The average compressive and tensile strengths were 15.1 and 3.2 MPa at 23 days old and 23.5 and 4.0 MPa at 32 days old, respectively. The corresponding maximum stresses were respectively, 17.8 and 3.3 MPa at 23 days and 30.7 and 4.6 MPa at 32 days. From the compression tests of 150 mm cubes, an average compressive strength of 23.8 MPa, was obtained at 32 days, with a maximum value of 27.2 MPa. The standard deviation of the results was 2.6. It should be noted that all ash beams were tested between days 31 and 33 after pouring.

2.4. Test Procedure

Figure 3 shows the symmetric loading diagram adopted for the beam tests. The beams were simply supported, and the supports were placed at 50 mm from each end. Two vertical symmetrical loads P/2 were placed 450 mm apart, leading to pure flexion. A central zone of the beam was then under a constant bending moment. The self-weight was not considered in the diagrams shown in this figure.

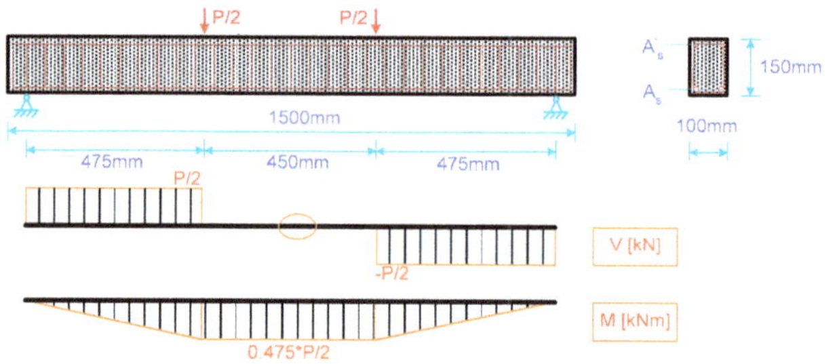

Figure 3. Load set-up and shear and bending diagrams.

A beam under test is shown in Figure 4. The evaluation of the deformation of the beams (the deflection) was carried out using nine Linear Variable Differential Transformers (LVDTs). Three of them were placed on each support (to measure vertical and horizontal displacements) and three others were placed in the central zone (to measure the mid span deflection), as shown in Figure 4.

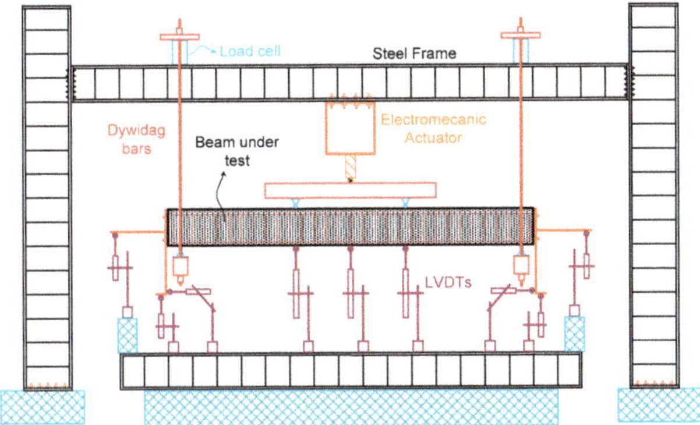

Figure 4. Beam under test.

The readings of the load cells and LVDTs were recorded by a Data Logger TML TDS-602, with a frequency of 2 Hz.

3. Test Results

The comparison of the behavior of the beams is based on key aspects of the evolution of the beam with load, such as the cracking point, Pcr, the yielding of steel, Py, the maximum load, Pmax, the ultimate strength, Pult, among others. The stiffness in each zone of the P-d graph is also an important aspect for comparison. The P-d diagram will be the basis for this analysis (P corresponds to the total load applied at each instant, and d to the displacement in the mid-span of the beam). The load P is obtained from the sum of the loads registered in the four load cells responsible for measuring the value of the support reactions. The value of the displacement in the mid-span, d, is obtained by subtracting the value of the LVDTs at the supports from the value measured of the LVDT placed in the mid-span of the beam.

The test results are presented in Figures 5–9. This type of beams generally presents three distinct phases in behavior, namely: State I, State II, and State III. State I, which corresponds to the Stiffness K_I, extends from the origin to the point Pcr, corresponding to the beginning of cracking. The second state, Stiffness K_{II}, goes from Pcr to the point where the reinforcement yields Py. Finally, the third and last State is the "plastic phase" and starts at Py and extends past the maximum load Pmax, and ending at the last load point Pult (Pult is assumed to be the load 15% lower than the maximum load, according to NP EN1998-1 [31]). In beams where the failure occurs due to reinforcement failure, this point generally coincides with the last point of the graph.

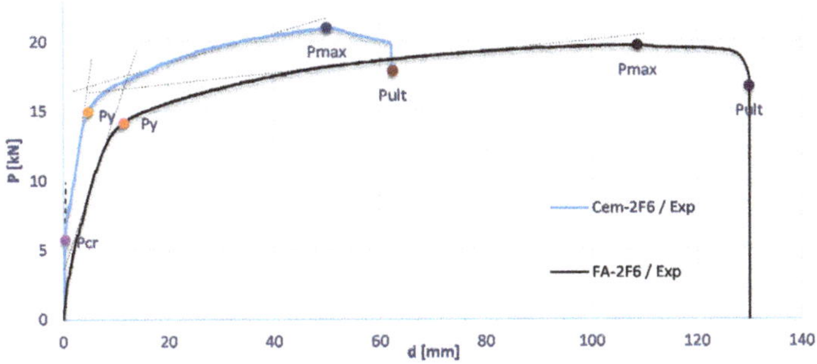

Figure 5. P-d experimental curves for beams Cem-2F6 and FA-2F6.

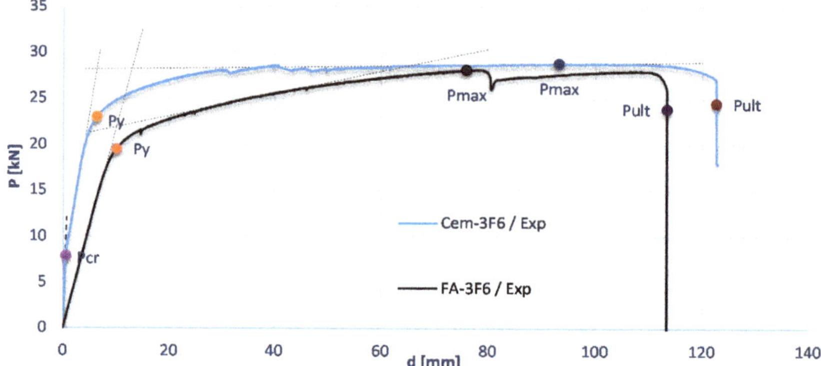

Figure 6. P-d experimental curves for beams Cem-3F6 and FA-3F6.

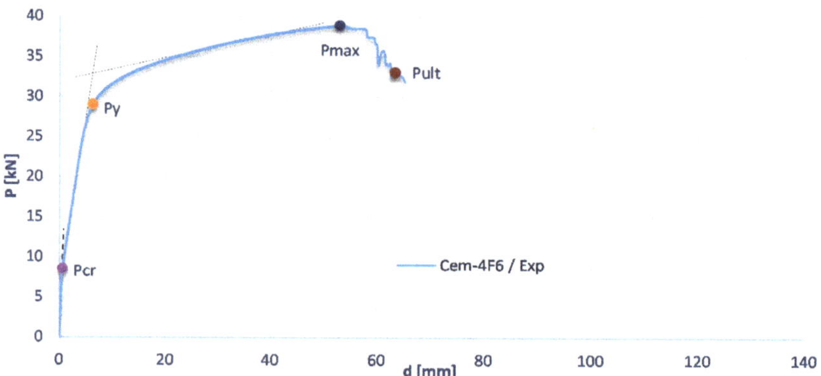

Figure 7. P-d experimental curve for beam Cem-4F6.

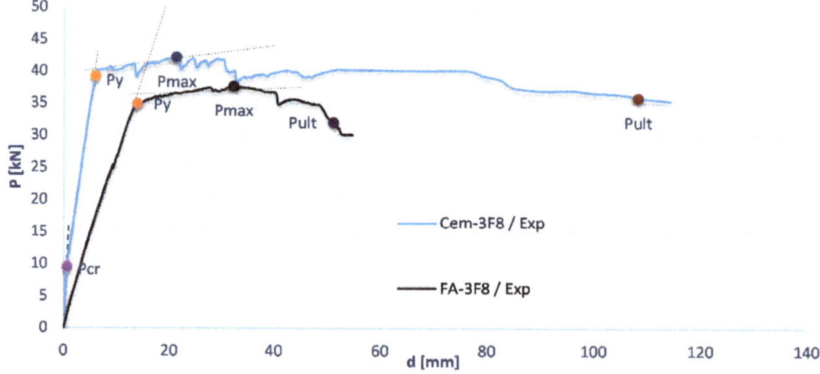

Figure 8. P-d experimental curves for beams Cem-3F8 and FA-3F8.

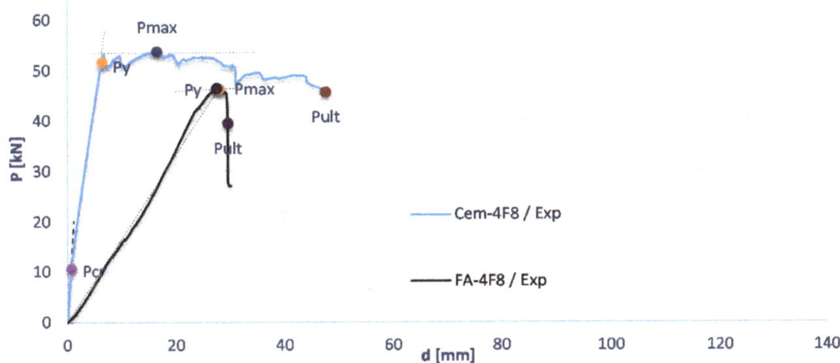

Figure 9. P-d experimental curves for beams Cem-4F8 and FA-4F8.

Figure 5 shows the P-d curves (load-deflection curves) for the insufficiently reinforced beams (ρ = 0.38%). The absence of State I in the fly ash beam curve is justified by the fact that these beams were cracked before the start of the test. This was provoked by the shrinkage due to the temperature of the curing time. This outcome was an assumed risk and could not be ruled out before casting. Since it did occur, this is an issue that needs to be further investigated. To cast these fly ash structures only under an optimized curing temperature of 60 to 70 degrees centigrade seems to be too restrictive to practical construction on site. Apart from this, it appears that there are no significant differences between both types of beams in terms of the stiffness of State II, nor in the values of Py and Pmax loads. In this case, the fly ash beam proved to be much more flexible and more ductile when compared to Cement (Cem) beam.

Figure 6 shows the P-d curves of the lightly reinforced beams (ρ = 0.57%). In State II, both types of beams have similar behavior. In terms of ductility, the beams are equivalent, showing both a long level of ductility, despite the type of steel used for reinforcement (cold hardened steel).

Figure 7 shows the P-d curve for beam CEM-4F6 (bellow normally reinforced: ρ = 0.75). Unfortunately, a rare malfunction of the data logger caused the loss of the recorded values of FA-4F6 beam. The relatively low value of the ductility of the beam stands out in this experimental curve.

Figure 8 shows the P-d diagrams of the normally reinforced beams (ρ = 1.01%). Looking at the Py and Pmax points, it is clear that the ductility of the cement beams is much higher than that of the fly ash beam.

Figure 9 shows the P-d curve of above normally reinforced beams (ρ = 1.34%). In this case, the ash beam revealed an absence of ductility. The cement beams also show a low level of ductility, which could be problematic for hyperstatic structures under seismic actions, for instance. The values of Py and Pmax are not very different when both beams are compared. However, the modulus of elasticity in State II is noticeably smaller for the fly ash beam when compared to cement beam. This means that the first beam is more flexible than the second one.

Table 4 shows some further information related to the tests. The "curing" refers to the age of the beam in the day of testing, ΔT is the duration of the test, and "failure by" means the type of failure shown in the tests, whether by the tensioned reinforcement, "As", by crushing the compressed material, "fc", or by failure of the stirrups, "V". Thus, all beams were tested in the time ranging from 31 to 37 days. In the case of insufficiently reinforced or lightly reinforced beams, the failure occurred due to the failure of the tensioned reinforcement, as expected. In the other cases, the failures occurred due to crushing of the compressed material, except for the FA-3F8 beam, where the failure occurred in a shear crack, in this case due to insufficient connection of the stirrup to the surrounding material (this was observed during the visual inspection of the zone after the test).

Table 4. Further information related with tests.

Beam	2F6		3F6		4F6		3F8		3F8	
	Cem	FA	Cem	FA	Cem	FA	Cem	FA	Cem	FA
Curing [days]	34	32	34	32	36	32	37	31	35	31
ΔT (min)	230	191	146	142	370	147	193	102	219	83
Failure by:	As	As	As	As	fc	fc	fc	fc	fc	V

ΔT = duration of test; As = failure by longitudinal steel bars; fc = failure by crushing of concrete; V—failure by shear (stirrups).

Table 5 shows the values obtained from the curves of the tested beams. Some conclusions are obvious. For example, yield and maximum loads increase as the maximum pulling force on steel increases (due to higher values of the area of cross section of the sum of the longitudinal bars). However, there are some other important trends in this table. For example, although there are no significant variations in the rigidity of State I, at point Pcr, the crack load increases significantly with As, as it was theoretically demonstrated in a previous publication [32].

Table 5. Parameters associated with the behavior of the beams.

Beam	2F6		3F6		4F6		3F8		3F8	
	Cem	FA	Cem	FA	Cem	FA	Cem	FA	Cem	FA
P_{cr} [kN]	5.73	–	7.87	–	8.55	–	9.47	–	10.6	–
d_{cr} [mm]	0.34	–	0.49	–	0.60	–	0.59	–	0.73	–
K_I [kN/mm]	19.7	–	17.2	–	16.7	–	15.9	–	16.8	–
P_y [kN]	15.0	13.8	23	19.2	29.0	–	39.2	34.9	51.6	45.5
d_y [mm]	4.70	12.1	6.29	9.54	6.21	–	5.68	15.3	6.42	24.3
K_{II} [kN/mm]	2.50	1.36	3.48	2.25	4.27	–	5.86	2.37	7.41	1.91
P_y/P_{cr}	2.61	–	2.92	–	3.40	–	4.14	–	4.86	–
K_I/K_{II}	7.87	–	4.93	–	3.83	–	2.7	–	2.26	–
P_{max} [kN]	21.0	19.7	28.8	21.5	38.9	~33.0	42.2	37.7	53.7	46.4
d_{max} [mm]	49.9	110	93.1	75.7	52.9	–	21.3	33.6	16.5	25.4
K_{III} [kN/mm]	0.11	0.04	0.01	0.12	0.15	–	0.11	0.04	0.00	0.05
P_{max}/P_y	1.40	1.43	1.26	1.12	1.34	–	1.08	1.08	1.04	1.02
d_{ult} [mm]	62.4	131	123	113	63.4	~81	108	52.4	47.5	27.4
dult/dy	13.3	10.9	19.5	11.9	10.2	–	19.0	3.43	7.41	1.13

Another important aspect is the general idea that the ductility of a structure is related to the ductility of the steel bars. Assuming the dult/dy as an indicator of the ductility of a beam, it appears that a high ductility of the beams could occur with low ductility steel, as that used in this experimental program. As explained before [33,34], ductility depends on other concrete confinement conditions, which could be much more important.

Another important aspect concerns the K_I/K_{II} quotient. Eurocode 2, EC2 [35] proposes a value of 3 for this quotient. According to the results obtained for these five cement beams, this quotient can be related to the percentage of reinforcement, ρ, or to the mechanical percentage of reinforcement, ω, through equations 1. The equations result from an adjustment of the results to an exponential curve (R^2 = 0.98).

As previously mentioned, the ash beams did not show State I (not cracked). In addition, the deformability was generally higher than that of the cement beams (due to lower K_{II} stiffness values). The ductility of the fly ash beams was always lower than that of the cement beams.

$$\frac{K_I}{K_{II}} = 0.0292\rho^{0.995}; \frac{K_I}{K_{II}} = 0.495\omega^{0.914} \quad (1)$$

Table 6 shows the deviations of the values of the fly ash beams in relation to those of the cement beams. In this study, it is important to start by mentioning that the value of the compressive strength of the fly ash (~23.5 MPa; 23.8 MPa in cubes) is much lower (−29%) than that found for the cement specimens (~33 MPa). In any case, the loads at the Py points are about 10 to 15% lower. The behavior tends to be more similar for Pmax.

Table 6. Deviations from the key values of the strength of the beams.

Viga	2F6	3F6	3F8	4F8
	(Cem-FA)/Cem	(Cem-FA)/Cem	(Cem-FA)/Cem	(Cem-FA)/Cem
P_y [kN]	−8.0%	−17%	−11%	−12%
d_y [mm]	157%	52%	169%	279%
K_{II} [kN/mm]	−46%	−35%	−60%	−74%
P_{max} [kN]	−6.2%	−25.3%	−11%	−14%
d_{max} [mm]	120%	−19%	58%	54%
P_{max}/Py [1]	2.1%	−11%	0%	−2%
d_{ult} [mm]	110%	−8%	−51%	−42%
dult/dy	−18%	−39%	−82%	−85%

The great difference between the behavior of ash beams when compared to that of the cement beams lies in the flexibility throughout State II. As a consequence, deformations at the Py point are the ones with the greatest deviations. However, deformations at the Pult point are lower for ash beams, which indicates lower ductility of ash beams when compared to that of the cement beams.

4. Theoretical Analysis

To complete this study, the authors have decided to use a nonlinear analysis algorithm [17,33,34,36] that they have already applied in other reinforced concrete (RC) beams before. For applying to the beams of the current investigation, as the input of the numerical procedure the experimental stress-strain curves of the materials were considered, both for the steel and for the mortar materials. However, the key parameters values of each stress-strain curve were deduced also by using the information of the tests of the beams.

For the fly ash beams, the tension part was not considered because the beams were cracked before tests, as explained before. However, the predicted behavior of un-cracked fly ash beams was also computed for added information.

Figures 10–14 show the curves for the 5 sets of beams. In these figures, the experimental curves are in continuous line and the theoretical curves are dashed. The red color represents the curves for the cement beams, and the grey color the curves for the fly ash beams. The blue dashed line (theoretical 2) represents the curve for the ash beams, in case the retraction problem could be minimized (beams with no cracks before testing).

These figures show a very good approximation of the theoretical curves to the experimental ones for the cement beams. The least successful approach occurred for beam Cem-3F6. Despite the attention put in these simulations it is not possible to simulate some deficiencies not detected in the beams, or some details that favor their strength.

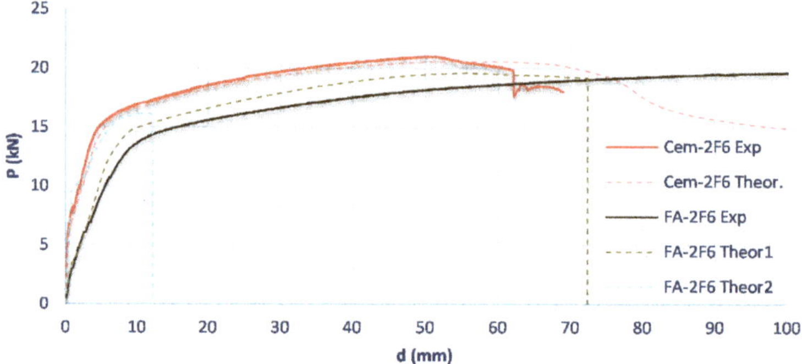

Figure 10. P-d curves for beams Cem-2F6 and FA-2F6.

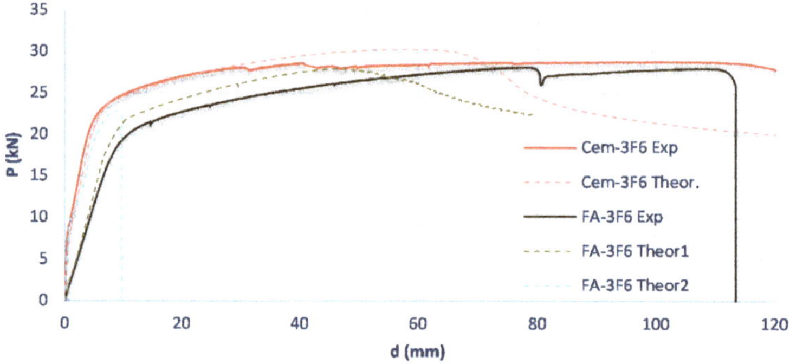

Figure 11. P-d curves for beams Cem-3F6 e FA-3F6.

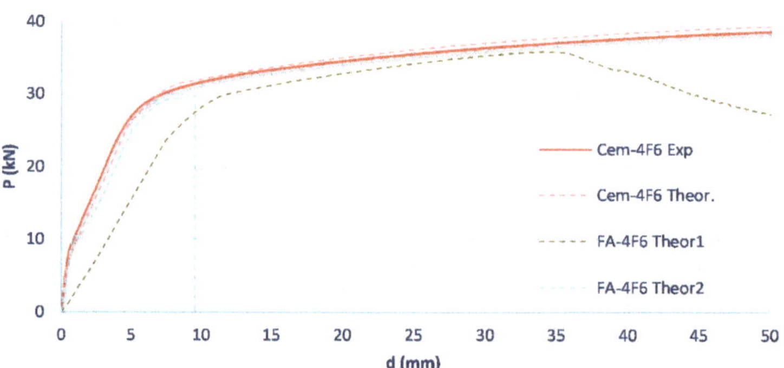

Figure 12. P-d curves for beams Cem-4F6 e FA-4F6.

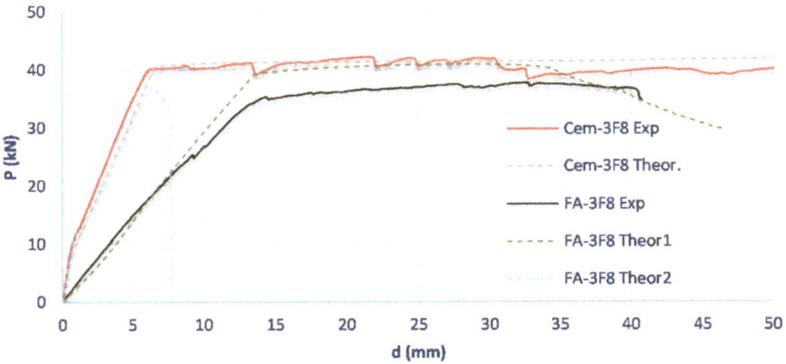

Figure 13. P-d curves for beams Cem-2F8 e FA-2F8.

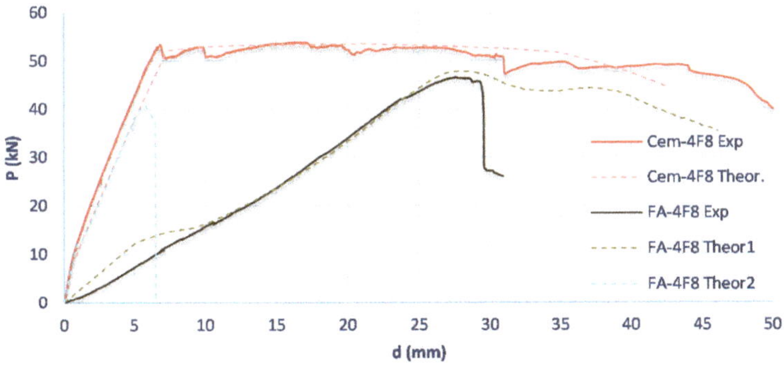

Figure 14. P-d curves for beams Cem-3F8 e FA-3F8.

In the case of fly ash beams, the discrepancies between the theoretical curves and the experimental curves are more significant. Assuming that the behavior of the steel bars is correct (the reinforcing bars of cement beams were the same as those of fly ash beams), these discrepancies can only be originated from the variability of the mortar material (fly ash), which in fact was confirmed in the laboratory. Beam FA-4F8 was the one that showed the highest degree of initial cracking.

In Figure 12, for beam 4F6, there is no experimental curve for the ash beam, as previously mentioned. However, the experimental curves can be presented on the assumption that the fly ash material was similar to the remaining fly ash beams.

The curves called "Theor. 2" corresponded to the behavior of the beams if the behavior of the material was the one verified in the tests of the 40 × 40 × 160 mm prisms, and of the cubes, namely in terms of compressive strength, tensile strength, and modulus of elasticity (as mentioned before, the beams were already cracked before testing). There are important aspects to this prediction. Firstly, for low levels of loading, the curves of the fly ash beams are close to those of the cement beams. In a second aspect, it appears that the beams show lower maximum loads. However, the worse aspect of their behavior when compared to the cement beams, is a very low ductility, since their ultimate load (85% Pmax) occurs for very small deformations (L/dult ~ 140). For comparison, all the cement beams were at least 5 times more ductile.

5. Conclusions

As previously stated, the number of studies on beams made from fly ash geopolymers are very scarce. This was probably the first study with curing at ambient temperature. The (very few) past experimental studies on beams of this material were carried out using special equipment to cure the material at 60 to 70 °C, which is rather difficult to implement outside a laboratory or a precast plant.

The risk of effects due to shrinkage was real, but it was not visible in small specimens, such as prisms or cubes. In the larger masses, such as the test beams, probably due to the presence of the reinforcing bars, the internal constraints led to some cracking during the hardening of the beams. Possibly, with other types of reinforcing, such as Fiber Reinforced Plastics, FRPs, this problem could be avoided, depending on the modulus of elasticity of such material.

In addition, three other aspects were found to be important in fly ash geopolymers beams when compared to cement beams: The low modulus of elasticity, the relatively low value of the maximum compressive strength, and the reduced values of the ductility.

Regarding the low modulus of elasticity of the fly ash beams, it does not prevent this material from being used in structures. Regarding the reduced value of the maximum compressive stress, this has natural influence on the ultimate and service loads. Cross sections need to be bigger than those of cement beams and this makes the inertia higher and, consequently, this can compensate the low modulus of elasticity when deformations have to be limited to certain values (L/400, for instance).

The reduced deformation after the peak load penalizes the ductility of this type of beams. This could be problematic when ductility is important, such as in seismic zones or in hyperstatic structures when redistribution of moments is expected.

In general, this material still needs further developments in order to reduce shrinkage for castings at ambient temperature and to increase the compressive strength. A potential use might be in railway sleepers for instance.

Author Contributions: All authors were involved in all parts of the work, but with different weights, as described next. For Conceptualization, methodology, analysis and writing: A.V.L., S.M.R.L., and I.P.; experimental work: A.V.L. and I.P.; computing of data: A.V.L.; final editing: S.M.R.L. All authors have read and agreed to the published version of the manuscript.

Funding: This research was sponsored by FEDER funds through the program COMPETE—Programa Operacional Factores de Competitividade—and by national funds through FCT—Fundação para a Ciência e a Tecnologia—, under the project UIDB/00285/2020.

Acknowledgments: The authors would like to express their gratitude to L.C. of the Department of Earth Sciences of the University of Coimbra and to A.T.-P. of the Technical Institute of Tundavala for the collaboration in this study. They also want to thank EDP—G.d.P.d.E. S.A., represented by H.F., for the donation of fly ashes and FAPRICELA, S.A., represented by N.B., for the donation of steel bars. Finally, they want to express their gratitude to the students N.M.d.S.C., A.S.G., and J.A.S.R., who have also collaborated in this investigation.

Conflicts of Interest: The authors declare no conflict of interest.

References

1. Lenne, J.; Preston, P. *Making Concrete Change—Innovation in Low-Carbon Cement and Concrete*; Chatham House, the Royal Institute of International Affairs: London, UK, 2018; p. 138.
2. Cembureau. *The Role of Cement in the 2050 Low Carbon Economy*; Cembureau: Brussels, Belgium, 2013; p. 64.
3. Teixeira-Pinto, A. Sistemas Ligantes Obtidos por Activacao Alcalina do Metacaulino (Binder Systems Obtained by Alkaline Activation of Metakaolin). Ph.D. Thesis, University of Minho, Guimaraes, Portugal, 2004. (In Portuguese).
4. Ling, Y. Proportion and Performance Evaluation of Fly Ash-Based Geopolymer and its Application in Engineered Composites. Ph.D. Thesis, Iowa State University, Ames, IA, USA, 2018; p. 151.
5. Criado, M.; Palomo, A.; Fernandez-Jimenez, A.M.; Banfill, P.F.G. Alkali activated fly ash: The effect of admixtures on paste rheology. *Rheol. Acta* **2009**, *48*, 447–455. [CrossRef]
6. Ridtirud, C.; Chindaprasirt, P.; Pimraksa, K. Factors affecting the shrinkage of fly ash geopolymers. *Int. J. Miner. Metall. Mater.* **2011**, *18*, 100–104. [CrossRef]
7. Vijayakumar, R.M. Evaluating Shrinkage of Fly Ash—Slag Geopolymers. Master's Thesis, University of Illinois at Urbana-Champaign, Champaign, IL, USA, 2013; p. 55.
8. Wattimena, O.K.; Antoni; Hardjito, D. A review on the effect of fly ash characteristics and their variations on the synthesis of fly ash based geopolymer. In *Green Construction and Engineering Education for Sustainable Future AIP Conference Proceedings*; Petra Christian University: Kota SBY, Indonesia, 2017; pp. 020041:1–020041:12.

9. Abraham, R.; Raj, S.D.; Abraham, V. Strength and behaviour of geopolymer concrete beams. In Proceedings of the International Conference on Energy and Environment-2013 (ICEE 2013), Putrajaya, Malaysi, 5–6 March 2013; pp. 158–163.
10. Kumaravel, S.; Thirugnanasambandam, S. Flexural behaviour of geopolymer concrete beams. *Int. J. Adv. Engg. Res. Stud. III I* **2013**, *6*, 1–4.
11. Monfardini, L.; Minelli, F. Experimental study on full-scale beams made by reinforced alkali activated concrete undergoing flexure. *Materials* **2016**, *9*, 739. [CrossRef] [PubMed]
12. Ahmed, H.Q.; Jaf, D.K.; Yaseen, S.A. Comparison of the flexural performance and behaviour of fly-ash-based geopolymer concrete beams reinforced with CFRP and GFRP bars. *Adv. Mater. Sci. Eng.* **2020**, *2020*, 1–15. [CrossRef]
13. Fang, G.; Bahrami, H.; Zhang, M. Mechanisms of autogenous shrinkage of alkali-activated fly ash-slag pastes cured at ambient temperature within 24 h. *Constr. Build. Mater.* **2018**, *171*, 377–387. [CrossRef]
14. Samantasinghar, S.; Singh, S.P. Fresh and hardened properties of fly ash–slag blended geopolymer paste and mortar. *Int. J. Concr. Struct. Mater.* **2019**, *13*, 47. [CrossRef]
15. Biondi, L.; Perry, M.; Vlachakis, C.; Wu, Z.; Hamilton, A.; McAlorum, J. Ambient cured fly ash geopolymer coatings for concrete. *Materials* **2019**, *12*, 923. [CrossRef] [PubMed]
16. Lopes, A.V.; Lopes, S.M.R.; Pinto, M.I.M. Influence of the composition of the activator on mechanical characteristics of a geopolymer. *Appl. Sci.* **2020**, *10*, 3349. [CrossRef]
17. Lopes, A.V.; Lou, T.J.; Lopes, S.M.R. Numerical Evaluation of the Non Linear Behaviour of Reinforced Concrete Beams. In Proceedings of the 14th International Conference on Civil, Structural and Environmental Engineering Computing, Cagliari, Sardinia, Italy, 3–6 September 2013; Civil-Comp Press: Stirlingshire, UK, 2013; p. 115.
18. Kim, J.K.; Yi, S.T.; Kim, J.H.J. Effect of specimen sizes on flexural composite strength of concrete. In *Fracture Mechanics of Concrete Structures*; Borst, Ed.; Smets & Zeitlinger: Lisse, Holland, 2001; pp. 697–704.
19. Carpinteri, A.; Chiaia, B. Embrittlement and decrease of apparent strength in large-sized concrete structures. *Sadhana* **2002**, *27*, 425–448. [CrossRef]
20. Kim, N.; Lee, J.; Chang, S. Equivalent multi-phase similitude law for pseudodynamic test on small scale reinforced concrete models. *Eng. Struct.* **2009**, *31*, 834–846. [CrossRef]
21. Fapricela. Available online: https://www.fapricela.pt/ (accessed on 27 May 2020).
22. NP EN10002-1. *Materiais Metálicos. Ensaios de Tracção—Parte 1: Método de Ensaio à Temperature Ambiente (Tensile Testing of Metallic Materials. Method of Test at Ambient Temperature)*; Instituto Português da Qualidade: Caparica, Portugal, 2006.
23. NP EN450-1. *Cinzas Volantes Para Betão—Parte 1: Definição, Especificações e Critérios de Conformidade (Fly Ash for Concrete—Part 1: Definition, Specifications and Conformity Criteria)*; Instituto Português da Qualidade: Caparica, Portugal, 2012.
24. Granizo, M. Activation Alcalina de Metacaolin: Desarrolllo de Nuevos Materials Cementantes (Alkaline Metakaolin Activation: Development of New Binding Materials). Ph.D. Thesis, Universidad Autonoma de Madrid, Madrid, Germany, 1998; p. 325.
25. Granizo, M. Activation Alcalina de Metacaolin: Desarrolllo de Nuevos Materials Cementantes (Alkaline Metakaolin Activation: Development of New Binding Materials). PhD Thesis, Universidad Autonoma de Madrid, Madrid, Germany, 1998; p. 325.
26. Davidovits, J. Chemistry of Geopolymeric Systems, Terminology. In *Proceedings of 2nd International Conference Géopolymère*; Geopolymer Institute: Saint-Quentin, France, 1999; pp. 9–40.
27. Pinto, A. *Introdução ao Estudo dos Geopolímeros (Introduction to the Study of Geopolymers)*; Sector Editorial dos SDE-UTAD: Vila Real, Portugal, 2006; p. 79.
28. Neville, M. *Properties of Concrete*; John Wiley and Sons: New York, NY, USA, 1973.
29. NP EN196-1. *Métodos de Ensaio de Cimentos—Parte 1: Determinação das Resistências Mecânicas (Methods of Testing Cement. Determination of Strength)*; Instituto Português da Qualidade: Caparica, Portugal, 2006.
30. NP EN12390-5. *Ensaios do Betão Endurecido—Parte 5: Resistência à Flexão de Provetes (Testing Hardened Concrete. Flexural Strength of Test Specimens)*; Instituto Português da Qualidade: Caparica, Portugal, 2009.
31. NP EN 1998-1. 2009. *Eurocódigo 8: Projecto de Estruturas Para Resistência Aos Sismos. Parte 1: Regras Gerais, Acções Sísmicas e Regras Para Edifícios (Design of Structures for Earthquake Resistance. General Rules, Seismic Actions and Rules for Buildings)*; Instituto Português da Qualidade: Caparica, Portugal, 2010.

32. Lopes, A.V.; Lopes, S.M.R. Importance of a rigorous evaluation of the cracking moment in RC beams and slabs. *Comput. Concr.* **2012**, *9*, 275–291. [CrossRef]
33. Lopes, S.M.R.; Lopes, A.V.; Lou, T. Numerical Estimation of RC Beams. In Proceedings of the International Conference on Recent Advances in Nonlinear Models, Structural Concrete Applications, CoRAN 2011, Coimbra, Portugal, 24–25 November 2010; University of Coimbra: Coimbra, Portugal, 2011; pp. 243–253.
34. Lopes, A.V.; Lou, T.; Lopes, S.M.R. Nonlinear flexural behaviour of RC beams. In *Civil Engineering and the Concrete Future, Proceedings of the Twin Covilhã International Conferences, 5th Int'l Conference on the Concrete Future, Covilha, Portugal, 26–29 May 2013*; CI-Premier Pte Ltd.: Covilha, Portugal, 2013; pp. CF113–CF122.
35. NP EN 1992-1-1:2010. *Eurocódigo 2: Projeto de Estruturas de Betão; Parte 1: Regras Gerais e Regras Para Edifícios (Eurocode 2: Design of Concrete Structures. General Rules and Rules for Buildings)*; Instituto Português da Qualidade: Caparica, Portugal, 2010.
36. Lou, T.; Lopes, S.M.R.; Lopes, A.V. A finite element model to simulate long-term behavior of prestressed concrete girders. *Finite Elem. Anal. Des.* **2014**, *81*, 48–56. [CrossRef]

Article

Influence of the Composition of the Activator on Mechanical Characteristics of a Geopolymer

Adelino V. Lopes [1], Sergio M.R. Lopes [2,*] and Isabel Pinto [2]

[1] Institute for Systems and Computers Engineering, Department of Civil Engineering, University of Coimbra, 3030-788 Coimbra, Portugal; avlopes@dec.uc.pt

[2] Centre for Mechanical Engineering, Materials and Processes, Department of Civil Engineering, University of Coimbra, 3030-788 Coimbra, Portugal; isabelmp@dec.uc.pt

* Correspondence: sergio@dec.uc.pt

Received: 1 April 2020; Accepted: 8 May 2020; Published: 12 May 2020

Featured Application: This work is a contribution to better knowledge of the behavior of a new family of materials obtained by alkali activation of a binder, which belongs to the geopolymers group. The possibility of its use in structures with finishes of different colors and textures is a feature that would be of great interest in some innovative architectural applications.

Abstract: Geopolymer materials are characterized by their high durability and low carbon dioxide emissions, when compared with more traditional materials, like concrete made from ordinary Portland cement. These are interesting advantages and might lead to a more sustainable construction industry. The aim of this study is the characterization of the mechanical behavior of the materials obtained by the activation of metakaolin. The activator is a mixture of sodium hydroxide with sodium silicate in different proportions. The influence of the composition of activator is studied. For the analysis of the mechanical properties of the different mixtures two different types of tests were performed, bending tensile strength tests and compressive strength tests. The results show that an activator with not less than 300 g of sodium hydroxide and not exceeding 600 g of sodium silicate per 750 g of metakaolin gives the best results, for both tensile strength and compressive strength.

Keywords: Geopolymer; Alkali activated; compressive strength; tensile strength; deformability

---

1. Introduction

Some old structures, such as the Egyptian pyramids or Roman constructions have lasted more than 2000 years and are still in good condition, showing a remarkable durability. Glukhovsky argued that many of them were made of aluminosilicate calcium hydrates similar to those of Portland cement and also of crystalline phases of analcite, a natural rock. These constructions have been shown to have outstanding behavior, not only concerning the mechanical performance, but also in terms of durability and resistance to weathering [1–3].

The (re)discovery of these materials could be very important for the construction industry. In fact, in addition to the higher strength properties shown, these materials might solve some important limitations of ordinary Portland cement (OPC): environmental problems with the release of carbon dioxide (CO_2) during the production process, and also performance due to aggressive environmental conditions [4–7]. Some authors claim that the technology reduces CO_2 emissions caused by the cement and aggregates by 80% to 90% when compared to OPC [8–10]. Other authors found that the difference in CO_2 emissions is not so high, but they confirm that the technology has clear advantages in such aspect [11]. Therefore, the need to promote research for materials alternative to OPC seems clear, and such material must be also competitive in its mechanical behavior. A possible solution seems to be the use of a binder obtained by alkaline activation. In general, when compared to OPC, these materials have shown acceptable mechanical strength (sometimes the strength is higher than that of normal concrete structures, made from OPC). The development of the strength can be very quick, which might speed up the construction rate. Durability and stability are also acceptable. Understandably, these characteristics depend on the mix proportions of the components of the concrete mass.

In general, the alkali activation is a hydration reaction of aluminum silicate with alkali or alkaline-earth type materials, such as: hydroxides (ROH, $R(OH)_2$), weak acid salts (R_2CO_3, R_2S, RF), strong acid salts (Na_2SO_4, $CaSO_4.2H_2O$) or silicate salts such as $R_2(n)$ SiO_2, where R is an alkaline ion of Na, K or Li type, or alkali earth such as Ca [12,13]. Basically, the process consists of mixing an alkaline activator with a binder rich in alumino-silicates. First, the alumino-silicates undergo a heat treatment, involving loss of water, and modification of the coordination of the aluminum ion with oxygen. This helps to enhance the results since the material shows a great susceptibility to combine chemically [13,14]. Slags from blast furnaces, fly ash from coal burning in thermoelectric power plants, volcanic ash with natural heat treatment, tile or brick powder, and metakaolin obtained from kaolin (natural alumino-silicates), are examples of potential materials for the alkali activation.

Alkali activators are obtained by the use of alkali metals such as sodium and potassium. Within a relatively large group of alkaline activators, two can be distinguished: a simple activation and a combination of activators. A simple activation is obtained by a single activator, i.e., just one chemical compound, for example sodium hydroxide (NaOH) or potassium hydroxide (KOH). Combination of activators is also possible and might involve hydroxide and silicate of the alkali metal in appropriate proportions.

According to A. Pinto [13], geopolymer mass activated with a simple activator is very plastic and shows low workability due to the high viscosity of the activator. This viscosity increases considerably when the concentration of hydroxide increases. The composite activators are slightly less viscous and, therefore, they would lead to higher workability. Using an activator based on sodium or potassium eventually leads to similar results [15]. Simply use of sodium silicate might be sufficient depending on the curing temperature.

According to Granizo [16], the molar concentration 12 M sodium hydroxide is the lowest concentration able to promote the alkaline reaction. Davidovits [17] suggested that the maximum concentration would be 20 M from which the identification of the alkaline reaction would no longer be possible.

In his study of the alkaline activation of metakaolin by using an alkaline solution with sodium silicate, Pinto [13,18] reported an increase in mechanical strength both for compression, with values from 30 to 60 MPa, and for tensile bending, with values between 5 and 7 MPa. Fernandez-Gimenez and Palomo [19] reported that the use of a solution consisting of sodium hydroxide (NaOH) and sodium silicate (Na_2SiO_3) as the alkali activator, rather than only NaOH caused an increase of 40 to 90 MPa after curing for one day.

Altan and Erdogan [15] studied the effect of temperature (at room temperature and at 80 °C), and the strength development of slag mortars alkali-activated with Na + and K + hydroxide and with sodium silicate. To obtain a high strength, these authors found that both activators are required for curing at room temperature. At 80 °C sodium silicate is essential and sufficient as the NaOH solution is not needed to the gain in strength. This is because at this temperature the dissolution of the slag is sufficiently high. At room temperature, NaOH concentration affects directly the rate of increase of the strength, as it affects the rate and amount of dissolution of the slag. At 80 °C, the amount of KOH was shown to be more effective than NaOH.

The behavior of each component of the activator is not yet fully understood. In the alkaline activation process, the sodium silicate acts as a binder while sodium hydroxide acts more in the dissolution process of the raw materials [20]. However, this proposal does not pose the possibility of Na_2SiO_3 playing both roles: providing at first the dissolution of aluminum silicate (breakdown of siloxane bonds) and afterwards acting as a binder, rearranging the structure [7].

Also interesting are recent studies on some mechanical properties of geopolymer composites reinforced by ultra-high-molecular-weight polyethylene fibers and on the strength of powder-based three dimensional (3D) printed geopolymer samples [21,22].

Past research was concentrated on material studies. However, if the geopolymers are to replace to a certain extent the cement-based materials, the mechanical behavior of such types of material needs to be fully understood. Structural members made from geopolymers need to be constructed and tested. For instance, the deformability of members is of great importance in the structural design of continuous beams and prestressing systems are often used to overcome structural concrete difficulties [23,24]. In normal concrete, the modulus of elasticity does not vary too much with the strength. Moduli of elasticity of geopolymers can be very different from those of normal concrete and this will pose new structural problems that need to be studied. Cracking behavior of members is of great importance too, because of its influence on the corrosion of the steel bars of the reinforced elements and also because of its influence on the deformability of the members [25].

The combination of geopolymers with other materials, such as steel and timber for structural applications also need to be studied. These studies might follow similar laboratorial or computational steps as those applied to combinations of concrete with steel or with timber [26,27].

There are already some few studies on the structural behavior of members made with geopolymers. Some reports of tests on columns and beams were presented in Chapter 26 of the book *Concrete Construction Engineering Handbook* [10]. Tests on columns were also performed by Maranan et al. [28]. More structural examples are reported by Mo et al. [29]. The bond between steel bars and geopolymers was studied by Dahou et al. [30]. The aspects that were studied in the above works need to be further developed and many other aspects of the structural behavior need to be studied. For instance, it is expected that the geopolymers can have a positive performance regarding the protection of the reinforcing bars and regarding the bond between bars and concrete. The relatively high tensile strength can be of great importance to limit the cracking extension of the reinforced members.

The authors of this article have already performed some tests on geopolymer-reinforced beams and they found it very important to study the basic mechanical characteristics of the material and to study the influence of the composition of the activator on such characteristics. This article presents some of the findings that would be very useful for the structural design of structural members.

2. Materials and Testing Procedures

The material under study, a geopolymer mortar, is a mixture of metakaolin with an activator and sand. After a period of curing of 20 d, bending tests were performed to study the mechanical behavior of specimens. At the beginning, the load was not very high before the elastic behavior could be observed and, in a second stage, the load was increased up to failure. Compression tests were also carried out. Bending tests give the information on the modulus of elasticity and indirectly on the tensile strength. The compression tests are the most common tests for concrete. However, the tensile strength and the modulus of elasticity are also very important because they determine the way the structure behaves under service loading. The durability of the structure itself or the risk of malfunction (cracking) of partition walls, for instance, derives mostly from such characteristics.

2.1. Materials

The metakaolin used in this work is identified as HRM MetaMAx® [31] and was supplied by Engelhard, an American company from Iselin, New Jersey. This is a high-reactivity metakaolin and the grain particles are of smaller size than the usual metakaolin. In order to evaluate the chemical composition of this material, an elemental analysis was made (see Table 1) and this analysis indicated that the main components were silica (SiO_2) and alumina (Al_2O_3).

Table 1. Elemental composition of metakaolin.

Element	SiO_2	Al_2O_3	K_2O	TiO_2	Fe_2O_3	MgO	P_2O_5	SO_3	FeO
%	57	36.97	0.15	2.73	0.40	1.26	0.42	0.54	0.36

The sand was collected in Coimbra, Centre Region, Portugal. From the particle size analysis, the sand can be classified as a poorly graded sand according to the Unified Soil Classification System, ASTM D 2487-06 [32]. The density of the sand particles is G = 2.64 g/cm^3.

During this work an activator has been used resulting from the combination in suitable proportions of sodium hydroxide (NaOH) (10 M) with sodium silicate (Na_2SiO_3). Table 2 shows the composition of the Base Mix of the studied mixtures. Six specimens (4 × 4 × 16 cm) were made for each batch.

Table 2. Composition of the Base Mix of the geopolymers.

Metakaolin (g)	Sand (g)	Activator (g)	=	Hydroxide (g)	+	Silicate (g)	Ratio Hydroxide:Silicate
750	1875	900	=	300	+	600	1:2

Table 3 presents the 8 compositions used with the correspondent relative concentration of the activator (sodium hydroxide and sodium silicate). The results of these mixtures are, thereafter, compared to the ordinary Portland cement (OPC) mortar with no additives (see Table 4).

Table 3. Other compositions of the geopolymers.

Mix Type	Metakaolin (g)	Sand (g)	Activator (g)	=	Hydroxide (g)	+	Silicate (g)	Ratio Hydroxide:Silicate
A	750	1875	750	=	300	+	450	1:1.15
A	750	1875	1050	=	300	+	750	1:2.5
B	750	1875	900	=	450	+	450	1:1
B	750	1875	900	=	360	+	540	1:1.5
B	750	1875	900	=	327	+	573	1:1.75
B	750	1875	900	=	277	+	623	1:2.25
B	750	1875	900	=	257	+	643	1:2.5
B	750	1875	900	=	180	+	720	1:4

Table 4. Composition of the ordinary Portland cement (OPC) mortar.

Cement	Sand (kg)	Water (kg)
9	39.9	6

The analysis of the influence of the activator composition is based on experimental results, namely the modulus of elasticity E, tensile strain ε_t, tensile strength σ_t and compressive strength σ_c. EN 196-1 [33] was followed. This European standard presents the procedure for the determination of both the tensile bending tests and compression tests for cement mortars and establishes the use of prismatic 4 × 4 × 16 cm specimens. EN 12390-parts 1, 3 and 5 [34–36] was also followed for the strength tests on mortar specimens.

Mixing was carried out mechanically, using a mixer for this purpose with the dimensions and characteristics specified in EN 196-1 [33].

2.2. Tensile Bending Tests Under Elastic Behaviour

The initial modulus of the specimens, E0, was determined by the tensile bending tests under elastic behaviour and Hooke´s law ($\sigma = \varepsilon.E$). The σ_t is related to the load applied, F, while ε_t is measured using an extensometer stuck in the center of the lower face of test specimens.

The determination of the modulus of elasticity is not described in EN 196-1 [33]. However, the general procedure suggested by the standard was adapted. The load model presented in EN 12390-5 [36], where the specimens are subjected to two F/2 loads placed at thirds spans, was adopted. During the test, the vertical load was applied with increments of 20 N, by using standard weights. The strain for each load level was recorded.

Four charge/discharge cycles were applied with the maximum load of 80 N (less than 1/20 of the failure load). The tensile strength σ_t (MPa) is given by the following equation:

$$\sigma_t = \frac{6M}{b \times h^2} = F \times \frac{L}{2} \times \frac{6}{b \times h^2} = 3\,\frac{F \times L}{b \times h^2} \quad (1)$$

where M (N mm) is the moment applied to the specimen; b (mm) and h (mm) are geometrical properties of the specimen (width and height respectively), F (N) is the applied load; L (mm) is the horizontal distance between the load F/2 and the nearest support.

2.3. Tensile Bending Tests

The tensile bending tests described herein were performed by using the same load model described in EN 12390-5 and used for the initial modulus of elasticity. However, for this test, the load increments were applied up to failure by hydraulic means. At first, a vertical deflection was applied with a maximum preload of about 0.2 kN (about 1/10 of the failure load) and after that the load was increased progressively to achieve a deformation speed of 0.003 mm/sec until failure.

2.4. Compressive Strength Tests

The pair of specimens obtained from each tensile bending test were subject to compression loading, in two stages. In the first stage of the test, a small preload of 5 kN was applied and afterwards the load was increased until failure, at a deformation rate of 0.01 mm/sec. The load at failure (F) was recorded for the computation of the compression strength σ_c (MPa) of the mix through the following Equation (2):

$$\sigma_c = \frac{F}{A_{specimen}} = \frac{F}{b \times h} \quad (2)$$

Figure 1 shows the specimens after the compressive tests. It is clear that the specimens present the double pyramid shape after failure and therefore, according to EN 12390-3 [36], this type of failure can be classified as satisfactory.

Figure 1. Typical failure shapes resulting from the compressive tests.

3. Test Results

After the tests, the results were analyzed. The tensile bending tests under elastic behavior were the first tests carried out in order to evaluate the static modulus of elasticity. After these, two other types of tests were performed up to failure to evaluate the strength characteristics: the bending tensile testing, and the compressive tests. For each composition of the activator, a total of 6 specimens were tested for the bending tensile test and 12 specimens for the compressive tests. All these tests were performed after 20 d of curing.

3.1. Tensile Strength

For the tensile bending test, the increment of the applied load was measured while the strain was controlled. The maximum load before failure was recorded. As should be expected, failure occurred mostly in the central area of the specimens, except for a few cases where there was a rupture slightly shifted from this central zone. After bending tests each specimen was thus divided into two prisms. Figure 2 shows the statistical tensile strength σ_t for the 9 different compositions of the activator. For example, TB-1:2.25 means Type B mixes (see Table 3) with 1:2.25 ratio of activator composition.

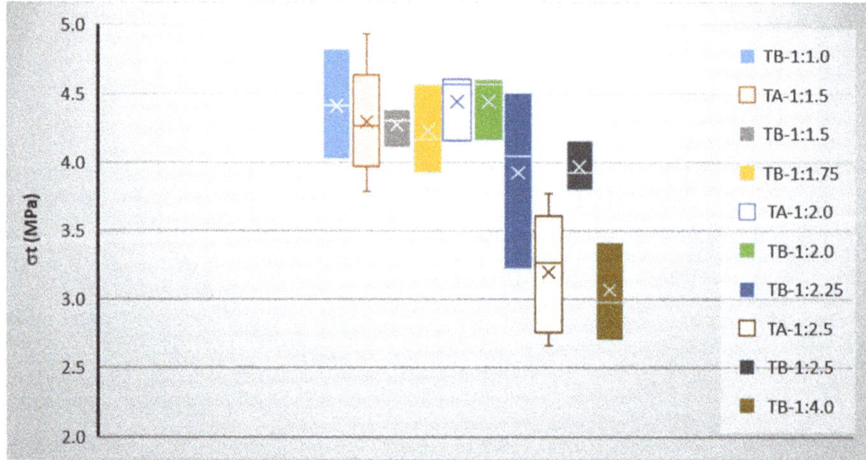

Figure 2. Statistical tensile strength for Type A and Type B mixes.

As can be seen, for Type A mixes (see Table 3) the average value of tensile strength values is maximum for the Base Mix (1:2). In fact, these mixes have been shown to have some excess of liquid or imperfections, which might have influenced the strength. As the amount of the activator varies relative to the Base Mix, the workability also changes. It was found that 1:1.5 ratio is less workable and 1:2.5 is more fluid. In fact, for the former mix it was necessary to add an extra 50 mL of water at the time of preparation of the batches. The results suggest that the more liquid the material has, the less strength it has.

As far as Type B mixes are concerned, it is found that the composition with the highest tensile strength corresponds to 1:1 (4.5 MPa) and that of the lowest value corresponds to 1:4 (3.1 MPa), i.e., only about 2/3 of the highest value. Excepting this latter mix (1:4), the tensile strength values range between about 4.0 and 4.5 MPa, and it seems it is not possible to see a direct relationship between σ_t and activator composition. The mixes of various compositions exhibit strengths not higher than the Base Mix. For Type B mixes and for the OPC mortar specimens, the computed standard deviation was found to be always lower than 0.7 MPa, which indicates some variability of results that should not be neglected. Failure surfaces always occurred in the central area of the specimens.

The OPC mortar show a tensile strength of 3.64 MPa after 28 d of cure and 3.66 MPa after 60 d. Taking into consideration the usual gain of compressive strength with time, it would be expected that the tensile strength would accompany this evolution as expressed in EC2 [37]. However, the strength values for 28 and 60 d are very similar. Phenomena such as the shrinkage of mortar may be the reason for the non-increase of this strength, but no macroscopic shrinkage was detected in the specimens.

The OPC mortar strength is generally about 15% lower than that obtained for most of the geopolymer mixes, which is not a dramatic difference in practical terms.

The results seem to show that the tensile strength of geopolymers is not significantly influenced by the composition of the mixes. The biggest variation is at 1:2.5 for Type A mixes and at 1:4 for Type B mixes, although the variation from the Base Mix is never too sharp. It was also found that the tensile strength of geopolymers is not significantly higher than that of the cement mortar, but it still generally shows a moderate gain. This aspect can significantly modify the behavior of structural members as far as phenomena such as cracking or bond properties of the steel reinforcement are concerned.

3.2. Tensile Strain

Some specimens were instrumented with strain gauges to evaluate the strains up to failure, during the bending tensile tests. It is assumed that failure would correspond to the point of maximum load applied to the specimen. The results are shown in Figure 3. The presented margin of error of the data is computed from the standard error of the mean (or alternatively from the product of the standard deviation of the data and the inverse of the square root of the sample size, which is the same) and it is typically approximately twice the standard deviation—the half-width of a 95% confidence interval.

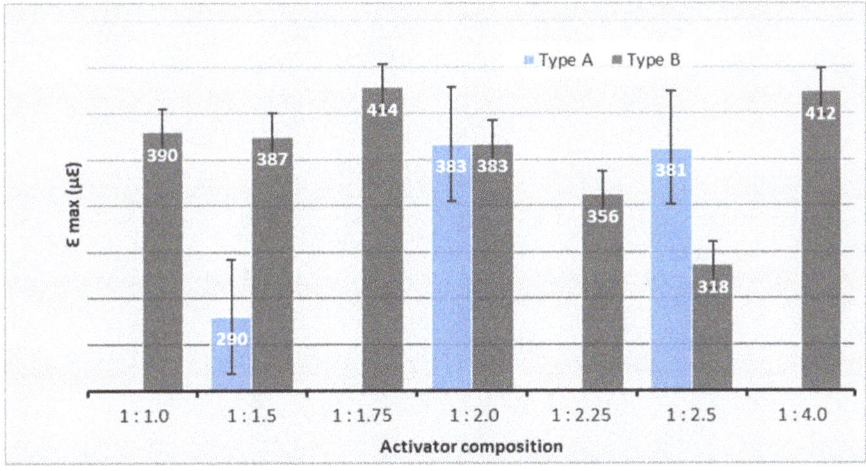

Figure 3. Strains at failure for Type A and Type B mixes.

As can be seen, Type A mixes have, on average, slightly lower values than those of Type B, except for 1:1.5, for which the strain value is far below the others (290 με). The possible reason given before for the explanation of low tensile strength values, i.e., some excess of liquid, does not seem to fully explain the difference in this case. The variability of the results would help to explain some relatively high variations.

For Type B mixes, the composition with the highest strain corresponds to 1:1.75 (414 με) and the lowest to 1:2.5 (318 με). It also appears that most of the mixes have strains higher than that of the Base Mix, except for 1:2.25 and 1:2.5. However, the values did not show any direct relationship to the activator composition. For Type B mixes, the standard deviation shows a maximum value of 108 με, revealing the significant variability of the results.

For OPC mortars, the tensile strain measured at failure was 118 με, which is clearly lower than those obtained for the different compositions of geopolymer specimens (the maximum standard deviation was found to be 21 με). On average, OPC mortars have values approximately one third of those of geopolymers. This means that the geopolymers seem significantly more deformable at tensile failure. These results are very important and have great impact on the structural behaviour of the members. The structures could be more adaptable to the imposed deformations (for instance differential soil settlements), but the deformations of the slabs could be a critical aspect in the structural design (as known, the deformation of the slabs must not exceed some values specified in standards to prevent cracking of non-structural components of the construction, such as partition walls, for instance).

3.3. Modulus of Elasticity

The modulus of elasticity (MOE) can be obtained from the tensile bending tests on specimens with bonded strain gauges. Two test methods can be used: the static and the quasi-static. The modulus of elasticity calculated from the tensile bending test up to failure is named the quasi-static modulus of elasticity (QSMOE) and the modulus obtained from the elastic tests is named the static modulus (SMOE).

During the first type of tests, the strain values were recorded after stabilization of the readings for each load. The maximum strain has never exceeded 100 $\mu\varepsilon$ and readings in the loading part of the curve should coincide with the readings in the unloading part down to the initial zero. This is an essential condition for the behavior to be considered elastic. Typically, the readings stabilized 2 to 4 s after loading/unloading. For the QSMOE tests, the strain rate was defined in such a way that failure should never occur in less than 120 s. This was considered an essential condition for the test to be considered quasi-static, that is, with negligible dynamic effects. Comparison of the results was carried out considering always the same maturity for the test specimens (20 d).

Figure 4 shows the results for Type A and Type B mixes. The bars for QSMOE and SMOE are light- and dark-colored respectively.

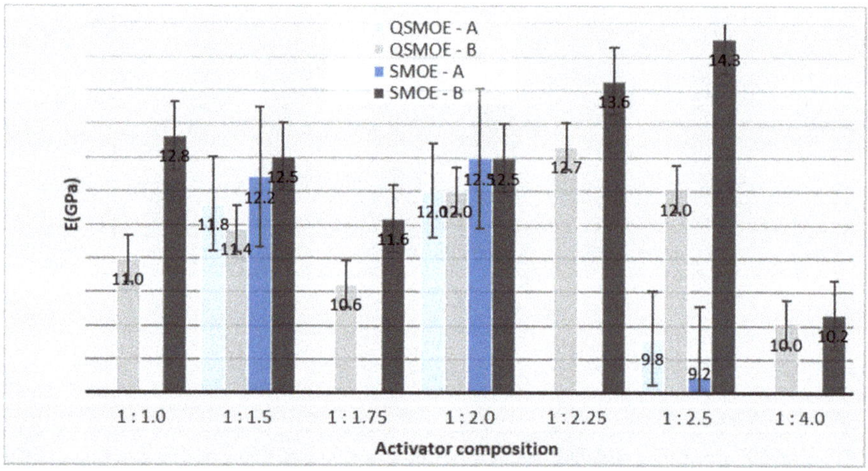

Figure 4. Quasi-static and static modulus of elasticity (QSMOE/SMOE) for Type A and Type B mixes.

The results show that QSMOE values for Type A mixes are in general equivalent to those of Type B. As mentioned before, excess of liquid in Type B mixes or perhaps lack of liquid in Type A mixes might help to explain some variability of the values.

The standard deviation for each composition for Type B mixes was always less than 4.0 GPa which, in percentage, is of the same order of variability as that for the maximum strain. Curiously, the maximum standard deviation for the OPC mortar was less than 2.3 GPa for values of moduli of elasticity much higher than those of geopolymers (at least the double). The highest value of QSMOE for Type B mixes was for 1:2.25 (12.7 GPa) whereas the lowest value was for 1:1.4 (10 GPa). This latter value is significantly low when compared with other values of Type B mixes, which shows the low rigidity of this ratio. With the exception of 1:1.75 and 1:4, it can be seen that for Type B mixes the proportion of the components of the activator does not influence significantly the results, as these are very close, between 11 and 13 GPa.

Comparing the values for the QSMOE and SMOE, it appears that the former are slightly higher than the latter, with the exception of one case: Type A Mix 1:2.5.

For the QSMOE of OPC mortars, a value of 28.3 GPa was obtained. This value was 19% lower than that of SMOE. In general, the modulus of elasticity of OPC mortars was about 2.3 times higher than that of geopolymers of different compositions. This confirms the higher deformability of geopolymers when compared with OPC mortars, which is an important conclusion, very much in line with the outcome of tensile strain results presented above.

As mentioned before, some specimens were instrumented with strain gauges, leading to readings of, not just the load at failure but also the overall evolution of the tensile strain with loading. Therefore, it is possible to plot diagrams with strain (ε) versus tensile bending stress (σ) for each instrumented specimen. In this situation, the QSMOE corresponds to the tangent of the ε-σ curve.

Figure 5 shows three stress–strain diagrams corresponding to 2 Type B mixes and to the OPC mortar. It can be seen that the behavior is essentially linear up to the maximum load, which was taken as the failure load. Failure can be classified as a brittle failure.

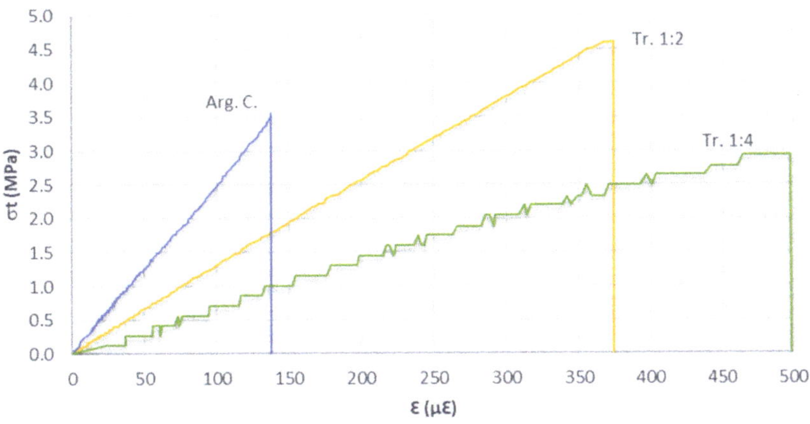

Figure 5. Stress–strain diagrams for tensile bending tests.

3.4. Compressive Strength

Figure 6 shows the statistical compressive strength, σc, for the 9 different activator compositions. As can be seen, for Type A mixes the highest compressive strength is that of the Base Mix (1:2). As for the other mechanical characteristics, the excess of liquid might be the reason for the reduction of the compressive strength for 1:1.5 and 1:2.5. This was an important reason that leads to the change of composition of the following mixes, Type B mixes, which were obtained by maintaining constant the activator weight.

For Type B mixes, the highest strength corresponds to 1:1.75 (42.8 MPa) and the lowest to 1:4 (20.5 MPa). Because the composition had small amounts of sodium hydroxide, and therefore low ability to dissolve the constituent materials, the compressive strength would be low. The standard deviation for Type B mixes was always less than 3.1 MPa, which can be considered a good value.

For OPC mortars the compressive strength measured for 28 days of curing was 21.7 MPa and for 60 days was 23.5 MPa, which corresponds to an 8% increase in strength, as would be expected. From the results, OPC mortar presents lower compression strength when compared to the geopolymers. For the OPC mortars, the maximum standard deviation was 2.6 MPa.

In order to validate the methodology described in EN 196-1 [33] for the calculation of the tensile bending strength and of the compression strength through 4 × 4 × 16 cm OPC mortar specimens, some compressive tests were also performed on 15 cm-wide cube specimens. Such cubes were loaded up to failure, following EN 12390-3 [35]. An average value of 24.1 MPa was obtained for the strength of 28 d old cubes (this value was 11.1% higher than that of small specimens), and therefore was not very different.

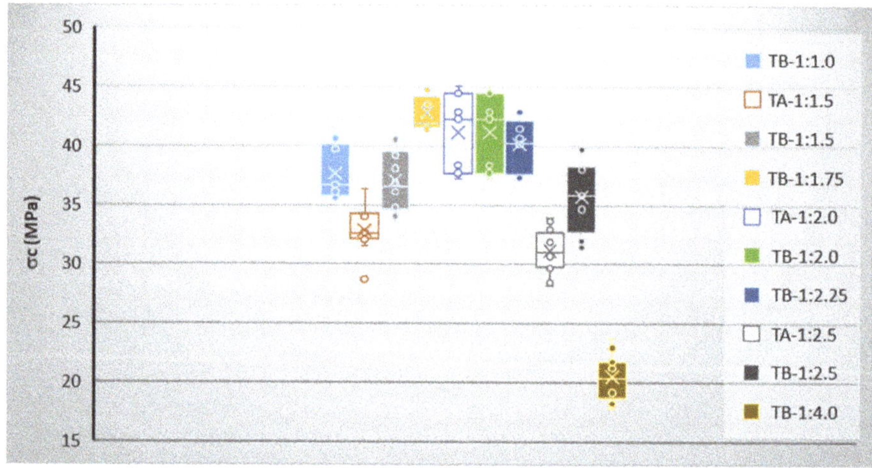

Figure 6. Statistical compressive strength for Type A and Type B mixes.

4. Comparisons of Results

One of the objectives of this study was to find the most efficient composition for the activator, but the results does not give any clear conclusion on this. Nevertheless, it seems that the Base Mix, 1:2 is a good solution, and compositions between 1:1.5 and 1:2.5 are equally acceptable in practical terms. When compared to Type A mixes, the great majority of the specimens of Type B mixes present better results for all the mechanical properties that were studied. For both compression and tensile strength the worst results correspond to 1:4, and the best performance appears to be between the ratios 1:1.75 and 1:2.

It is known that sodium hydroxide (NaOH) is responsible for breaking down the internal connections of the binder at the beginning of the process. Afterwards, sodium silicate (Na_2SiO_3) activates the necessary connections between the binder and the aggregates. Therefore, for a certain amount of metakaolin, an optimal balance between the quantities of NaOH and of Na_2SiO_3 would be expected for the best performance of the geopolymer. To find out this optimal balance, each of the activators were separated and some graphs for the individual influence of each one of the activators were plotted for the tensile and the compressive strength. As a consequence, 4 graphs were produced (Figures 7–10). Figures 7 and 8 show the evolution of the tensile strength σ_t, with the amount of NaOH and Na_2SiO_3, respectively.

For Figure 7, the first value corresponds to composition 1:4, whereas the last one corresponds to 1:1. In this analysis, Type B mixes were the only ones considered for the trend line, since Type A mixes had different amounts of activator and, therefore, comparisons should be too complex. This figure might suggest two different possible trends: the linear increase of σ_t with the amount of NaOH (the black trend line correspond to this possibility); or an increase of σ_t up to 300 g of NaOH (corresponding to 1:2) and a constant value thereafter (this trend line is not presented in the figure). This latter possibility is not as simple as the first, but seems to fit better the points of the graph.

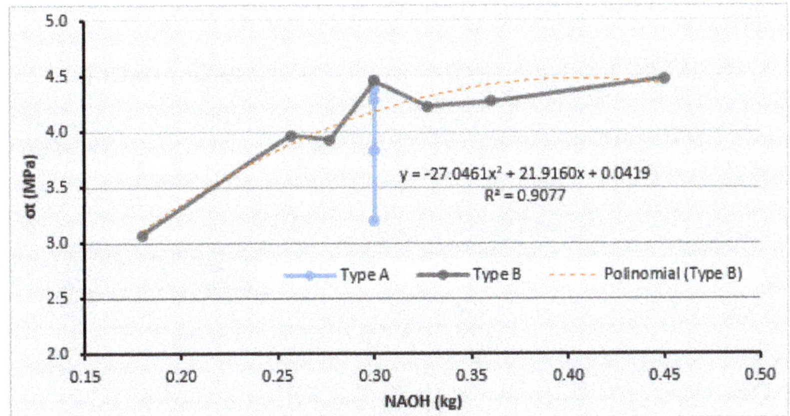

Figure 7. Influence of sodium hydroxide (NaOH) on the tensile strength.

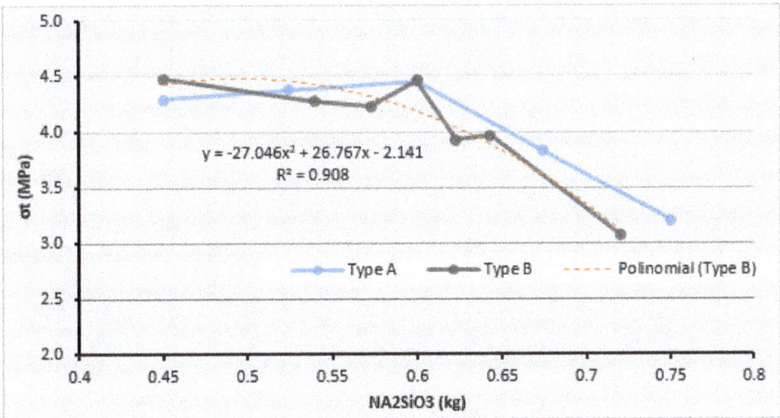

Figure 8. Influence of sodium silicate (Na2SiO3) on the tensile strength.

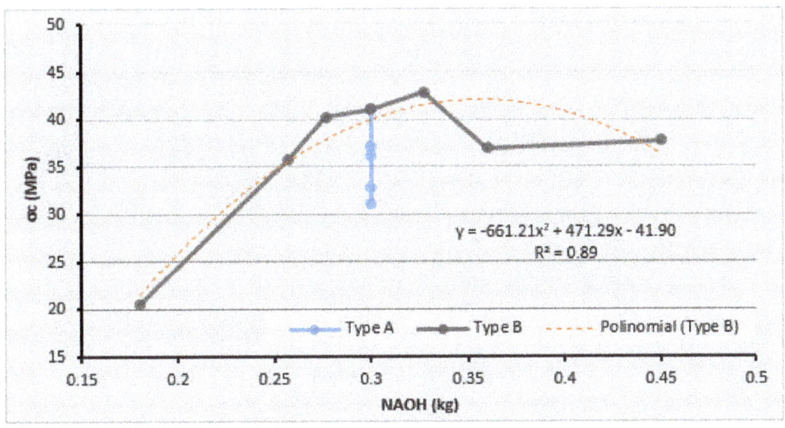

Figure 9. Influence of sodium hydroxide (NaOH) on the compressive strength.

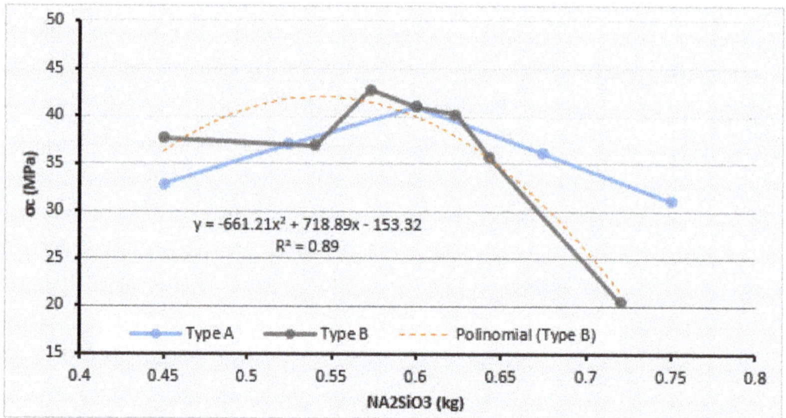

Figure 10. Influence of sodium silicate (Na_2SiO_3) on the compressive strength.

In Figure 8 the influence of this activator seems to have a different trend when compared with Figure 7. This is understandable because the sum of correspondent values of both figures gives the same constant value, the amount of activator (900 g). However, in Fig. 8 the values for Type A mixes are reasonably close to those of Type B mixes. Quantities of sodium silicate between 450 and 600 g seem to lead to values of σ_t that are approximately the same.

Figures 9 and 10 show the variation of the compressive strength, σ_c, with the amount of NaOH and Na_2SiO_3, respectively. For this type of strength, and by combining both figures, it can be concluded that 390 g of NaOH and 514 g of Na_2SiO_3 would be the optimal combination.

Finally, the results seem to show that combination 1:2 (Base Mix) is the most adequate to obtain a material with good properties for structure applications. However, the activator composition of 300 g of NaOH, 450 g of Na_2SiO_3 and 150 g of water is more economic than the Base Mix and does not seem to lead to mechanical characteristics of inferior quality.

5. Conclusions

The main objective of this work was to evaluate the variation of the mechanical properties, mainly the tensile and compressive strengths with the variation of the composition of geopolymers. Different activator compositions were tried in order to conclude about the consequences in terms of mechanical properties and the possibility of using these materials in structural members and ultimately in the construction industry.

The values obtained for the tensile and compressive strength of geopolymers are good in comparison with those of OPC mortar. Some changes in the composition proved not to affect the mechanical properties significantly. This is a good indication because it shows that there is room for some adaptation of the composition when some practical problems need to be solved. One of the most important issues is the workability of the fluid mass before pouring into the molds. OPC mortars or concretes are well developed in this aspect and nowadays self-compacting concretes are easily produced, even at large scale [38,39]. Geopolymers still need more research on this aspect.

The values of the tensile strain and of the modulus of elasticity indicate that the geopolymers are more deformable than OPC mortars. This has some advantages and some drawbacks. For instance, differential soil settlements will have less impact on the internal forces of a geopolymer structure when compared to the impact of similar settlements in an OPC concrete structure. On the other hand, higher deformability of geopolymer structures when compared to that of OPC concrete structures could be a problem when maximum deformations are imposed by design standards. This could lead to thicker cross sections, therefore, with impact on the price. It should be noted that the higher deformability at the elastic phase of the mechanical behaviour of geopolymers could be somehow counterbalanced by their higher tensile strength. Therefore, many particularities of the structural behaviour of geopolymers need to be investigated in the near future. For instance, the bond of reinforcement to geopolymer might be different (possibly higher) when compared to OPC structures. This would lead to shorter anchorage lengths, for instance. The bond between geopolymers and OPC should also require special attention from researchers since rehabilitation works of old OPC structures are in increasing demand.

Focusing on the particular results of this investigation, for mixes with a constant amount of activator, the highest efficiency of the geopolymers appears to take place for ratios (hydroxide: silicate) between 1:1.75 and 1:2. Geopolymers within this ratio range are shown to have the best strength capacities, either on tensile or compression and also better results in terms of modulus of elasticity (better results here means results closer to those of OPC mortars). Furthermore, if the composition of the activator falls within this range, it does not affect significantly any of the mechanical characteristics.

Finally, this article describes an exploratory work on the study of mechanical characteristics of geopolymers and the authors believe that the findings could help other researchers in defining new lines of research. Future work should take different directions suggested earlier. Given some interesting advantages, the authors argue that this material deserves to be further investigated.

Author Contributions: Author Contributions: Conceptualization, methodology, analysis and writing: A.V.L., S.M.R.L. and I.P., Experimental work: A.V.L. and I.P., computing of data: A.V.L., final editing: S.M.R.L. All authors have read and agreed to the published version of the manuscript.

Funding: This research was sponsored by FEDER funds through the program COMPETE–Programa Operacional Factores de Competitividade– and by national funds through FCT–Fundação para a Ciência e a Tecnologia–under the project UID/EMS/00285/2020.

Acknowledgments: The authors would like to express their gratitude to Amandio Teixeira-Pinto for the valuable discussions and to Mario Oliveira for his help on experimental work and on some analyses.

Conflicts of Interest: The authors declare no conflict of interests.

References

1. Majidi, B. Geopolymer technology, from fundamentals to advanced applications: A review. *Mater. Technol. Adv. Perform. Mater.* **2009**, *24*, 79–87. [CrossRef]
2. Pacheco-Torgal, F.; Castro-Gomes, J.; Jalali, S. Alkali-activated binders: A review. Part 1. Historical background, terminology, reaction mechanisms and hydration products. *Constr. Build. Mater.* **2008**, *22*, 1305–1314. [CrossRef]
3. Skvara, F.; Svoboda, P.; Dolezal, J.; Smilauer, V.; Kopecky, L.; Sulc, R. Geopolymer concrete—An ancient material too? *Ceram. Silik.* **2008**, *52*, 296–298.
4. Attanasio, A.; Pascali, L.; Tarantino, V.; Arena, W.; Largo, A. Alkali-Activated Mortars for Sustainable Building Solutions: Effect of Binder Composition on Technical Performance. *Environments* **2018**, *5*, 35. [CrossRef]
5. Wang, Y.; Hu, S.; He, Z. Mechanical and Fracture Properties of Fly Ash Geopolymer Concrete Addictive with Calcium Aluminate Cement. *Materials* **2019**, *12*, 2982. [CrossRef]
6. Rapazote, J.G.; Laginhas, C.; Teixeira-Pinto, A. Development of Building Materials through Alkaline Activation of Construction and Demolition Waste (CDW)—Resistance to High Temperatures. In Proceedings of the International Conference on Durability of Building Materials and Components, Porto, Portugal, 12–15 April 2011.
7. Teixeira-Pinto, A.; Fernandes, P.; Jalali, S. Geopolymer Manufacture and Application—Main problems When Using Concrete Technology. In Proceedings of the Geopolymers International Conference, Melbourne, Australia, 28–29 October 2002.
8. Davidovits, J. Global warming impact on the cement and aggregates industries. *World Resour. Rev.* **1994**, *6*, 263–278.
9. Davidovits, J. *Geopolymer Cement—A Review*; Institut Geopolymere: Saint-Quentin, France, 2013.
10. Rangan, B.V. Low-Calcium, Fly-Ash-Based Geopolymer Concrete. In *Concrete Construction Engineering Handbook*, 2nd ed.; Nawy, E.G., Ed.; Taylor & Francis Group: New York, NY, USA, 2008; Chapter 26; pp. 1–26.

11. Singh, N.B.; Middendorf, B. Geopolymers as an alternative to Portland cement: An overview. *Constr. Build. Mater.* **2020**, *237*, 117455. [CrossRef]
12. Pacheco-Torgal, F.; Castro-Gomes, J.; Jalali, S. Alkali-activated binders: A review. Part 2. About materials and binders manufacture. *Constr. Build. Mater.* **2008**, *22*, 1315–1322. [CrossRef]
13. Teixeira-Pinto, A. Sistemas Ligantes Obtidos por Activacao Alcalina do Metacaulino (Binder Systems Obtained by Alkaline Activation of Metakaolin). Ph.D. Thesis, University of Minho, Guimaraes, Portugal, 2004. (In Portuguese)
14. Provis, J.L.; Bernal, S.A. Geopolymers and Related Alkali-Activated Materials. *Annu. Rev. Mater. Res.* **2014**, *44*, 299–327. [CrossRef]
15. Altan, E.; Erdogan, S.T. Alkali activation of a slag at ambient and elevated temperatures. *Cem. Concr. Compos.* **2012**, *34*, 131–139. [CrossRef]
16. Granizo, M.L.; Blanco, M.T. Alkaline activation of metakaolin an isothermal conduction calorimetry study. *J. Ther. Anal.* **1998**, *52*, 957–965. [CrossRef]
17. Davidovits, J. Chemistry of Geopolymeric Systems, Terminology. In Proceedings of the Second International Conference Geopolumère, Saint-Quentin, France, 30 June–2 July 1999; pp. 9–40.
18. Komnitsas, K.; Zaharaki, D. Geopolymerisation: A review and prospects for the minerals industry. *Miner. Eng.* **2007**, *20*, 1261–1277. [CrossRef]
19. Fernandez-Jimenez, A.; Palomo, A. Composition and microstructure of alkali activated fly ash binder: Effect of the activator. *Cem. Concr. Res.* **2005**, *35*, 1984–1992. [CrossRef]
20. Fernandez-Jimenez, A.; Palomo, A.; Sobrados, I.; Sanz, J. The role played by the reactive alumina content in the alkaline activation of fly ashes. *Microporous Mesoporous Mater.* **2006**, *91*, 111–119. [CrossRef]
21. Nematollahi, B.; Sanjayan, J.; Qiu, J.; Yang, E.H. High ductile behavior of a polyethylene fiber-reinforced one-part geopolymer composite: A micromechanics-based investigation. *Arch. Civ. Mech. Eng.* **2017**, *17*, 555–563. [CrossRef]
22. Nematollahi, B.; Xia, M.; Sanjayan, J. Post-processing Methods to Improve Strength of Particle-Bed 3D Printed Geopolymer for Digital Construction Applications. *Front. Mater.* **2019**, *6*, 1–10. [CrossRef]
23. Lou, T.; Lopes, S.M.R.; Lopes, A.V. Nonlinear and time-dependent analysis of continuous unbonded prestressed concrete beams. *Comput. Struct.* **2013**, *119*, 166–176. [CrossRef]
24. Lopes, S.M.R.; Carmo, R.N.F. Deformable strut and tie model for the calculation of the plastic rotation capacity. *Comput. Struct.* **2006**, *84*, 2174–2183. [CrossRef]
25. Lopes, A.V.; Lopes, S.M.R. Importance of a rigorous evaluation of the cracking moment in RC beams and slabs. *Comput. Concr.* **2012**, *9*, 275–291. [CrossRef]
26. Lou, T.; Lopes, S.M.R.; Lopes, A.V. Numerical modeling of externally prestressed steel-concrete composite beams. *J. Constr. Steel Res.* **2016**, *121*, 229–236. [CrossRef]
27. Monteiro, S.R.S.; Dias, A.M.P.G.; Lopes, S.M.R. Transverse distribution of internal forces in timber–concrete floors under external point and line loads. *Constr. Build. Mater.* **2016**, *102*, 1049–1059. [CrossRef]
28. Maranan, G.B.; Manalo, A.C.; Benmokrane, B.; Karunasena, W.; Mendis, P. Behavior of concentrically loaded geopolymer-concrete circular columns reinforced longitudinally and transversely with GFRP bars. *Eng. Struct.* **2016**, *117*, 422–436. [CrossRef]
29. Mo, K.H.; Alengaram, U.J.; Nicolas, R.S.; Jumaat, M.Z. Structural performance of reinforced geopolymer concrete members: A review. *Constr. Build. Mater.* **2016**, *120*, 251–264. [CrossRef]
30. Dahou, Z.; Castel, A.; Noushini, A. Prediction of the steel-concrete bond strength from the compressive strength of Portland cement and geopolymer concretes. *Constr. Build. Mater.* **2016**, *119*, 329–342. [CrossRef]
31. Engelhard. Basic Concrete Materials & Methods. *Chemical and Physical Characteristics of MetaMax®*. Available online: http://newenglandresins.com/wp-content/uploads/2012/01/raw-material-library/BASF%20-Kaolin-Clay/Metamax%20TD%20sheet.pdf (accessed on 23 January 2019).
32. American Society for Testing Materials. *Standard Classification of Soils for Engineering Purposes: ASTM D 2487-06*; American Society for Testing Materials: West Conshohocken, PA, USA, 2006.
33. NP EN 196-1. *Available in English as Methods of Testing Cement. Part 1. Determination of Strength: BS EN 196-1*; British Standards Institution (BSI): London, UK, 2016. (In Portuguese)
34. NP EN 12390-1. *Available in English as Testing Hardened Concrete. Shape, Dimensions and Other Requirements for Specimens and Moulds: BS EN 12390-1*; British Standards Institution (BSI): London, UK, 2012. (In Portuguese)

35. NP EN 12390-3. *Available in English as Testing Hardened Concrete. Compressive Strength of Test Specimens: BS EN 12390-3*; British Standards Institution (BSI): London, UK, 2019. (In Portuguese)
36. NP EN 12390-5. *Available in English as Testing Hardened Concrete. Flexural Strength of Test Specimens: BS EN 12390-5*; British Standards Institution (BSI): London, UK, 2019. (In Portuguese)
37. NP EN 1992-1-1. *Available in English as Design of Concrete Structures. General Rules and Rules for Buildings: BS EN 1992-1-1*; British Standards Institution (BSI): London, UK, 2004. (In Portuguese)
38. Nepomuceno, M.; Oliveira, L.; Lopes, S.M.R. Methodology for mix design of the mortar phase of self-compacting concrete using different mineral additions in binary blends of powders. *Constr. Build. Mater.* **2012**, *26*, 317–326. [CrossRef]
39. Nepomuceno, M.; Oliveira, L.; Lopes, S.M.R. Methodology for the mix design of self-compacting concrete using different mineral additions in binary blends of powders. *Constr. Build. Mater.* **2014**, *64*, 82–94. [CrossRef]

Article

Experimental Study on the Torsional Behaviour of Prestressed HSC Hollow Beams

Luís Bernardo [1,*], Sérgio Lopes [2] and Mafalda Teixeira [1]

1 Department of Civil Engineering and Architecture, Centre of Materials and Building Technologies (C-MADE), University of Beira Interior, 6201-001 Covilhã, Portugal; mafalda.m.teixeira@ubi.pt
2 Department of Civil Engineering, Center for Mechanical Engineering, Materials and Processes (CEMMPRE), University of Coimbra, 3030-788 Coimbra, Portugal; sergio@dec.uc.pt
* Correspondence: lfb@ubi.pt

Received: 23 December 2019; Accepted: 13 January 2020; Published: 16 January 2020

Abstract: This article describes an experimental program developed to study the influence of longitudinal prestress on the behaviour of high-strength concrete hollow beams under pure torsion. The pre-cracking, the post-cracking and the ultimate behaviour are analysed. Three tests were carried out on large hollow high-strength concrete beams with similar concrete strength. The variable studied was the level of longitudinal uniform prestress. Some important conclusions on different aspects of the beams' behaviour are presented. These conclusions, considered important for the design of box bridges, include the influence of the level of prestress in the cracking and ultimate behaviour.

Keywords: concrete structures; beams & girders; torsion; high-strength concrete; prestressing

1. Introduction

Pure torsion does not really occur too often in concrete structures; it usually occurs together with other internal forces such as shear, bending and axial forces. However, in some structures, such as the case of curved box bridges, the torsional action can be very important for design.

The application of prestress usually increases the cracking and ultimate resistances of concrete structures. Prestress is particularly important in High-Strength Concrete (HSC) structures. In general, HSC structures are expected to be more flexible than Normal-Strength Concrete (NSC) structures because of the smaller cross-section area of members. This high flexibility could be problematic for the serviceability limit states. The use of the prestress technique can also help to solve such problems associated with high flexibility.

The application of longitudinal prestress in members under high torsional loads is a common situation, as for instance in curved bridges. Many of such structures are also built with HSC and use hollow cross-sections for the girders (Figure 1) because they present some advantages when compared to solid cross-sections. In large cross-sections under high torsion, the internal shear flow is mainly absorbed by the outer concrete shell. Thus, the concrete at the centre zone of the cross-section is redundant and can be removed. As a consequence, hollow cross-sections allow for a high reduction in weight and concrete consumption, without compromising the torsional strength.

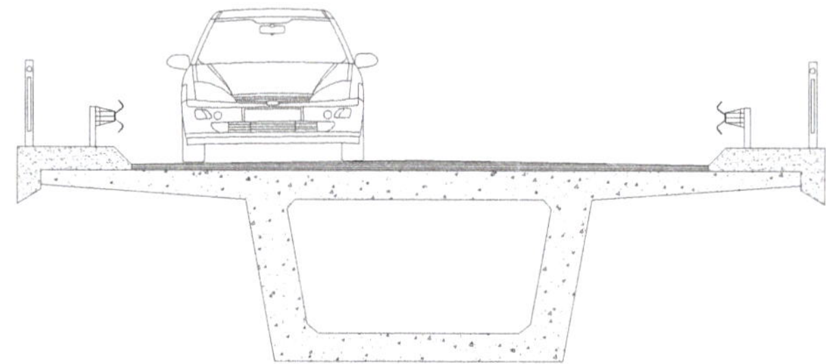

Figure 1. Example of a hollow (box) cross-section for the girder of a bridge deck.

Since the end of the last century, it is well known that the uniaxial stress-strain curve for HSC is quite different from that for NSC [1]. Therefore, it is not obvious that the computing and design models for HSC members can be directly extrapolated from NSC. Nowadays, many codes of practice already include HSC range. However, some aspects of the structural behaviour of HSC members still need to be studied in order to check if the physical models accepted for NSC can be adopted for HSC. The response of HSC beams to torsional loads constitutes an example of such cases.

There are still few experimental studies on hollow beams under pure torsion reported in the literature and most of them involve only a small number of NSC beams [2–5]. Only some few recent works report new results for both NSC and HSC hollow beams [6,7]. When compared with NSC beams, such studies demonstrate some of the advantages of HSC, namely, to increase the cracking and ultimate torsional strengths, as well as the torsional stiffness of the beams. However, some disadvantages of using HSC are also pointed out, in particular related to the torsional ductility which constitutes an important property to be considered for design. By using the experimental results reported in the previously referred experimental studies and also by using numerical models, other recent studies show that some ductility can be observed in NSC beams under torsion, namely for a certain range of the torsional reinforcement. However, almost no torsional ductility in HSC beams under torsion is observed [8–10]. Such studies also show that the level of the observed torsional ductility is normally low when compared with flexural ductility [11,12].

Few previous studies leading with testing of prestressed concrete rectangular beams under torsion can be found in the literature, namely—Mitchell and Collins in 1974 [4], El-Degwy and McMullen in 1985 [13], Hsu and Mo in 1985 [14] and Wafa et al. in 1995 [15]. Among the beams tested in the referred studies (twenty-seven beams), only three where hollow and built with NSC. This is because building hollow beams for testing is more complicated when compared with solid beams. Furthermore, due to the complexity of the experimental program, only a constant longitudinal prestressing was considered in all the referred studies. The authors did not find any previous study on the experimental behaviour of HSC prestressed hollow beams under torsion.

From the foregoing, experimental studies on the behaviour of prestressed hollow beams under torsion are needed, in particular for HSC beams. This article presents an experimental study on the global behaviour of HSC hollow beams with uniform longitudinal prestress. The beams were loaded under pure torsion and tested up to failure. As for the previous referred studies, a uniform longitudinal prestress was applied. This was considered to be sufficient to give some indications on the influence of prestress on the behaviour of HSC hollow beams under torsion.

2. Experimental Program

2.1. Test Specimens

For this study, three hollow beams were tested up to failure. The beams had a squared cross-section and were 5.90 m long. During testing, the beams had an extremity fixed to the strong floor of the laboratory and the load was applied to the other extremity by an electromechanical actuator. In this extremity, a special device transformed the linear point load applied by the actuator into a torque applied to the beam extremity. The dimensions of the test models are presented in Figure 2. The geometry and dimensions of the adopted cross-section are in line with some of the works previously referred and leading with hollow beams.

External prestressing was applied through four 0.6″ wires (0.6 inches or 1.52 cm diameter) centred in the cross-section (Figure 2). The torsional reinforcement ratio was kept constant for the three beams and it corresponds to an average value which was defined accounting for the range of values that were used in a previous experimental work on similar beams with no prestressing [6]. The concrete strength was approximately constant, varying between 77.8 and 80.8 MPa. The average level of stress in concrete induced by prestress (f_{cp}) varied between 0 MPa (beam with no prestress) and 3.08 MPa, after short term losses. The maximum value is not very high because the losses were somewhat important. This is because the effect of anchorage slip becomes important when the length of the wires is relatively small.

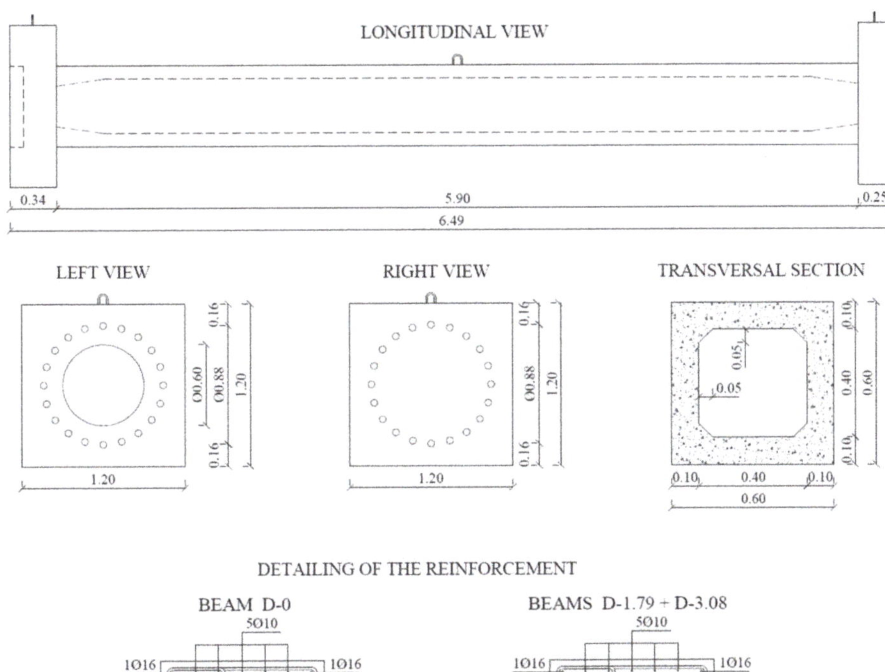

Figure 2. Geometry and detailing of test beams.

Table 1 summarizes the characteristics of each test beam, namely—the real thickness of the walls of the cross-section (t), the distance between parallel branches of stirrups, x_1 and y_1, the total area of ordinary longitudinal reinforcement (A_{sl}), the area of one branch of the transverse reinforcement (A_{st}), the longitudinal spacing of the stirrups (s), the total area of prestress longitudinal reinforcement (A_{sp}), the ratios of longitudinal reinforcement ($\rho_l = A_{sl}/A_c$, where $A_c = xy$ and $x = y = 60$ cm) and transverse reinforcement ($\rho_t = A_{st}u/A_c s$, where $u = 2(x_1 + y_1)$), the total ratio of reinforcement

($\rho_{tot} = \rho_l + \rho_t$), the balanced ratio of the longitudinal to transverse reinforcement ($m_b = A_{sl}s/A_{st}u$) and the average values for the compressive concrete strength obtained from cylindrical specimens (f_c).

The beams are named according to the series to which they belong (series D) and to the average stress (in MPa) in concrete induced by prestress, f_{cp}, after short term losses.

Table 1. Properties of test beams.

Beam.	t cm	x_1 cm	y_1 cm	A_{sl} cm^2	A_{st} cm^2	A_{sp} cm^2	s cm	ρ_l %	ρ_t %	ρ_{tot} %	m_b	f_c MPa	f_{ctm} MPa	E_c GPa
D-0	10.9	53.5	53.7	23.75	0.79	0	7.0	0.66	0.67	1.33	0.99	77.8	4.33	40.7
D-1.79	11.4	54.3	54.2	23.75	0.79	5.60	7.0	0.66	0.68	1.34	0.97	80.8	4.43	41.1
D-3.08	11.5	55.0	54.6	23.75	0.79	5.60	7.0	0.66	0.68	1.34	0.96	78.8	3.66	37.4

2.2. Materials Properties

The average value of the compressive concrete strength used in each tested beam was obtained from 5 cube specimens, casted and tested at the same time of the corresponding beam. The equivalent cylindrical values were computed by following the indications from Eurocode 2 [16]. Table 2 presents the concrete mix design used to produce the concrete.

Table 1 also presents, for each tested beam, the average values for the tensile concrete strength (f_{ctm}) and for the concrete Young's Modulus (E_c). These two parameters were computed from f_c also following the indications from Eurocode 2 [16].

As example, Figure 3 illustrates a stress (σ)–strain (ε) curve recorded during the test of one of the concrete samples. The initial part of the graph shows the influence of the adjustment due to the existent gap between the loading plates of the test machine and the concrete specimen.

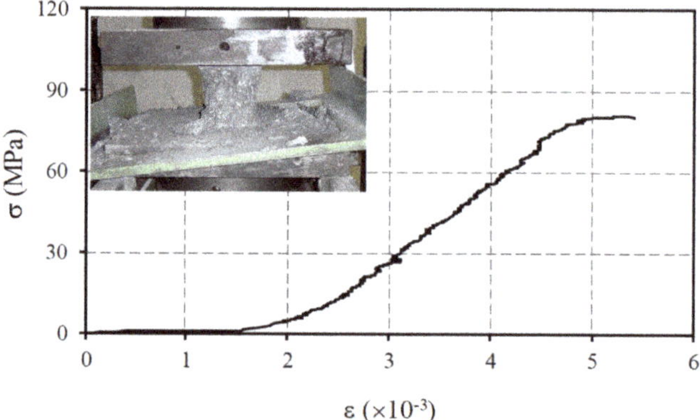

Figure 3. Uniaxial σ–ε curve for concrete.

Table 2. Concrete mix design (contents per m^3).

Components	Dosage
Thin sand	164 kg
Thick sand	908 kg
Crushed Granit 5–11 mm	734 kg
Normal Portland Cement (C): Type I/42.5R	375 kg
Admixture—Rheobuild 1000	4.8 ℓ
Silica Fume (Sikacrete HD)	41 kg
Water (A)	145 ℓ
A/(C + Additions)	0.35

The ordinary reinforcement used in the beams consisted of hot rolled ribbed steel bars (with 10 and 16 mm diameters) sold commercially as Class A500. In order to know the average values of the yield stress and corresponding strain of the bars (f_y and ε_y, respectively), tensile tests on steel samples were carried out (6 samples of each of the diameters that were used in the beams). The following average values were obtained—$f_y = 686$ MPa and $\varepsilon_y = 3.43 \times 10^{-3}$. For the steel Young´s Modulus, the typical value set in codes of practice was assumed, $E_s = 200$ GPa [16].

As example, Figure 4a illustrates some σ–ε curves recorded during the tensile tests of steel specimens (ordinary reinforcement).

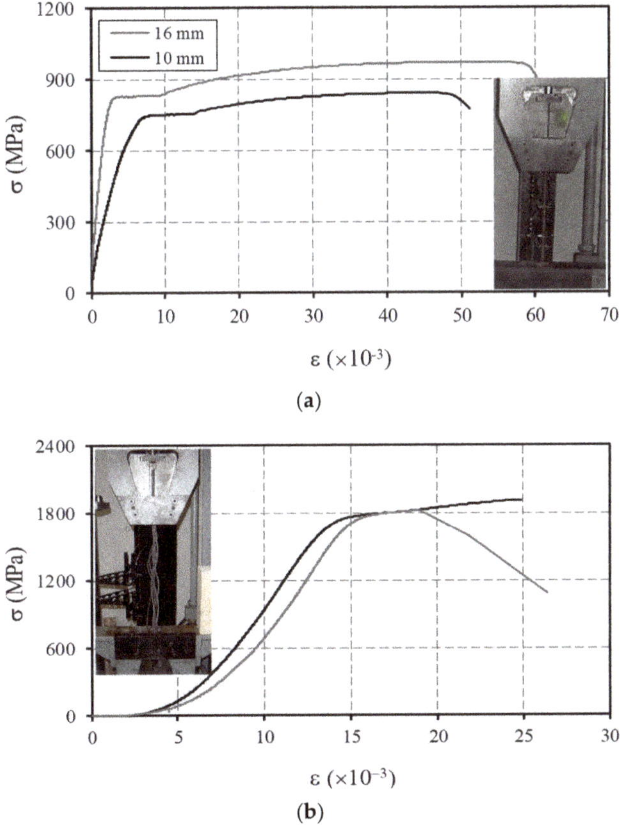

Figure 4. Uniaxial σ–ε curves for reinforcement: (**a**) ordinary and (**b**) prestress.

The prestress reinforcement used in Beams D-1.79 and D-3.08 consisted of four 0.6″ wires (0.6 inches or 1.52 cm diameter) belonging to Class S1670/1860. Tests on 6 prestress wire specimens were also carried out and the average values of the 0.1% limit proportional stress ($f_{p0.1\%}$) and the ultimate stress (f_{pu}) of the wires were obtained—$f_{p0.1\%} = 1670.5$ MPa and $f_{pu} = 1867.1$ MPa, respectively. The corresponding strain at 0.1% was computed by assuming a linear relationship between strains and stresses, which led to $\varepsilon_{p0.1\%} = 8.567 \times 10^{-3}$. The Young's Modulus was assumed to be the one indicated by the supplier, $E_p = 195$ GPa.

As example, Figure 4b illustrates σ–ε curves recorded during the tensile test of two prestress steel specimens (with 4 × 0.6 inches wires each).

It should be referred that some slip was observed between the steel specimens and the claws of the test machine. This explains the apparent different initial stiffness between the curves for 10 and 16 mm bars in Figure 4a and also the initial part of the curves in Figure 4b. This problem had no implication to the previously presented strain values at the end of the elastic stage. The strains were computed from Hooke's law by knowing the stresses and assuming the Young's Modulus.

2.3. Testing Procedure

The test device is made of three main components:

- a test frame where the mechanical actuator is fixed;
- a device that receives the load from the mechanical actuator and applies a torsional moment in one end of the test beams;
- a device that fixes the test beam at the other end, restricting its transversal rotation (twist) and allowing its longitudinal deformation (elongation).

Figure 5 illustrates the global test device with a beam in its test position.

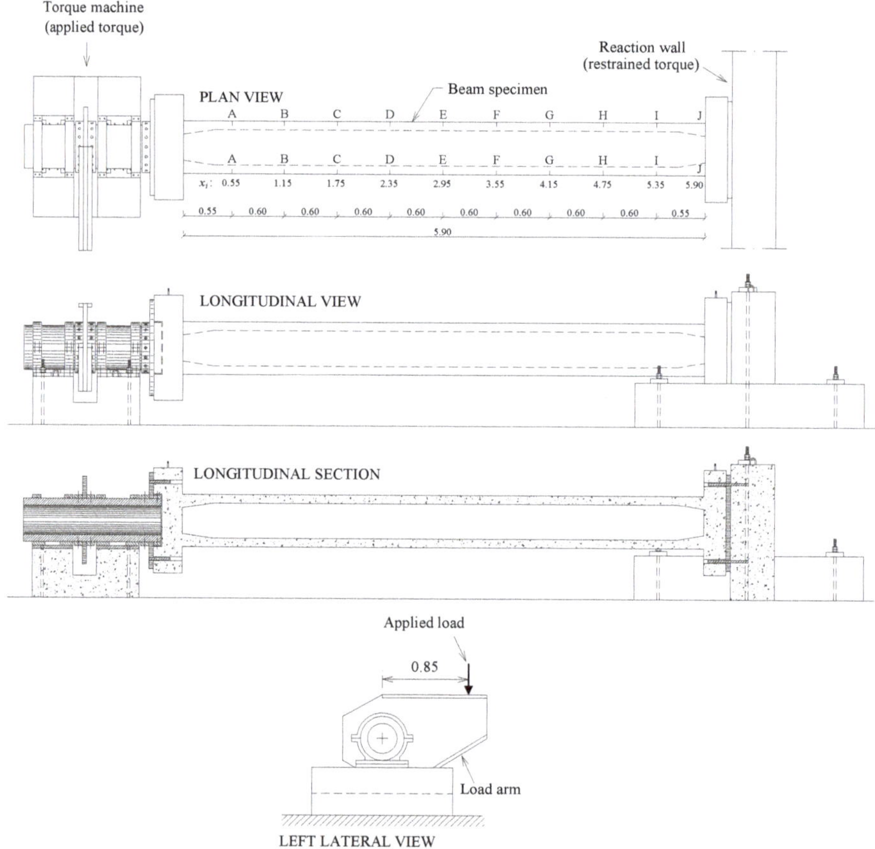

Figure 5. Test setup.

The load was applied by imposing a low deformation rate with an electromechanical actuator. Several load cells were placed in different points of the global test device in order to record at any time the general loading state of the beams.

The transversal rotations (twists) were read in 10 sections uniformly spaced along the length of the beams (Sections A-A to J-J), as illustrated in Figure 5. For this, 10 pairs of displacement transducers were placed at the top face of the beam (they were fixed to an external anchored referential). Special care was taken to allow free horizontal relative movements between the beams and the LVDT arms. Because of this, the horizontal projection of the distance between pairs of transducers remained constant during the test and the rotation angles at each section could be easily computed from the transducers' readings.

The torsional reinforcement bars were instrumented in three selected sections at quarters of the beams' length. Resistance strain gauges were stuck on the 4 longitudinal corner bars and on the 4 branches of one stirrup.

For Beams D-1.79 and D-3.08, in order to evaluate the effective longitudinal prestressing force after losses and also the changes in this force due to the application of increasing torque, a load cell was placed between the head of the beam and the head of anchorage. Figure 6 illustrates longitudinal cuts at the ends of the prestressed beams and shows the technical solutions adopted to prestress the beams and to read the prestress force during the tests.

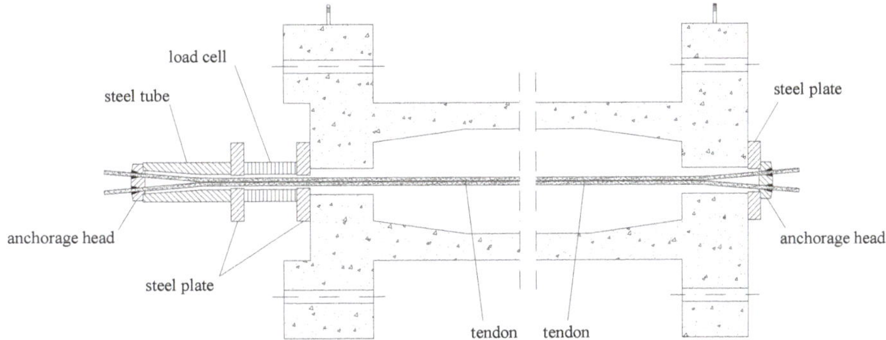

Figure 6. Anchorage zones at both ends of the prestressed beams.

A Data Logger was used to record all the readings. Figure 7 shows general views of a beam in its position with the instrumentation ready for testing.

Figure 7. Beam specimen in test position.

3. Global Analysis of the Experimental Results

3.1. Torsional Moments vs. Twists

Figure 8 presents the graphs of torque (T) *versus* the average twists (θ_m) for the tested beams. The torque, T, was obtained multiplying the load applied by the actuator by the horizontal projection of the level arm, 0.85 m, which remained constant (see Figure 5). The average twist, θ_m, was obtained by dividing the experimental angle measured in Section A-A to the distance between Sections A-A and J-J, 5.35 m (Figure 5). In each T–θ_m curve, identification marks were used to highlight the points corresponding to cracking () and to yielding of the transverse (□) and longitudinal (△) reinforcement. The yielding points were calculated from the experimental values of the strains recorded by the strain gauges stuck to the reinforcement bars.

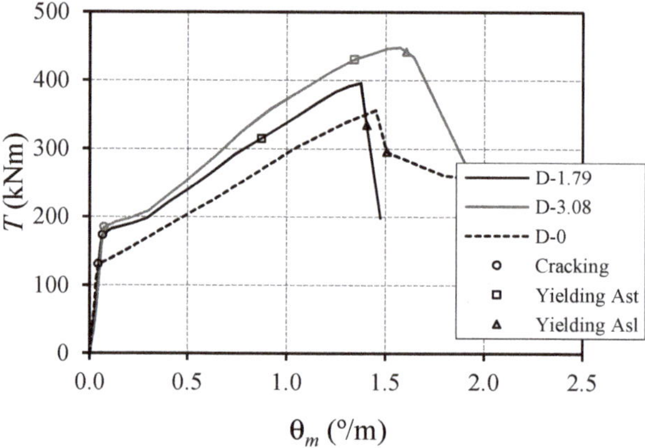

Figure 8. T–θ_m curves.

As expected, Figure 8 shows the high influence of prestress in the cracking torque. It is known that prestress delays the formation of cracking. For a moderate concrete stress (induced by prestress) of 1.79 MPa (Beam D-1.79) an increase of approximately 32.5% on the cracking torque is observed, when compared with the beam without prestress (Beam D-0). This shows the efficiency of uniform longitudinal prestress to delay the cracking in beams under torsion. This high influence of prestress can be explained because in torsion the concrete is under a lower and more uniform level of tensile stresses (in the whole section) when compared with the bending situation. Therefore, even for low levels of prestress, the cracking stage is delayed. It is also observed that, in State I (non-cracked state), the steel bars, including the prestressed wires, generally have little influence on the stiffness of the beams. In fact, the T–θ_m curves are almost coincident at this state.

In State II (cracked state), the T–θ_m curves are almost parallel to each other. This shows that the contribution of the longitudinal prestress for the stiffness of the beams is small at this state. This is due to the adopted prestressing technique (external longitudinal and centred prestress).

As expected, Figure 8 also shows that the use of prestress increases the resistant torque, T_r, of the beams. However, there is not a clear tendency with respect to the associated twist at the ultimate torque, θ_{T_r}. In fact, since prestress induces a compressive stress state in concrete, it would be expected that the deformation capacity of concrete in the compressed areas of the beam (namely in the struts) would decrease as the level of prestress increases. As a consequence, the twist corresponding to the ultimate torque should decrease as the stress induced by prestress increases. This is not the case for Beam D-3.08, which has the highest level of prestress and reaches a twist θ_{T_r} that exceeds the same one of the other beams. However, this observation can be explained due to the type of failure of Beams D-0 and D-1.79, which was fragile and somehow premature (failure by pull off of the concrete corners).

This subject will be discussed later. This aspect also justifies the different shape of the descending branches of T–θ_m curves that is observed and the absence of yielding points before the peak torque is reached for Beam D-0.

T–θ_m curves of Figure 8 also show that, before the peak torque is reached, the prestressed beams only present points corresponding to yielding of the transverse reinforcement. It is observed that the yielding of longitudinal reinforcement only occurs after the peak torque. After cracking, the longitudinal prestress reinforcement starts working as ordinary reinforcement under the torsional loading. Consequently, the beams with balanced longitudinal to transverse reinforcement ratios will lose this balance because of the influence of the prestress wires. Hence, the calculation of the balanced ratio of the longitudinal to transverse reinforcements should account for the area of the prestressed steel ($m_{b,tot} = (A_{sl} + nA_{sp})s/A_{st}u$ with $n = E_p/E_s$), which leads to an excess of longitudinal reinforcement of about 20%. Therefore, the transverse reinforcement should yield before the longitudinal reinforcement, as observed in Figure 8.

Table 3 presents, for each tested beam, the main properties of T–θ_m curves, namely—the cracking torque and correspondent twist (T_{cr} and θ_{cr}), the torsional stiffness in State I ($(GC)^{\mathrm{I}}$), the torsional stiffness in State II ($(GC)^{\mathrm{II}}$), the torque corresponding to the yielding of the transverse reinforcement and correspondent twist (T_{ty} and θ_{ty}), the resistant torque (peak torque) and correspondent twist (T_r and θ_{T_r}). Since the yielding of the longitudinal reinforcement occurs after the peak torque, the corresponding values are not presented.

Table 3. Properties of T–θ_m curves.

Beam	T_{cr} kNm	θ_{cr} °/m	$(GC)^{\mathrm{I}}$ kNm^2	$T^{\mathrm{II}}=a\theta^{\mathrm{II}}+b$	$(GC)^{\mathrm{II}}$ kNm^2	T_{ty} kNm	θ_{ty} °/m	T_r kNm	θ_{T_r} °/m
D-0	130.5	0.04	172,940	a = 170.87; b = 119.43	9790	-	-	355.9	1.45
D-1.79	172.9	0.07	152,188	a = 194.73; b = 142.95	11,157	314.9	0.87	396.0	1.38
D-3.08	184.7	0.08	141,823	a = 148.05; b = 132.34	14,212	430.0	1.34	447.7	1.57

The torsional stiffness in State I was calculated dividing T_{cr} by θ_{cr} (with θ_{cr} in radians unit). Prior to the calculation of the torsional stiffness in State II, the equation of the line of T–θ_m curve in the linear elastic stage was previously calculated from linear interpolation. For this calculation, the points of the T–θ_m curves located in the zone that can be identified as belonging to State II were selected. Only the zone of the curves that is approximately a straight line was considered. After the calculation of the equation, $T = a\theta + b$ (see Table 3), the stiffness $(GC)^{\mathrm{II}}$ is equal to the slope a of the line (with twists converted to radian units).

The analysis of the values displayed in Table 3 confirms the trends observed in the T–θ_m curves from Figure 8 and previously discussed.

3.2. Force in the Prestress Reinforcement vs. Twists

Figure 9 presents the graphs of the force in the longitudinal prestress reinforcement (F_{ps}) *versus* the average twist (θ_m). The force F_{ps} was obtained directly from the load cell placed in the anchorage zone of the prestressed wires. The evolution of the recorded values starts from the initial value of the applied prestress force (after short-term losses).

The curves of Figure 9 show the existence of a small horizontal zone where the force in the prestress reinforcement is almost constant and equal to the force due to initial prestress. This zone ends as the first crack appears in the beam. After this zone, the force in the prestressed wires increases gradually. In fact, before cracking, the internal reinforcement steel bars are also under very low levels of stress. The strains in the beam during the pre-cracking state are very small, as confirmed by the readings of the strains in the reinforcement bars (recorded from the strain gauges). Decompression of concrete takes place at a certain point of this initial horizontal zone (before this point, concrete is only in compression).

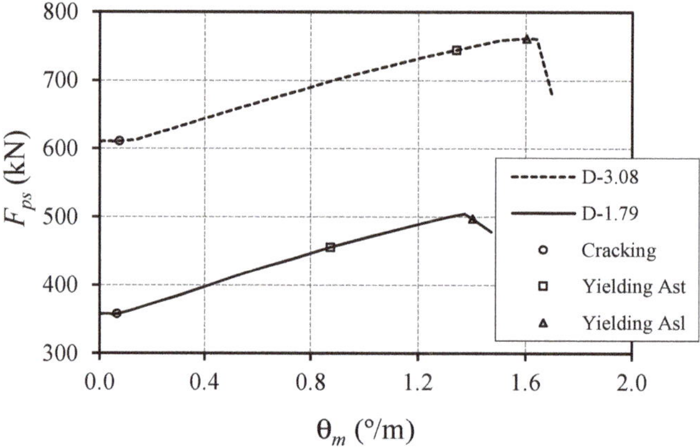

Figure 9. F_{ps}–θ_m curves.

After cracking, the torsional longitudinal reinforcement bars become effective and the tensile stress in the bars increases as the longitudinal deformation of the beam increases. The elongation of the beams also causes the increasing of the initial force in the prestressed reinforcement. This increasing occurs up to the failure of the beams.

4. Failure Modes and Cracking Patterns

Figures 10 and 11 show some photographic records of the failure zone of the tested beams. From these photos and from the analysis of the T–θ_m graphs at their final part it is possible to distinguish two failure modes.

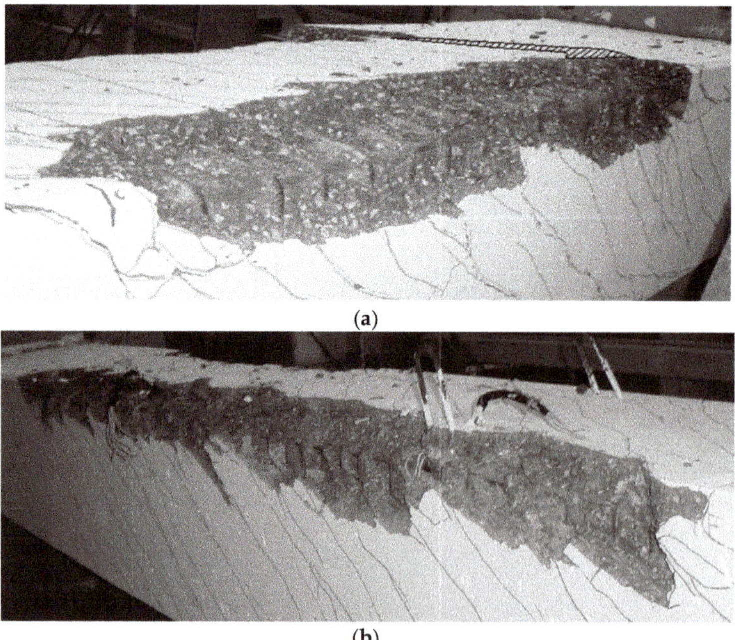

(**a**)

(**b**)

Figure 10. Fragile failure by concrete break off at the corners: (**a**) Beam D-0 and (**b**) Beam D-1.79.

Figure 11. Fragile failure by crushing of concrete: Beam D-3.08.

With a low level of prestress (1.79 MPa), Beam D-1.79 had the same failure mode as Beam D-0 (with no prestress), characterized by concrete break off at the corners (Figure 10). It should be mentioned that this particular fragile failure mode, somehow frequent in hollow beams, was observed and reported in previous studies [6,10]. For a moderate prestress level (3.08 MPa), Beam D-3.08 clearly shows a fragile failure by crushing of concrete in the struts due to compression. This failure mode was particularly destructive and explosive (Figure 11). These results seem to show that the prestress level influences the failure mode. The results seem to indicate that, as the initial prestress increases, the failure becomes more fragile. This reveals an unfavourable effect of the prestress, which is observed for moderate levels of prestress.

To analyse the cracking pattern of the tested beams, photographs were taken and posterior graphic tools were used to produce graphic type images for the three visible faces (lateral faces and top face) for all the beams. Such illustrations are presented in Figures 12–14.

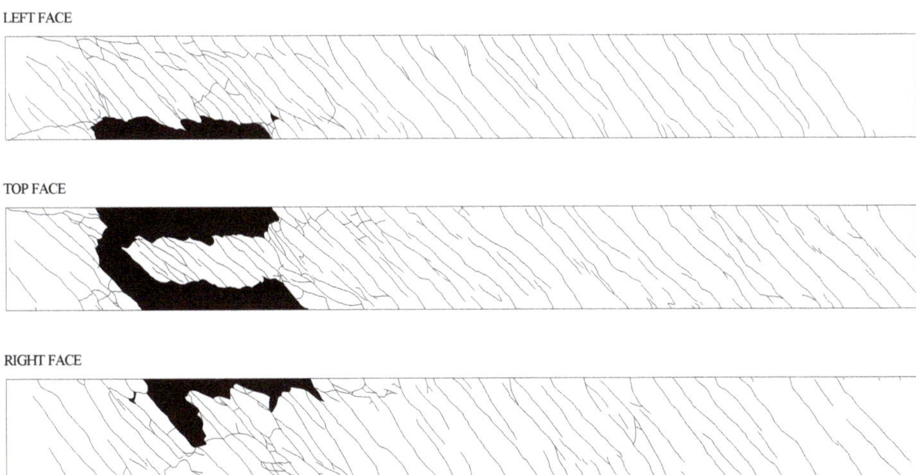

Figure 12. Cracking pattern: Beam D-0.

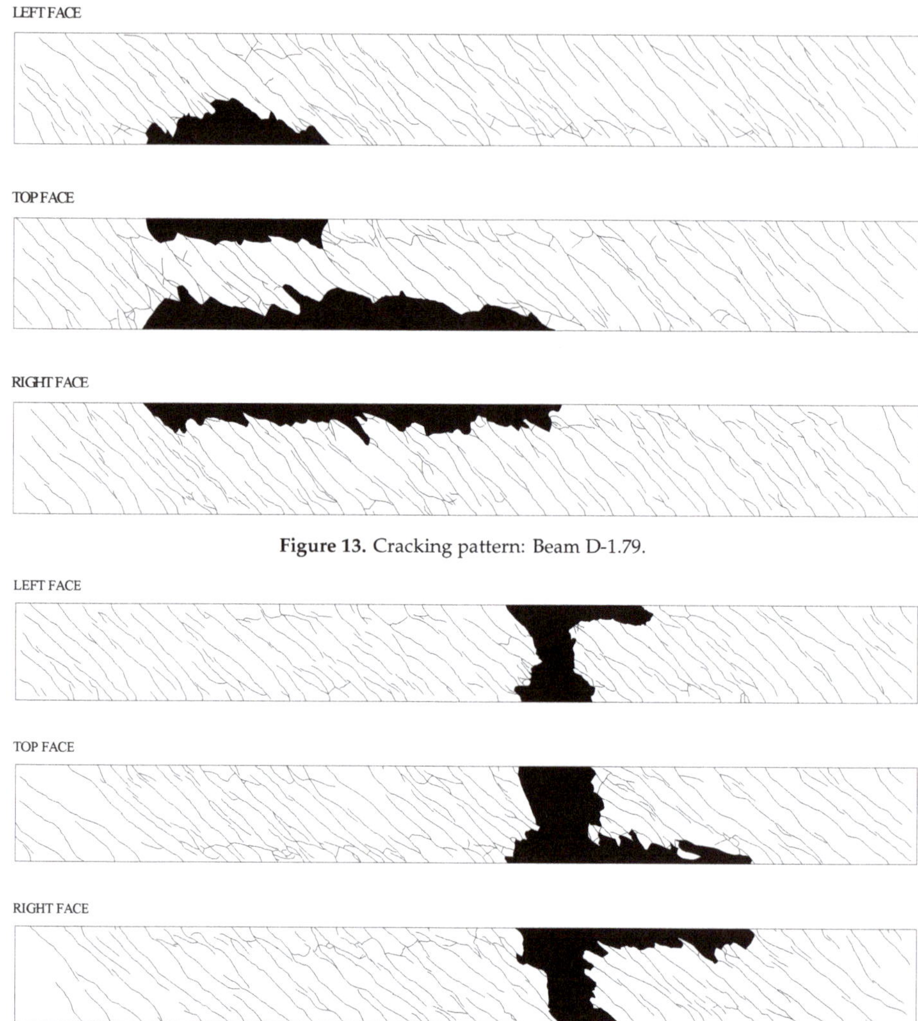

Figure 13. Cracking pattern: Beam D-1.79.

Figure 14. Cracking pattern: Beam D-3.08.

Figures 12–14 do not seem to show a significant influence of the prestress level in the distribution of cracking.

Table 4 presents, for each beam, the average longitudinal distance between cracks (d_m) and the average angle of such cracks to the longitudinal axis of the beams (α_m). Only diagonal cracks that cross the entire width of the faces were considered to compute d_m. Minors cracks or those associated to localized concrete spalling were not accounted. The angles α_m were taken at the mid-width of the faces.

Table 4. Cracking parameters.

Beam	d_m cm	α_m °
D-0	13.0	46.3
D-1.79	12.9	44.8
D-3.08	13.6	44.2

Table 4 confirms that prestress does not seems to influence the distribution of cracking, since the values of d_m are similar for the three beams. The values of α_m varied from 44.2° to 46.3°, not far from the expected theoretical value of 45° for beams with balanced reinforcement and without prestress. For beams with prestress, the referred angle is slightly smaller to the one of the similar beam without prestress. This shows a small influence of prestress (the prestress levels were not high either). It is known that the existence of an axial compressive force in beams symmetrically reinforced under torsion causes the angle of the diagonal cracks to decrease [17].

5. Conclusions

The experimental results obtained with prestressed HSC hollow beams tested in this study showed the effectiveness of longitudinal prestress to delay the cracking and to increase their resistance to torsion. After cracking, the longitudinal prestress reinforcement starts to effectively behave as an ordinary reinforcement, contributing for the internal equilibrium state of the beams.

Despite the limited number of tested beams, the tests showed that the level of longitudinal prestress can influence the failure mode. It was observed that, as the level of prestress increases, the risk of a fragile failure by concrete crushing in the struts becomes higher. Therefore, the level of compressive stresses in the concrete struts must be carefully controlled and the contribution of the prestress reinforcement must not be neglected. This problem becomes more important in HSC beams when compared to NSC beams, because HSC is more fragile when compared to NSC. In fact, the range of the reinforcement ratios that leads to ductile hollow beams under torsion is larger in NSC beams when compared to HSC beams [6].

The use of longitudinal prestress did not result in high differences with respect to the cracking behaviour of the tested beams.

Despite the limited number of tests presented in this study, the reported results show that there are noticeable differences in the behaviour of the tested beams. After this research program, it became now obvious that some particular aspects of the behaviour of prestressed HSC hollow beams need to be further investigated.

Author Contributions: S.L. supervised the experimental program. L.B. supervised the construction of the test beams, performed the experimental tests and analysed the recorded data with the help of M.T.; L.B. wrote the article with review by S.L. and M.T. All authors have read and agreed to the published version of the manuscript.

Funding: No funding supported this work.

Conflicts of Interest: The authors declare no conflict of interest.

References

1. Dahl, K.K.B. *Uniaxial Stress-Strain Curve for Normal and High Strength Concrete*; Report No. 182; Department of Structural Engineering, Technical University of Denmark: Kgs. Lyngby, Denmark, 1992; p. 58.
2. Hsu, T.T.C. *Torsion of Structural Concrete—Behavior of Reinforced Concrete Rectangular Members*; Torsion of Structural Concrete, SP-18; American Concrete Institute: Detroit, MI, USA, 1968; pp. 261–306.
3. Lampert, P.; Thürlimann, B. *Essais de Poutre en Béton Armé Sous Torsion Simple et Flexion Combinées (Tests on Reinforced Concrete Beams under Pure Torsion and Combined Bending)*; Torsion, Bulletin d'Information N° 71, Comitée Européen du Béton: Brussels, Belgium, 1969; pp. 177–207.
4. Mitchell, D.; Collins, M.P. *Behavior of Structural Concrete Beams in Pure Torsion*; Civil Engineering Publication No.74-06; Department of civil Engineering, University of Toronto: Toronto, ON, Canada, 1974; p. 88.
5. Leonhardt, F.; Schelling, G. *Torsionsversuche na Stahlbetonbalken (Torsion Tests on Reinforced Concrete Beams)*; Bulletin No. 239; Deutscher Ausschuss für Stahlbeton: Berlin, Germany, 1974; p. 122.
6. Bernardo, L.F.A.; Lopes, S.M.R. Torsion in HSC Hollow Beams: Strength and Ductility Analysis. *ACI Struct. J.* **2009**, *106*, 39–48.
7. Jeng, C.H. Unified Softened Membrane Model for Torsion in Hollow and Solid Reinforced Concrete Members: Modeling Precracking and Postcracking Behavior. *J. Struct. Eng.* **2015**, *141*, 04014243. [CrossRef]

8. Bernardo, L.F.A.; Lopes, S.M.R. Behaviour of Concrete Beams under Torsion—NSC Plain and Hollow Beams. *Mater. Struct.* **2008**, *41*, 1143–1167. [CrossRef]
9. Bernardo, L.F.A.; Lopes, S.M.R. Theoretical Behaviour of HSC Sections under Torsion. *Eng. Struct.* **2011**, *33*, 3702–3714. [CrossRef]
10. Lopes, S.M.R.; Bernardo, L.F.A. Twist Behaviour of High-Strength Concrete Hollow Beams—Formation of Plastic Hinges along the Length. *Eng. Struct.* **2009**, *31*, 138–149. [CrossRef]
11. Lopes, S.M.R.; Bernardo, L.F.A. Plastic Rotation Capacity of High-Strength Concrete Beams. *Mater. Struct.* **2003**, *36*, 22–31. [CrossRef]
12. Bernardo, L.F.A.; Lopes, S.M.R. Plastic Analysis of HSC Beams in Flexure. *Mater. Struct.* **2009**, *42*, 51–69. [CrossRef]
13. El-Degwy, W.M.; McMullen, A.E. Prestressed Concrete Tests Compared with torsion Theories. *PCI J.* **1985**, *30*, 96–127.
14. Hsu, T.T.C.; Mo, Y.L. Softening of Concrete in Torsional Members—Prestressed Concrete. *J. Am. Concr. Inst.* **1985**, *82*, 603–615.
15. Wafa, F.F.; Shihata, S.A.; Ashour, S.A.; Akhtaruzzaman, A.A. Prestressed High-Strength Concrete Beams Under Torsion. *J. Struct. Eng.* **1995**, *121*, 1280–1286. [CrossRef]
16. *NP EN 1992-1-1. Eurocode 2: Design of Concrete Structures—Part 1: General Rules and Rules for Buildings*; European Committee for Standardization—CEN: Brussels, Belgium, 2010.
17. Hsu, T.T.C. *Torsion of Reinforced Concrete*; Van Nostrand Reinhold Company: New York, NY, USA, 1984; p. 516.

Article

Flexural and Shear Performance of Prestressed Composite Slabs with Inverted Multi-Ribs

Sun-Jin Han [1], Jae-Hoon Jeong [1], Hyo-Eun Joo [1], Seung-Ho Choi [1], Seokdong Choi [2] and Kang Su Kim [1,*]

1 Department of Architectural Engineering, University of Seoul, 163 Seoulsiripdaero, Dongdaemun-gu, Seoul 02504, Korea; sjhan1219@gmail.com (S.-J.H.); ghostwise@naver.com (J.-H.J.); hyoyjoo@gmail.com (H.-E.J.); ssarmilmil@gmail.com (S.-H.C.)

2 Yunwoo Structural Engineers Co., Ltd., 4F, Tower B, 128, Beobwon-ro, Songpa-gu, Seoul 05854, Korea; csd@yunwoo.co.kr

* Correspondence: kangkim@uos.ac.kr; Tel.: +82-2-6490-2762; Fax: +82-2-6490-2749

Received: 28 October 2019; Accepted: 13 November 2019; Published: 17 November 2019

Abstract: Half precast concrete slabs with inverted multi-ribs (Joint Advanced Slab, JAS), which enhance composite performance between slabs by introducing shear keys at connections between the slabs and improve structural performance by placing prestressing tendons and truss-type shear reinforcements, have recently been developed and applied in many construction fields. In this study, flexural and shear tests were performed to verify the structural performance of JAS members. Towards this end, two flexural specimens and four shear specimens were fabricated, and the presence of cast-in-place concrete and the location of the critical section were set as the main test variables. In addition, the flexural and shear performance of the JAS was quantitatively evaluated using a non-linear flexural analysis model and current structural design codes. Evaluation results confirmed that the flexural behavior of the JAS was almost similar to the behavior simulated through the non-linear flexural analysis model, and the shear performance of the JAS can also be estimated appropriately by using the shear strength equations presented in the current design codes. For the JAS with cast-in-place concrete, however, the shear strength estimation results differed significantly depending on the way that the shear contributions of the precast concrete unit and cast-in-place concrete were calculated. Based on the analysis results, this study proposed a design method that can reasonably estimate the shear strength of the composite JAS.

Keywords: inverted multi tee; prestressed concrete; precast concrete; structural performance; flexural analysis; shear strength

1. Introduction

The precast concrete (PC) slab has many advantages in terms of the quality control of concrete through prefabrication at the factory, reduction of construction costs by shortening the construction period, and reduction of waste on construction sites [1–3]. Accordingly, various types of half-PC slab members, such as the flat precast slab, hollow-core slab, multi-rib slab, and double-tee slab, have been developed, and related research on them has been actively done. The flat precast slab has the advantage of ensuring easy mold operation and rebar placement. However, because of the thin flanges, it requires temporary work, such as the installation of shores, to resist the construction load at the time of casting topping concrete [4,5]. The hollow-core slab can reduce the lifting weight because of the presence of hollows in the cross-section and can greatly increase the thickness of the cross-section under design conditions in which a load is quite large [3,6–8]. However, the hollow-core slab manufactured using an extruder is very vulnerable to web shear at the end of the member because shear reinforcement cannot be placed on the web [3,9,10]. Furthermore, since the ACI 318-14 code [11] prescribes that the web shear strength of a one-way prestressed member that exceeds 315 mm in thickness and has less than the minimum shear reinforcement shall be reduced by half, the application of the hollow-core slab produced by the extrusion method can provide non-economical design results [3,10]. Meanwhile, the multi-rib and double-tee slabs have cross-sectional shapes that are very efficient for positive moment resistance because of the removal of an unnecessary concrete section on the tension side. However, there is a disadvantage in that the top flange is very thin, making it difficult to achieve continuity at the member ends [12–14].

In order to maximize the advantages of the half-PC system, a joint advanced slab (JAS) to minimize the temporary work process and lifting weight during construction, and facilitate the placement of shear reinforcement, has recently been developed, as shown in Figure 1. The JAS has advantages, in that it has greater flexural stiffness of the cross-section than that of the flat precast slab because of the presence of inverted multi-ribs, and it can minimize the on-site rebar placement process, because the shear keys introduced at both ends of the slab panel replace the reinforcing steel required for connections between PC slabs. In addition, the introduction of prestressing into the cross-section improves the flexural and shear performance of the member, and the N-type lattice reinforcement is placed in the ribs as a stirrup to resist horizontal and vertical shear forces effectively. Moreover, since the JAS ensures continuity at the end region of the slab, it can resist the entire static moment generated by the load applied after the placement of cast-in-situ concrete by dividing it into positive and negative moments.

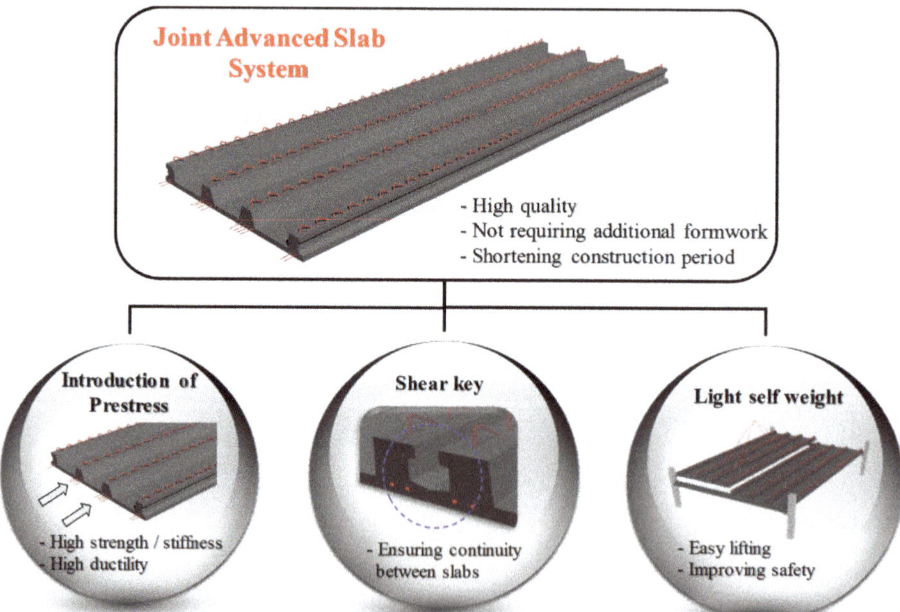

Figure 1. Characteristics of the joint advanced slab (JAS).

In this regard, experimental research was done to examine the flexural and shear performance of the newly developed JAS, where the presence of cast-in-place concrete and the location of the critical section were set as the main test variables. For the flexural specimens, the strain gauges were attached to the longitudinal tension reinforcement and concrete in a compression zone and were attached to the stirrups for the shear specimens. Then, the member behavior, according to the loads, was measured and analyzed in detail. In addition, this study quantitatively evaluated the flexural and shear performance of the JAS by comparing the test results and analysis results from the non-linear flexural analysis model and structural design codes. In particular, based on the shear test results, a method to reasonably estimate the shear contributions of the PC unit and cast-in-place concrete was proposed for the shear design of the JAS.

2. Experimental Program

2.1. Test Specimens

Table 1 and Figure 2 show the details of the test specimens. A total of six specimens, one PC unit flexural specimen, one PC flexural specimen with cast-in-place concrete two PC unit shear specimens, and two PC shear specimens with cast-in-place concrete, were fabricated to evaluate the structural performance of the JAS. The width (b) and length (L) of all specimens were 1100 mm and 7200 mm, respectively, and six prestressing strands with a diameter of 15.2 mm were placed on the tension side of the specimens. In addition, the N-type lattice reinforcement with a diameter of 10 mm and a spacing of 200 mm was placed in inverted multi ribs. The PC unit flexural and shear specimens were named UF(Unit-Flexural) and US(Unit-Shear), and the flexural and shear specimens with cast-in-place concrete were named CF(Composite-Flexural) and CS(Composite-Shear), respectively. Since the JAS is manufactured through the pretension method, a transfer length zone exists, and the stress of the strands in this region is smaller than the effective prestress (f_{se}). Therefore, for the US and CS specimens, shear tests were conducted in the transfer length zone and the strain plateau zone, respectively, as shown in Figure 3. To distinguish between the test specimens, 't' and 'f' were added to each of the specimen names. Here, the USf and CSf specimens were additionally strengthened with eight 19 mm diameter rebars on the tension side, in order to avoid flexural failure and to induce shear failure. The magnitude of the effective prestress (f_{se}) introduced in the specimen was 61% of the ultimate strength (f_{pu}) of the tendon. The concrete compressive strength (f_c') of the PC unit specimens was 27.0 MPa, and the PC compressive strength in the composite specimens was 36.0 MPa. In addition, the compressive strengths of the cast-in-place concrete were 18.5 and 21.3 MPa, respectively. It is noted that the designed compressive strengths of PC and cast-in-place concrete were 40 and 30 MPa, respectively; however, the compressive strengths were found to be smaller than the designed compressive strengths because the outdoor air temperature was below zero degree Celsius during the fabrication of the specimens [15].

Table 1. Details of the test specimens.

Specimen	Description	Test Region	h (mm) ***	d_p, d_s (mm) ***	A_{ps}, A_s (mm^2) ***		Stirrups **	f_c' (MPa) ***	
					Strands *	Rebar **		PC	Topping
UF	Flexure (without topping)	-	230	190	6-Φ15.2 (832.2)	-	D10@200 + D10@200 (inclined)	27.0	-
CF	Flexure (with topping)	-	330	290	6-Φ15.2 (832.2)	-		36.0	21.3
USt	Shear (without topping)	Transfer length zone	230	190	6-Φ15.2 (832.2)	-		27.0	-
USf	Shear (without topping)	Strain plateau zone	230	190	6-Φ15.2 (832.2)	8-D19 (2288)		27.0	-
CSt	Shear (with topping)	Transfer length zone	330	290	6-Φ15.2 (832.2)	8-D19 (2288)		36.0	18.5
CSf	Shear (with topping)	Strain plateau zone	330	290	6-Φ15.2 (832.2)	8-D19 (2288)		36.0	18.5

* Tensile strength of the prestressing strands (f_{pu}) was 1936 MPa. ** Yield strengths of the tensile reinforcement (f_y) and stirrup (f_{vy}) were 503 and 406 MPa, respectively. *** Notations: h: member height, d_s: effective depth, A_{ps}: sectional area of strands, A_s: sectional area of tensile reinforcement, f_c': compressive strength of concrete.

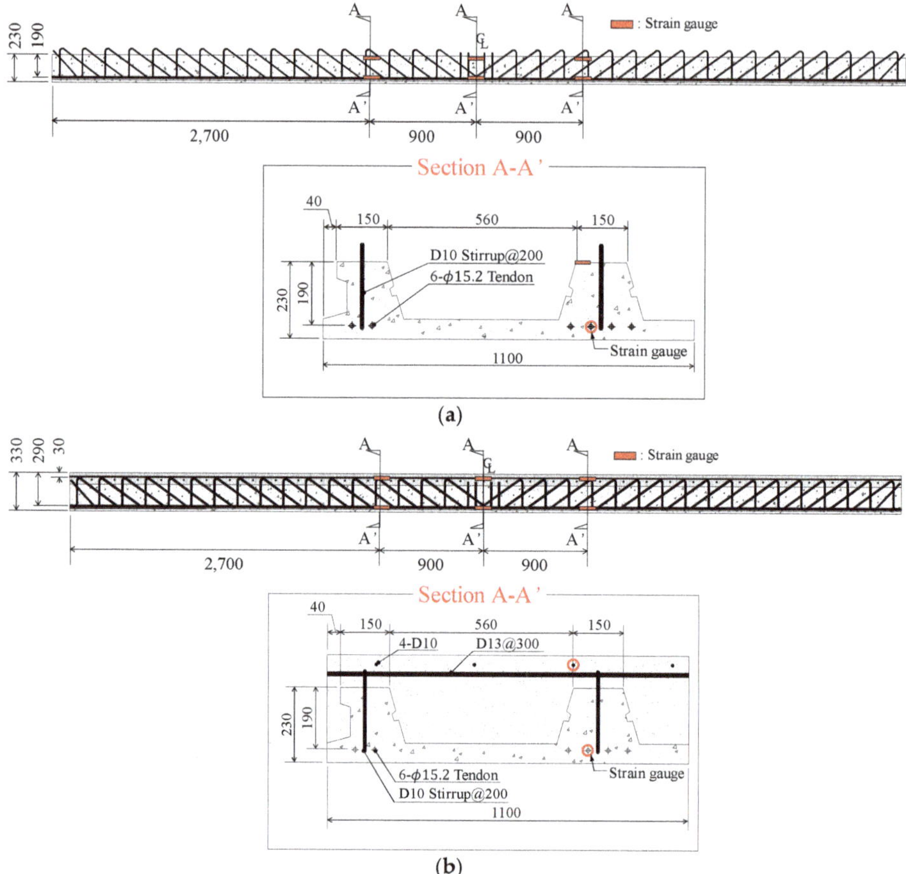

Figure 2. *Cont.*

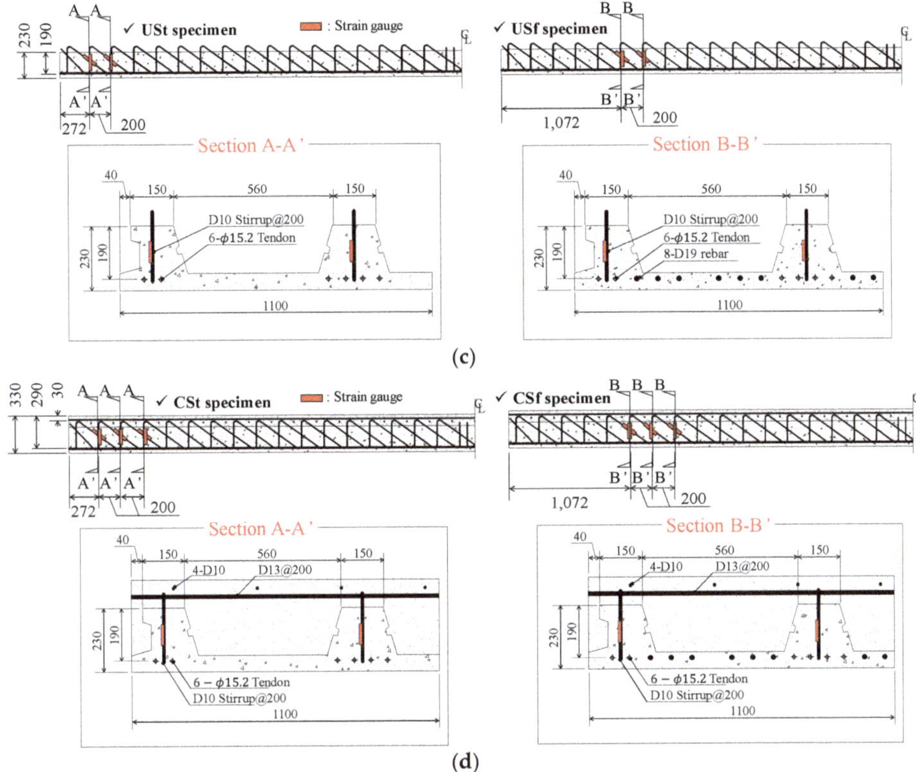

Figure 2. Details of test specimens (unit: mm); (**a**) UF specimen; (**b**) CF specimen; (**c**) USt and USf specimens; (**d**) CSt and CSf specimens.

As shown in Figure 2, in the UF specimen, strain gauges were attached to the strands on the tension side and concrete at the extreme compression fiber, whereas they were attached to the strands on the tension side and compression reinforcement in the CF specimen. Strain distributions in cross-sections according to loads were then measured in detail. For shear specimens, the shear contributions of the stirrups were measured by attaching gauges to the lattice bars placed in the test region.

2.2. Test Set-Up

Figure 3 shows the loading details of the test specimens. Two-point loading was applied to the simply supported UF and CF specimens, and one-point loading was applied to the US and CS specimens with a shear span to depth ratio (a/d_s) of 2.5. For the shear specimens, the left and right span lengths were set differently to induce shear failure in the target region. The deflections of the specimens were measured using linear variable differential transformers (LVDTs) installed at the center of the span of the flexural specimens, and the bottom of the cross-section located at the loading point of the shear specimens.

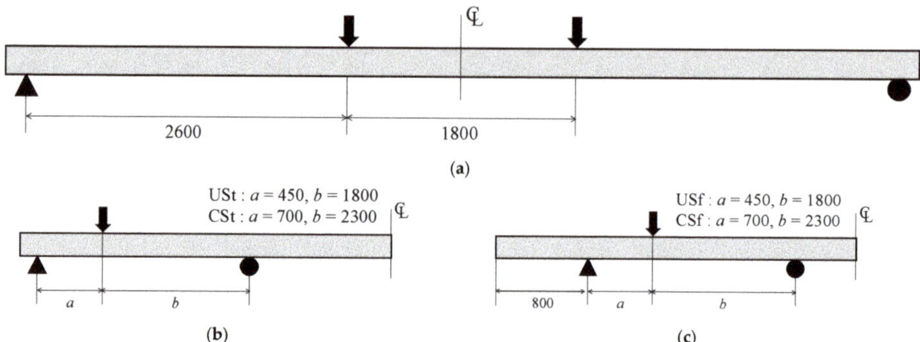

Figure 3. Test set-up (unit: mm); (**a**) UF and CF specimens; (**b**) USt and CSt specimens; (**c**) USf and CSf specimens.

3. Experimental Results

3.1. Failure Modes of the Test Specimens

Figure 4 shows the crack patterns of specimens at failure. The UF specimen, which is a PC unit slab without cast-in-place concrete, failed because of the sudden crushing of a rib on the compression side after flexural cracks had formed around the loading point. On the other hand, the CF specimen exhibited a typical flexural failure mode because of the crushing of concrete on the compression side after several flexural cracks propagated in the region of the maximum flexural moment. Meanwhile, both the USt and USf specimens underwent shear failure as web-shear cracks propagated at the ribs. However, as shown in Figure 4c,d, the shear crack angle generated in the USt specimen was steeper than that observed in the USf specimen, because the effective prestress (f_{se}) of the USt specimen is smaller than that of the USf specimen as the shear span of the USt specimen is located within the transfer length zone. The CSt and CSf specimens with case-in-place concrete showed a crack behavior different from that of the PC unit specimens. That is, inclined cracks occurred near the loading section with a relatively large flexural moment, and horizontal cracks towards the support were observed along with the interface between the PC unit slab and the cast-in-place concrete. In addition, the angle of the shear crack was about 60°, which was steeper than the shear crack angle ranging from 35° to 40° [16] that is generally observed in prestressed concrete (PSC) members. This is because the sectional area of cast-in-place concrete is considerably larger than that of the PC unit, and no prestressing is introduced to the cast-in-place concrete. Consequently, it is estimated that the effect resulting from the introduction of prestressing is lower than in the PC unit specimens.

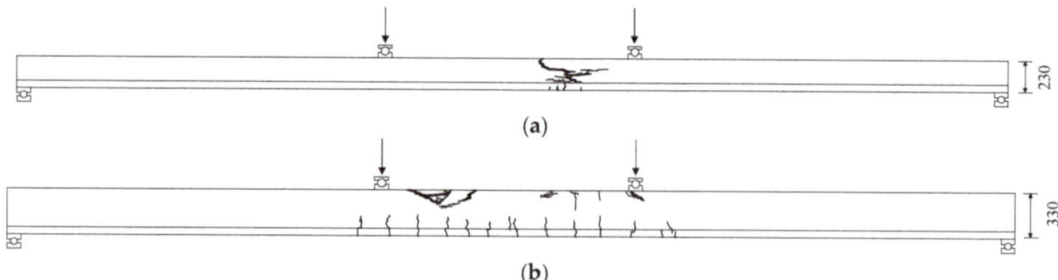

Figure 4. *Cont.*

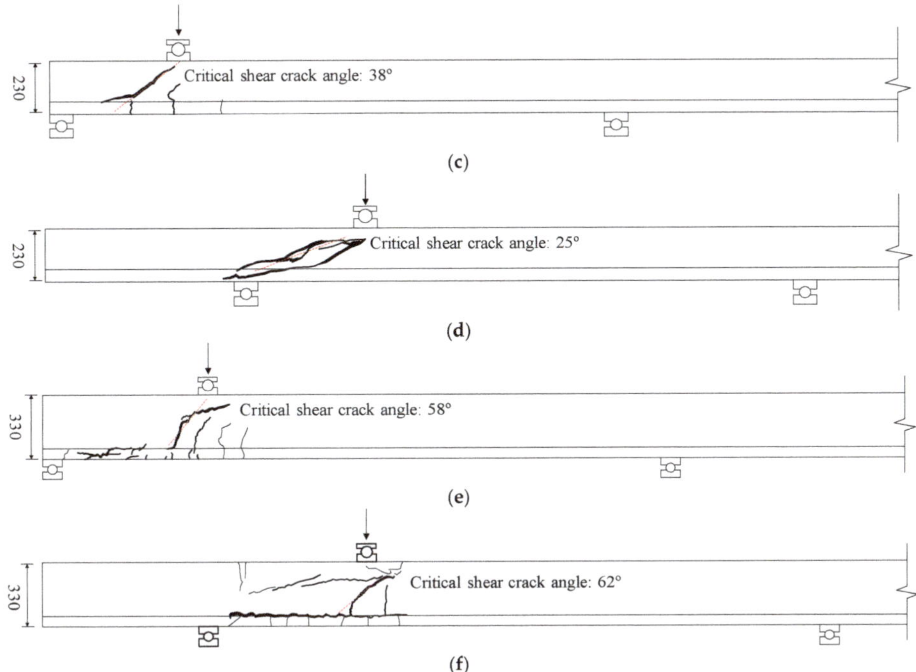

Figure 4. Crack patterns of test specimens at failure; (**a**) UF specimen; (**b**) CF specimen; (**c**) USt specimen; (**d**) USf specimen; (**e**) CSt specimen; (**f**) CSf specimen.

3.2. Load-Deflection Responses

Figure 5 shows the load-deflection curves of UF and CF specimens, and Table 2 summarizes the test results. In the UF specimen, which is a PC unit slab, the initial flexural crack was observed at a load of about 70 kN (M_{cr} = 100.0 kN·m). Later, flexural failure occurred as the concrete on the compression side in the upper part of the rib was crushed at a load of 96.2 kN (M_u = 125.1 kN·m). In the CF specimen, which is a composite member, flexural cracks were observed at a load of 147 kN (M_{cr} = 191.5 kN·m). Then, the stiffness of the specimens continued to be reduced, and flexural failure occurred at a load of 267.1 kN (M_u = 347.2 kN·m). The UF specimen showed slightly less ductility than the CF specimen; although the compressive force generated by the flexural moment should be resisted by the ribs in a UF specimen, the rib is very small narrow to the overall width of the member, as shown in Figure 2a, so that the position of the neural axis is relatively low.

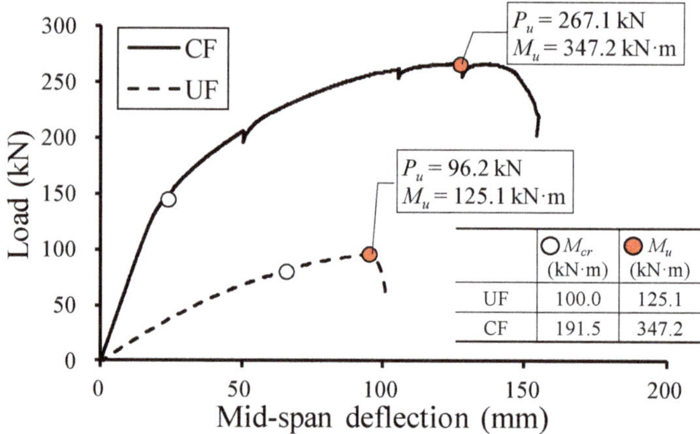

Figure 5. Load-deflection responses of flexural test specimens.

Table 2. Summary of test results.

Specimen	Failure Mode	Failure Loads (P_u, kN)	Flexural Strength (M_u, kN·m)	Shear Strength (V_u, kN)
UF	Flexure	96.2	125.1	-
CF	Flexure	267.1	347.2	-
USt	Shear	291.0	-	232.8
USf	Shear	325.6	-	260.5
CSt	Shear	548.0	-	420.1
CSf	Shear	848.8	-	650.7

Figure 6 shows the load-deflection curves of the USt, USf, CSt, and CSf specimens. In the USt specimen tested in the transfer length zone of the PC unit slab, shear cracks were observed at a load of about 151 kN (V_{cr} = 121 kN), and shear failure occurred at a load of 291 kN (V_u = 233 kN). In the USf specimen tested in the strain plateau zone of the PC unit slab, the initial shear crack occurred at a load of about 170 kN (V_{cr} = 136 kN). Even after that, new shear cracks were formed, as shown in Figure 4d, and shear failure occurred at a load of 326 kN (V_u = 261 kN). However, unlike the USt specimen, there was no decrease in the stiffness of the member, even after shear cracking, in the USf specimen. On the CSt specimen, which is a composite member, flexural cracks around the loading point developed into inclined shear cracks at a load of about 370 kN (V_{cr} = 284 kN), and shear failure took place as the horizontal crack towards the support along the interface between the PC unit slab and cast-in-place concrete occurred at a load of 548 kN (V_u = 420 kN). In the CSf specimen, the initial inclined crack was observed at a load of 420 kN (V_{cr} = 322 kN), and shear failure took place as the horizontal crack towards the support occurred at a load of 849 kN (V_u = 651 kN) similarly to that in the CSt specimen. As can be seen from the test results, the shear cracking strength (V_{cr}) and ultimate shear strength (V_u) of the USf and CSf specimens tested in the strain plateau zone, where the effective prestress (f_{se}) is fully developed, were found to be greater than those of the USt and CSt specimens, respectively.

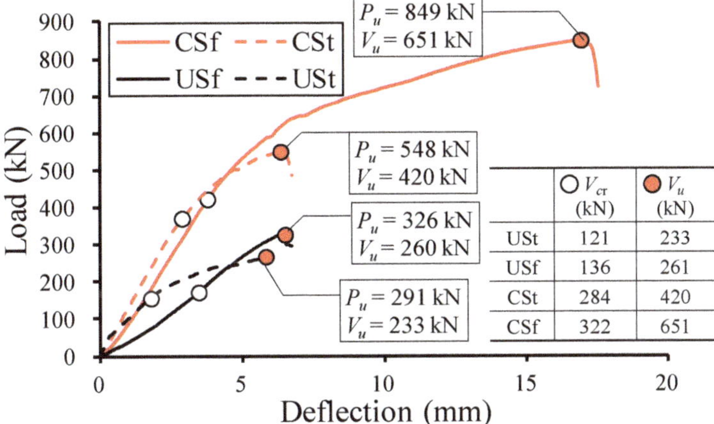

	V_{cr} (kN)	V_u (kN)
USt	121	233
USf	136	261
CSt	284	420
CSf	322	651

Figure 6. Load-deflection responses of shear test specimens.

3.3. Measured Strains

Figure 7a,b shows the strain distributions of the sections measured at the regions of the UF and CF specimens with maximum moment. In the UF specimen, the position of the neural axis was raised with increasing loads, and the strain in concrete at the compression edge was close to the ultimate strain ($\varepsilon_{cu} = 0.003$) at a load ($0.95P_u$), where P_u is the failure load. However, the strain in strands did not increase as much as did the strain in concrete at the compression edge, because the ribs are very narrow in the PC unit slab, and thus the neural axis position is low. Accordingly, the member was found to be not very ductile. On the other hand, the CF specimen, which has a sufficient compressive resistance area because of the presence of cast-in-place concrete, exhibited a large tensile strain of the strand at flexural failure and demonstrated highly ductile behavior.

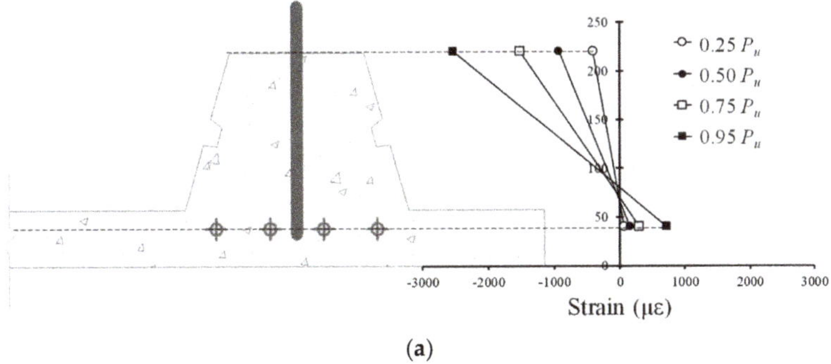

(**a**)

Figure 7. *Cont.*

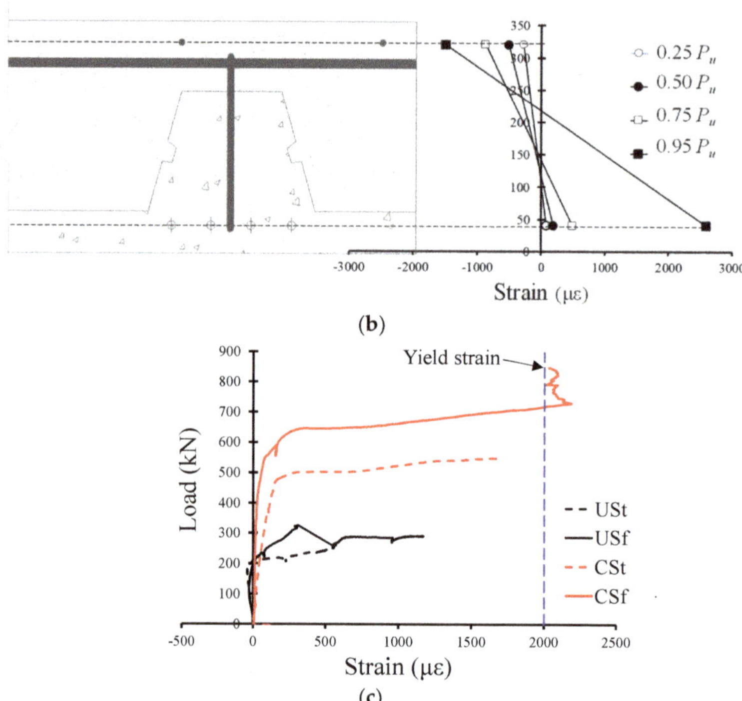

Figure 7. Measured strains from test specimens; (**a**) Strain distribution in the section located at maximum moment region (UF specimen); (**b**) Strain distribution in the section located at the maximum moment region (CF specimen); and (**c**) Stirrup strains measured from the shear test specimens.

Figure 7c shows the strains measured from stirrups placed in the shear specimens. In the PC unit specimens (USt and USf), the stirrup made a partial contribution to the shear resistance of the member but did not yield even at the maximum load. This is because the embedment length of the stirrup is short in the PC unit slab, and the hook at the top protrudes before the placement of cast-in-situ concrete. It should be noted that the shear reinforcement in the JAS member is designed not to secure the shear performance of the PC unit slab but to ensure the shear performance of composite members. On the other hand, composite specimens (CSt and CSf) with cast-in-place concrete, in which shear reinforcement is properly anchored in the cross-section, exhibited a strain close to the yield strain after shear cracking and effectively contributed to the shear resistance of the member.

4. Analysis of Flexural Behavior and Shear Strength

4.1. Non-Linear Flexural Analysis

In this study, a layered analysis [17,18] shown in Figure 8a was performed to evaluate the flexural behavior of the JAS. Since no damage at the interface between the PC unit slab and cast-in-place concrete was observed in the CF specimen, the PC and the cast-in-place concrete were considered to be fully composite in the flexural analysis. On the assumption of strain in concrete at the compression edge (ε_t) and neural axis depth (c), the strain of compression reinforcement (ε_s'), the strain of strands (ε_{ps}), and the concrete strain ($\varepsilon_{c,i}$) in each layer can be calculated through the compatibility conditions. Then the stresses corresponding to each strain can be obtained from the constitutive laws for steel and concrete materials shown in Figure 8b. In this study, the Collins model [19], the elasto-perfectly plastic model [20,21], and Ramberg–Osgood model [22,23] were adopted as the constitutive equations of the concrete, reinforcing bars, and strands, respectively. The compression force of the PC (C_{PC}), of the cast-in-place concrete (C_{RC}), and of the reinforcement (C_s), and tension force of strands (T_{ps}) can be calculated by multiplying the stress and areas of each element, and the moment-curvature response can be obtained from iterative calculations performed until the conditions for equilibrium of forces (i.e., $C_{PC} + C_{RC} + C_s + T_{ps} \approx 0$) are satisfied. In addition, the load-deflection response can be derived from the numerical integration of the curvature (ϕ_i) for each section, as shown in Figure 8c. A numerical calculation example can be found in Appendix A.

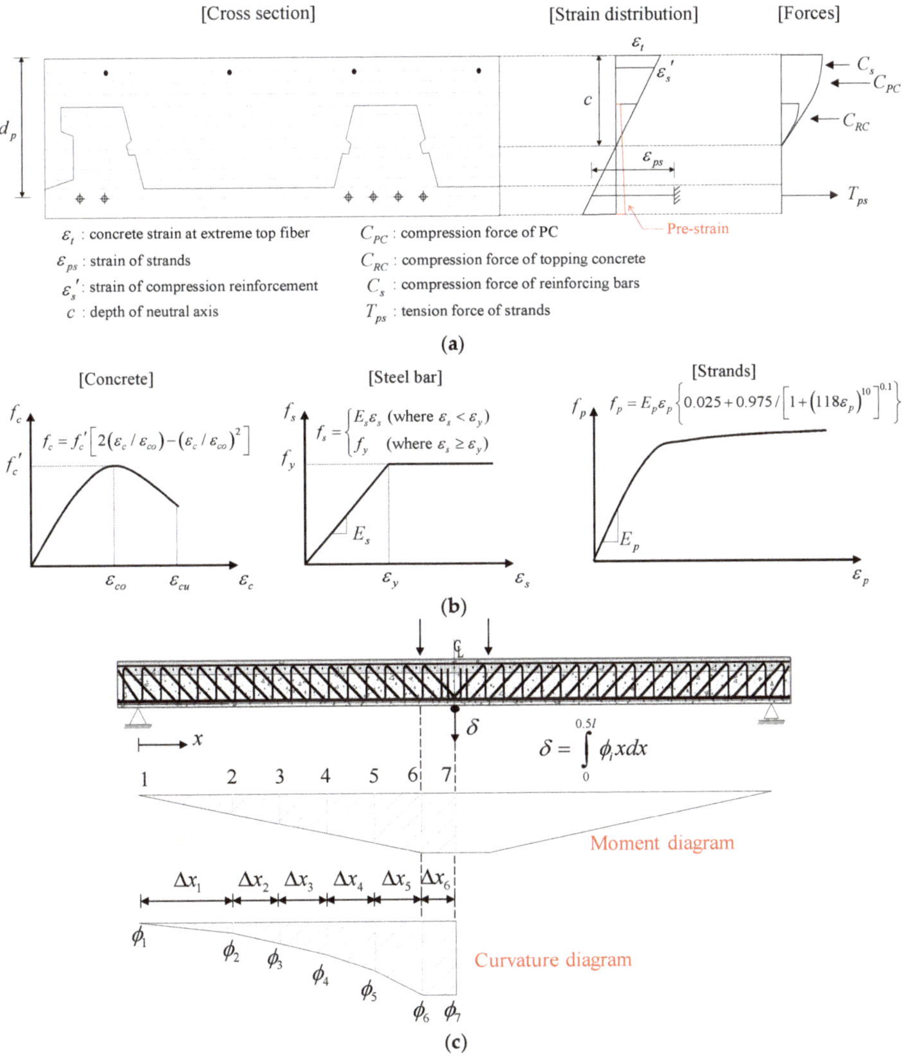

Figure 8. Non-linear flexural analysis; (**a**) Layered analysis model; (**b**) Constitutive laws of materials; (**c**) Numerical integration for estimating mid-span deflection.

4.2. Estimation of Shear Strength

Table 3 shows code equations for estimating the shear strength presented in ACI 318-14 [11], which were applied to evaluate the shear performance of the JAS. As mentioned earlier, shear reinforcement placed in the US series specimens without cast-in-place concrete did not contribute effectively to the shear resistance of the member. Therefore, only the contribution of the concrete (V_{cw}) was taken into account in estimating the shear strength of the PC unit members, without regard for the contribution of the shear reinforcement (V_s). Meanwhile, the compressive strengths of the PC unit and cast-in-place concrete are different in composite members, and ACI 318-14 [11] specifies that it is allowable to use the material properties of the elements that result in the most critical value of shear strength (V_n). However, in this case, it tends to be very conservative in evaluating the shear strength of the member. In addition, if the web shear strength equation for prestressed concrete members (V_{cw}) is applied to get the web shear strength of JAS members, it causes the fallacy that prestress is effective in all cross-sections including the cast-in-place concrete, while prestress force is introduced only to the PC unit in the JAS. Consequently, this approach provides an unsafe estimation on the shear strength of JAS members. For this reason, a way to estimate the shear strength by considering the entire composite PC slab to be reinforced concrete (RC) in order to achieve the safe side design has often been applied in practice [12].

Table 3. Code equations for estimating shear strength [11].

Term	Equations
Shear strength of reinforced concrete (V_c)	$V_c = 0.17\sqrt{f_c'}b_w d_s$
Web shear strength of prestressed concrete (V_{cw})	$V_{cw} = \left(0.29\sqrt{f_c'} + 0.3f_{pc}\right)b_w d_p$, where $f_{pc} = \frac{f_{se}A_{ps}}{A_{pc}}$
Shear contribution of stirrups (V_s)	$V_s = \frac{A_v f_{vy} d_s}{s_v}(\sin\alpha\cot\beta + \cos\alpha)$

* Notations: b_w: web width, d_s: effective depth of tension reinforcement, d_p: effective depth of prestressing strands, f_c': compressive strength of concrete, f_{se}: effective prestress, A_{pc}: sectional area of prestressed member, A_{ps}: sectional area of prestressing strands, A_v: sectional area of the stirrup, f_{vy}: yield strength of the stirrup, s_v: spacing of stirrups, α: inclined angle of the stirrup, β: angle of critical shear crack.

In this study, the shear strength (V_n) of the composite member was estimated by three methods (labeled Methods 1, 2, 3), as shown in Figure 9. Note that the area indicated in red is where the shear strength is estimated by regarding the composite member to be PSC, whereas the area indicated in blue is where the shear strength is estimated by regarding it as being RC. Method 1 is a method for calculating the shear contribution (V_{cw}) by defining the distance from the compression edge of the composite member to the center of the strands layer as d_p, and estimating the shear strength by considering the remaining part as RC. On the other hand, Method 2 estimates V_{cw} by considering the distance from the top of the PC unit rib to the center of the strands layer as d_p. Method 3 estimates the shear strength by regarding the cross-section as being RC. According to the test results, the stirrup effectively contributed to the shear resistance mechanism in the composite member. Therefore, the contribution of the shear reinforcement (V_s) was taken into consideration in the estimation of the shear strength. However, since the critical shear crack angle (β) observed in the test specimens was distributed in the range from 58° to 62°, β was applied at 60° in the estimation of V_s.

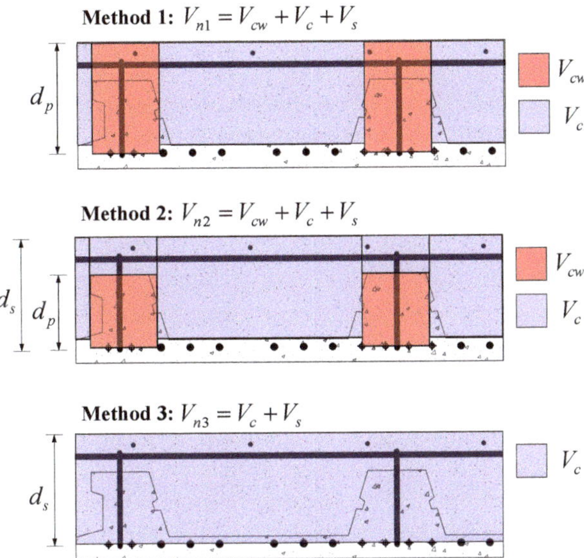

Figure 9. Shear strength estimation methods for composite member.

4.3. Comparison of Test and Analysis Results

Figure 10 compares the test and non-linear flexural analysis results of the UF and CF specimens. It was found that the analysis model provides very approximate predictions of the flexural strength and behavior of the specimens, and thus, can be applied to the flexural design of the JAS.

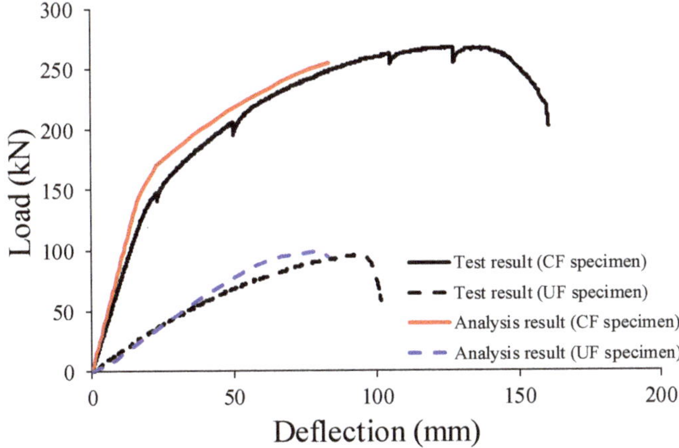

Figure 10. Comparison of the test and analysis results on the flexural specimens.

Figure 11 and Table 4 compare the test and analysis results of the shear specimens (US and CS series). As shown in Figure 11a, for the PC unit specimens, the shear strength estimation method, considering only the contribution of concrete (V_{cw}) without regard for the contribution of the shear reinforcement (V_s), provided the conservative calculation results. For the composite specimens, Method 1 (V_{n1}), which applies the effective depth (d_p) of the composite member, including the thickness of the topping slab in the upper part of the rib, provided the analysis results of the unsafe side, whereas Method 3 (V_{n3}), which estimates the shear strength by considering the gross-section to be RC, evaluated the shear strength of the composite specimens to be on the excessively safe side. Meanwhile, Method 2 (V_{n2}), which calculates V_{cw} by regarding the distance from the top of the PC unit rib to the center of the strands layer as d_p, slightly overestimated the shear strength of the CSt specimen, but provided a much more reasonable evaluation of the shear strength of composite members when compared to Methods 1 or 2. Considering that the safety factor and strength reduction factor are applied in the practical design, the most economical and reasonable design can be derived by using Method 2 when estimating the shear strength of the composite JAS. In this case, it is appropriate to apply 60° as the angle of the shear crack for estimating the contribution of the shear reinforcement (V_s).

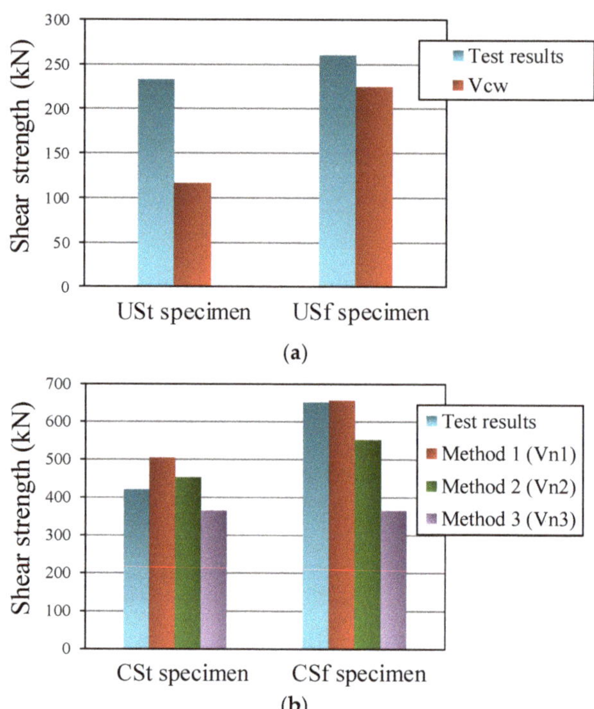

Figure 11. Comparison of the test and analysis results on the shear specimens; (**a**) Non-composite members; (**b**) Composite members.

Table 4. Comparison of shear strengths.

Specimen	Test Results (V_u, kN)	Calculation Results (V_n, kN)		
		V_{n1} (Ratio, V_u/V_{n1})	V_{n2} (Ratio, V_u/V_{n2})	V_{n3} (Ratio, V_u/V_{n3})
USt	232.8		115.9 (2.01)	
USf	260.4		223.9 (1.16)	
CSt	420.3	504.8 (0.83)	452.7 (0.93)	364.7 (1.15)
CSf	651.2	655.9 (0.99)	551.7 (1.18)	364.7 (1.78)

5. Conclusions

In this study, experimental and analytical research has been performed to evaluate the flexural and shear performance of the half-precast concrete slab with inverted multi-ribs (Joint Advanced Slab, JAS). The crack patterns, load-deflection responses, and strain behavior of the longitudinal and shear reinforcement of the specimens were measured and analyzed in detail, and the flexural and shear performance of the JAS was quantitatively evaluated using a non-linear flexural analysis method and ACI 318-14 code. In addition, a design method that can most reasonably estimate the shear strength of the JAS was proposed based on the test results. On this basis, the following conclusions can be drawn:

1. The UF specimen, a PC unit slab, exhibited a ductility slightly lower than that of the CF specimen with cast-in-place concrete; although the compressive force generated by the flexural moment should be resisted by the ribs, the rib is much narrower than the overall width of the member, and thus the position of the neural axis is relatively low.
2. The non-linear flexural analysis provided very approximate evaluations of the flexural behavior and strength of the JAS, regardless of the presence of cast-in-place concrete, and thus can be applied to the flexural design of the JAS.
3. The shear test results showed that the lattice reinforcement placed in composite specimens exhibited effective shear resistance performance, whereas the lattice reinforcement placed in PC unit specimens did not make a significant contribution to the shear resistance of the member. Therefore, the shear design results of the safe side can be obtained without considering the contribution of shear reinforcement (V_s) in the design of the PC unit slab.
4. The shear crack angle observed in composite specimens was about 60°, which was steeper than that observed in the general PSC member. In addition, various methods for estimating the shear contributions of the PC unit and cast-in-place concrete were examined based on the ACI 318-14 code. The results confirmed that a method of calculating the contribution of shear reinforcement (V_s) by applying the crack angle (β) of 60°, and estimating the PC unit shear contribution (V_{cw}) by regarding the distance from the top of the PC unit rib to the center of the strands layer as the effective depth (d_p) of the composite member, while estimating the shear strength by considering the remaining part as RC, provides the most reasonable analysis results.

Author Contributions: Original draft manuscript—S.-J.H.; Validation—J.-H.J. and H.-E.J.; Investigation—S.-H.C. and S.C.; Supervision and Review Writing—K.S.K.

Funding: This research received no external funding.

Acknowledgments: This work was supported by the National Research Foundation of Korea (NRF) grant funded by the Korea government (MSIT) (No. 2019R1A2C2086388).

Conflicts of Interest: The authors declare no conflict of interest.

Appendix A

In this section, an example of numerical calculations on the flexural behavior of composite JAS is presented. The sectional details and material properties of JAS are shown in Figure 2b and Table 1, respectively.

- Input parameters (Figure A1):

$$h_{pc} = 230 \text{ mm}, h_{top} = 100 \text{ mm}, d_p = 290 \text{ mm}, d'_s = 40 \text{ mm}, b = 1100 \text{ mm}, b_w = 150 \text{ mm}$$

$$f'_{c,pc} = 36.0 \text{ MPa}, f'_{c,rc} = 21.3 \text{ MPa}, f_y = 503 \text{ MPa}, f_{pu} = 1936 \text{ MPa}, f_{se} = 0.61 f_{pu}$$

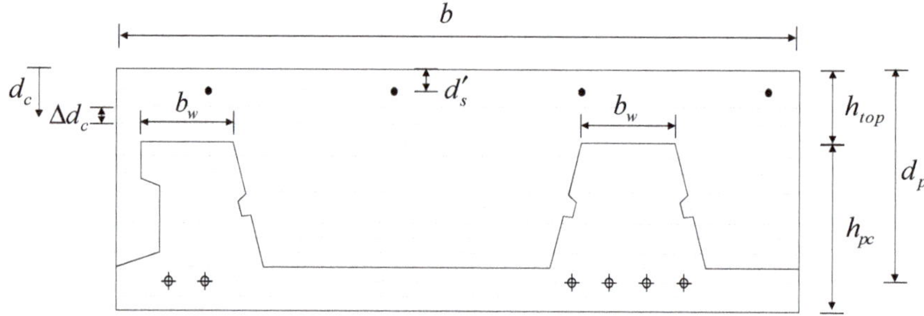

Figure A1. Section details of JAS.

-Moment-curvature response

Step 1: Select $\varepsilon_t = -0.001$

Step 2: Assume $c = 100$ mm

Step 3: Obtain the distribution of strains, as shown in Figure A2.

$\varepsilon_b = \frac{c-(h_{pc}+h_{top})}{c} \times \varepsilon_t = 0.0026$,

$\varepsilon'_s = \frac{c-d'_s}{c} \times \varepsilon_t = -0.00099$, and $\varepsilon_{ps} = \frac{f_{se}}{E_s} + \frac{c-d_p}{c} \times \varepsilon_t = 0.0076$

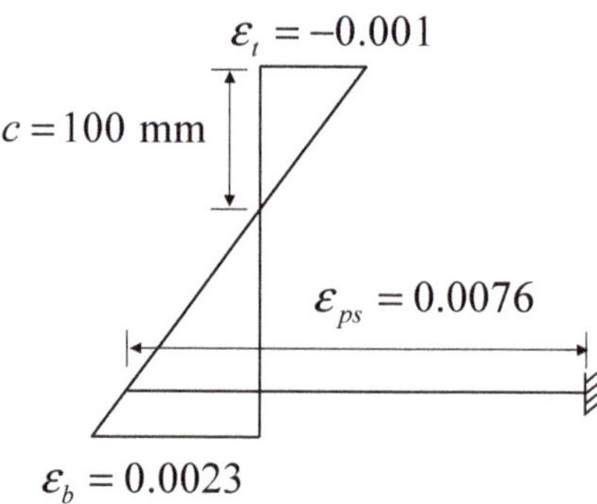

Figure A2. Strain distribution

Step 4: Calculate stresses of concrete layers, strands, and compressive reinforcement from constitutive laws of materials shown in Figure 8b.

Step 5: Calculate forces of concrete, strands, and compressive reinforcement.

$$C_{pc} = \sum f_{c,pc} \cdot b_i \cdot \Delta d_c = 15.1 \text{ kN}$$

$$C_{RC} = \sum f_{c,rc} \cdot b_i \cdot \Delta d_c = -732.4 \text{ kN}$$

$$C_s = f'_s A'_s = -56.5 \text{ kN}$$

$$T_{ps} = f_{ps} A_{ps} = 1258.8 \text{ kN}$$

Step 6: Check force equilibrium.

$C_{pc} + C_{RC} + C_s + T_{ps} \neq 0$ (Not satisfied)

Step 7: Go back to the *step 2*, and update c until the force equilibrium is satisfied.

If the neutral axis depth (c) is assumed to be 135 mm, the forces of concrete, strands and compressive reinforcement are calculated, as follows:

$$C_{pc} = -119.4 \text{ kN}, \ C_{RC} = -953.1 \text{ kN}, \ C_s = -56.6 \text{ kN}, \ T_{ps} = 1136.0 \text{ kN}$$

The equilibrium condition is also satisfied, $C_{pc} + C_{RC} + C_s + T_{ps} = 6.9 \text{ kN} \approx 0$.
Step 8: Calculate curvature (ϕ) and moment (M).

$$\phi = \frac{\varepsilon_t}{c} = 7.4 \times 10^{-6}$$

$$M = \sum f_{c,pc} \cdot b_i \cdot d_c \cdot \Delta d_c + \sum f_{c,rc} \cdot b_i \cdot d_c \cdot \Delta d_c + f_s' A_s' d_s' + f_{ps} A_{ps} d_p = 276.3 \text{ kN} \cdot \text{m}$$

Step 9: Repeat step 1 to 8 with an updated ε_t until ε_t reaches the crushing strain of concrete (−0.003). The moment-curvature response of composite JAS is presented in Figure A3:

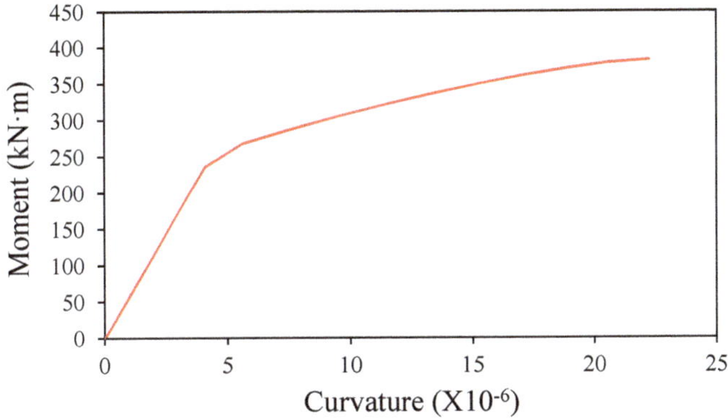

Figure A3. Moment-curvature response.

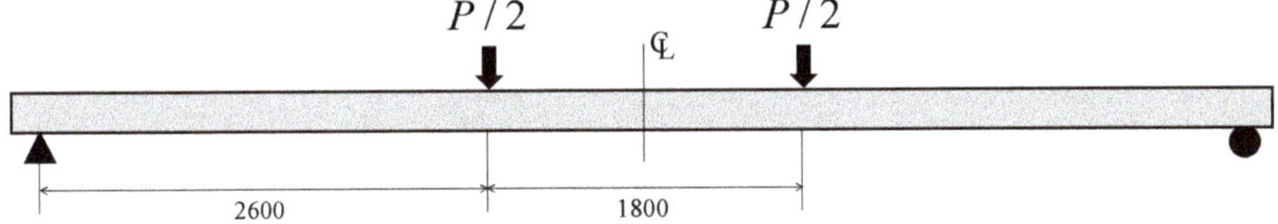

Figure A4. Loading details.

-Load-deflection response.
The loading details of JAS is shown in Figure A4.
Step 1: Select $P = 50$ kN.
Step 2: Obtain moment and curvature distributions from Figure A3.

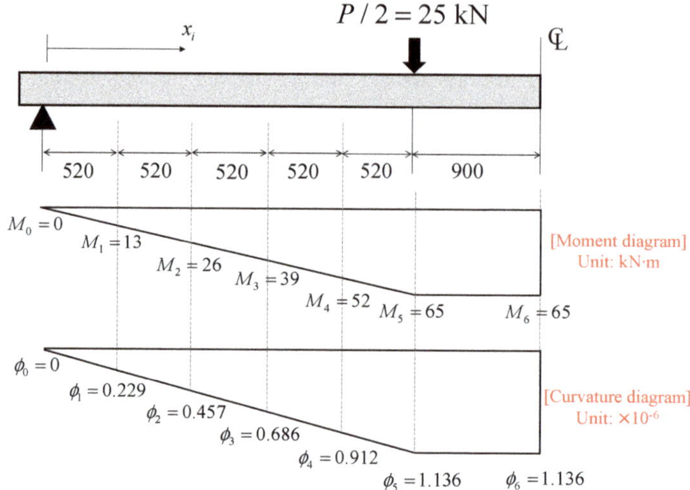

Figure A5. Moment and curvature distributions.

Step 3: Calculate mid-span deflection by numerically integrating curvatures shown in Figure A5, as follows:

$$\delta = \int_0^{0.5l} \phi_i x dx = \left(\frac{\phi_0 x_0 + \phi_1 x_1}{2}\right)\Delta x_0 + \left(\frac{\phi_1 x_1 + \phi_2 x_2}{2}\right)\Delta x_1 + \cdots + \left(\frac{\phi_5 x_5 + \phi_6 x_6}{2}\right)\Delta x_5 = 5.73 \text{ mm}$$

Step 4: Repeat step 1 to 3 with updating P, then the load-deflection response can be obtained, as shown in Figure 10.

References

1. Ju, H.; Han, S.J.; Choi, I.S.; Choi, S.; Park, M.K.; Kim, K.S. Experimental study on an optimized-section precast slab with structural aesthetics. *Appl. Sci.* **2018**, *8*, 1234. [CrossRef]
2. Kim, S.E.; Colin, J. *Multi-Story Precast Concrete Framed Structures*, 2nd ed.; Wiley-Blackwell: Hoboken, NJ, USA, 2014; Available online: https://www.wiley.com/en-us/Multi+Storey+Precast+Concrete+Framed+Structures%2C+2nd+Edition-p-9781405106146 (accessed on 8 March 2019).
3. Lee, D.H.; Park, M.K.; Oh, J.Y.; Kim, K.S.; Im, J.H.; Seo, S.Y. Web-shear capacity of prestressed hollow-core slab unit with consideration on the minimum shear reinforcement requirement. *Comput. Concr.* **2014**, *14*, 211–231. [CrossRef]
4. Newell, S.; Goggins, J.; Hajdukiewicz, M.; Holleran, D. Behaviour of hybrid concrete lattice girder flat slab system using insitu structural health monitoring. In Proceedings of the Civil Engineering Research in Ireland Conference (CERI 2016), Galway, Ireland, 29–30 August 2016.
5. Hou, H.; Liu, X.; Qu, B.; Ma, T.; Liu, H.; Feng, M.; Zhang, B. Experimental evaluation of flexural behavior of composite beams with cast-in-place concrete slabs on precast prestressed concrete decks. *Eng. Struct.* **2015**, *126*, 405–416. [CrossRef]
6. Yang, L. Design of prestressed hollow core slabs with reference to web shear failure. *J. Struct. Eng.* **1994**, *120*, 2675–2696. [CrossRef]
7. Palmer, K.D.; Schultz, A.E. Factors affecting web-shear capacity of deep hollow-core units. *PCI J.* **2010**, *55*, 123–146. [CrossRef]
8. Palmer, K.D.; Schultz, A.E. Experimental investigation of the web-shear strength of deep hollow-core units. *PCI J.* **2011**, *56*, 83–104. [CrossRef]
9. Pisanty, A. The shear strength of extruded hollow-core slabs. *Mater. Struct.* **1992**, *25*, 224–230. [CrossRef]

10. Im, J.H.; Park, M.K.; Lee, D.H.; Kim, K.S.; Seo, S.Y. *Evaluation of Web-Shear Capacity of Thick Hollow-Core Slabs*; The Regional Association of Architectural Institute of Korea: Daegu, Korea, 2014; Volume 16, pp. 139–145. Available online: https://aikra.jams.or.kr/co/com/EgovMenu.kci?s_url=/sj/search/sjSereClasList.kci&s_MenuId=MENU-000000000053000 (accessed on 21 March 2019). (In Korean)
11. ACI Committee 318. *Building Code Requirements for Reinforced Concrete and Commentary*; ACI 318-14; American Concrete Institute: Farmington Hills, MI, USA, 2014.
12. Ju, H.; Han, S.J.; Joo, H.E.; Cho, H.C.; Kim, K.S.; Oh, Y.H. Shear performance of optimized-section precast slab with tapered cross section. *Sustainability* **2019**, *11*, 163. [CrossRef]
13. Naito, C.; Cao, L.; Peter, W. Precast concrete double-tee connections, part 1: Tension behavior. *PCI J.* **2009**, *54*, 49–66. [CrossRef]
14. Cao, L.; Naito, C. Precast concrete double-tee connectors, part 2: Shear behavior. *PCI J.* **2009**, *54*, 97–115. [CrossRef]
15. Rubene, S.; Vilnitis, M. Impact of low temperatures on compressive strength of concrete. *Int. J. Theor. Appl. Mech.* **2017**, *2*, 97–101.
16. Collins, M.P.; Mitchell, D. *Prestressed Concrete Structures*; Prentice-Hall: Upper Saddle River, NJ, USA, 1991.
17. Lee, D.H.; Han, S.J.; Joo, H.E.; Kim, K.S. Control of tensile stress in prestressed concrete members under service loads. *Int. J. Concr. Struct. Mater.* **2018**, *12*, 38. [CrossRef]
18. Han, S.J.; Lee, D.H.; Oh, J.Y.; Choi, S.H.; Kim, K.S. Flexural responses of prestressed hybrid wide flange composite girders. *Int. J. Concr. Struct. Mater.* **2018**, *12*, 53. [CrossRef]
19. Vecchio, F.J.; Collins, M.P. The modified compression-field theory for reinforced concrete elements subjected to shear. *ACI J. Proc.* **1986**, *83*, 219–231. [CrossRef]
20. Scholz, H. Ductility, redistribution, and hyperstatic moments in partially prestressed members. *ACI Struct. J.* **1990**, *87*, 341–349. [CrossRef]
21. Rodriguez-Gutierrez, J.A.; Aristizabal-Ochoa, J.D. Partially and fully prestressed concrete sections under biaxial bending and axial load. *ACI Struct. J.* **2000**, *97*, 553–563. [CrossRef]
22. Mattock, A.H. Flexural strength of prestressed concrete sections by programmable calculator. *PCI J.* **1979**, *24*, 32–54. [CrossRef]
23. Park, H.; Cho, J.Y. Ductility analysis of prestressed concrete members with high-strength strands and code. *ACI Struct. J.* **2017**, *114*, 407–416. [CrossRef]

Article

Experimental Investigation on Compressive Properties and Carbon Emission Assessment of Concrete Hollow Block Masonry Incorporating Recycled Concrete Aggregates

Chao Liu [1,*], Chao Zhu [2], Guoliang Bai [2], Zonggang Quan [3] and Jian Wu [4]

1 College of Science, Xi'an University of Architecture and Technology, Xi'an 710055, China
2 College of Civil Engineering, Xi'an University of Architecture and Technology, Xi'an 710055, China; chaozhu@live.xauat.edu.cn (C.Z.); guoliangbai@126.com (G.B.)
3 Xi'an Research and Design Institute of Wall and Roof Materials, Xi'an 710061, China; xaqcy1965@163.com
4 Shaanxi Key Laboratory of Safety and Durability of Concrete Structures, Xijing University, Xi'an 710123, China; wujian2085@126.com
* Correspondence: chaoliu@xauat.edu.cn; Tel.: +86-180-9256-1062

Received: 17 October 2019; Accepted: 12 November 2019; Published: 14 November 2019

Featured Application: The compressive properties and carbon emission assessment of recycled aggregate concrete hollow block masonry were analyzed. The formula for calculating the compressive strength of hollow block masonry was presented. The global warming potential between natural and recycled aggregate concrete hollow blocks was compared. This study will contribute to the popularization and application of recycled aggregate concrete hollow block masonry.

Abstract: In this paper, the compressive strength experiment for three groups of recycled aggregate concrete (RAC) specimens with different replacement ratios of recycled aggregate (0%, 50%, 100%) was carried out. The mechanism of the block and mortar properties on the compressive strength of block masonry was investigated by means of a static loading test. The formula for calculating the compressive strength of a recycled concrete block was obtained based on experimental data. Moreover, the global warming potential (GWP) of recycled aggregate concrete (RAC) block masonry was evaluated by life cycle assessment (LCA) methodology. The feasibility for application of RAC block masonry was discussed combined with environmental impact data analysis. The results show that the strength of RAC blocks is the principal element affecting the compressive strength of block masonry, and the strength of mortar also has a certain impact for the compressive of RAC block masonry; the sub-coefficient of material performance should be enhancive appropriately for ensuring the construction quality of RAC block masonry; the total GWP of RAC block is lower than that of natural aggregate concrete (NAC) block. The environmental benefits of the promotion and application of RAC block masonry are inspiring.

Keywords: recycled aggregate concrete; block masonry; compressive strength; carbon emission; stress–strain curves

1. Introduction

In the last decade, the acceleration of urbanization has led to a sharp rise in the production of construction and demolition wastes (C&DW) in China, and the annual production of waste concrete is up to 20 billion tons. These ever-increasing C&DW pose serious challenges to environmental governance [1–3]. Meanwhile, China's construction raw materials, such as sand and stone, have been overexploited [4,5]. The preparation of recycled aggregate concrete (RAC) by replacing the natural aggregate (NA) with recycled aggregate (RA) from waste concrete can solve the dual problems of environment and resources effectively. The application of RAC is of great significance to the sustainable development of concrete [6–8].

Extensive research on mechanical and durability of RAC has been carried out in recent years [9–11]. Most scholars suggest that RAC is weaker than natural aggregate concrete due to the higher water absorption and crushing index of RA [12,13]. Some articles also reported that the complex and weak interface transition zone (ITZ) in RAC is also an important factor resulting in the performance deterioration of RAC [14]. However, the performance of RAC can be conformed to the specific requirements through reasonable design or adjustment of mixing ratio, and RAC has some applications in the construction industry [15–17].

The masonry structure is composed of blocks that are the oldest building materials, and this form of structure has been widely employed in the world [18]. The combination of RA with the masonry structure is absolutely inspiring. The RAC has certain advantages in block production compared with that in components, which is attributed to the block production technology. Several studies have reported that the process of vibrating and compacting can make up for the disadvantage of RAC slump, making the RAC block have a higher strength [19]. In addition, the reduced water content used in the block production process will significantly reduce the creep and shrinkage of RAC [20].

There have been some reports on recycled concrete bricks in recent years. Poon et al. [21] carried out the study on compressive of RAC bricks, and noticed that the compressive strength of the brick is almost constant with the RA replacement ratio of 50%, but the higher the replacement percentage, the lower the compressive strength of the paving slab. Poon et al. [22] investigated the effect of water–cement ratio and aggregate type on the performance of paving concrete blocks. They reported that the compressive strength of paving slabs decreases as the water–cement ratio increases and is proportional to the crush index of the aggregate. The effect of recycled aggregate on the compressive strength of precast concrete hollow blocks was reported by Matar et al. [19], which determined the content of RA in blocks with appropriate compressive strength. However, a few contrary conclusions have been proposed by Corinaldesi [23]. It can be found that from the above description, most of the current research involves the mechanical properties of RAC paving slabs. However, there are few reports on the use of RCA to manufacture blocks for buildings. This is disappointing for the promotion and application of recycled concrete block masonry structures.

In this paper, three groups of RAC specimens with different replacement ratios of recycled aggregate (0%, 50%, 100%) were prepared. The mechanism of the block and mortar properties on the compressive strength of RAC block masonry were investigated by means of static loading test. Moreover, the global warming potential (GWP) of RAC block masonry was evaluated by life cycle assessment (LCA) methodology. The feasibility of application of RAC block masonry was discussed. It can be inferred that the compressive strength of RAC block masonry increased regularly with the strength of the RAC block being enhanced, and the effect law of mortar performance was similar. The sub-coefficient of material performance should be enhancive appropriately for ensuring the construction quality of RAC block masonry. The environmental benefits of the promotion and application of RAC block masonry were inspiring. These results have a certain positive impact on the development of RAC block masonry.

2. Experimental Program

2.1. Materials and Mixture Proportions

The ordinary Portland cement with a grade of 32.5 MPa was used as cementitious. The RA employed in the study was purchased from a franchiser in Xi'an, China. NCA was crushed limestone with a nominal size of 5–10 mm and the size range was the same as RCA. The river sand was used as FA with the water absorption and fineness modulus of 1.02% and 2.18, respectively. The size distributions of coarse aggregates are illustrated in Figure 1. The aggregates employed in this research demonstrate a continuous grain curve and conforms to Chinese standard JGJ 52-2006 [24]. The physical properties of the employed NCA and RCA were tested according to Chinese standard GB/T 14685-2011 [25] and are listed in Table 1. The results show that RCAs have a higher crushing value and water absorption compared to NCAs, which may be attributed to the old adhered mortar.

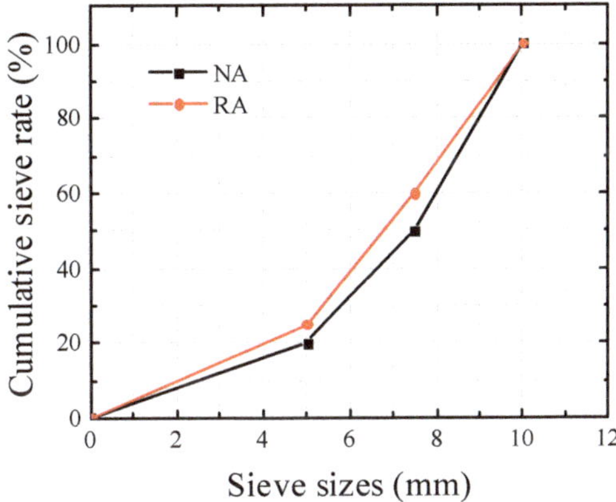

Figure 1. The size distribution of NA and RA used in this study.

Table 1. Physical properties of coarse aggregates.

Aggregate Type	Apparent Density (kg/m^3)	Crushing Value (%)	Water Adsorption (%)	Particle Size Range (mm)
NA	2802	6.0	0.4	5~10
RA	2522	24.7	3.5	5~10

The three types of mixture proportions were used in the study with the RA replacement ratios of 0%, 50%, and 100%, respectively (by weight). The workability of mixture was guaranteed by means of additional water. The mixture proportions for concrete hollow block are summarized in Table 2.

Table 2. Mixing proportions of concrete.

Replacement Ratio of RA (%)	Water Absorption of Mixed Aggregate (%)	Unit Weight (kg/m^3)				
		Water	Cement	FA	NA	RA
0	1.5	190	380	633	1232	-
50	3.5	239	380	633	616	616
100	5.0	276	380	633	-	1232

2.2. Specimens Preparation

Three types of RAC hollow blocks were fabricated with the RA replacement ratio of 0%, 50%, and 100%, respectively, and the strength grades of corresponding block were MU10, MU7.5, and MU5. The commonly employed block with dimensions of 390 mm × 240 mm × 190 mm and a half block with dimensions of 190 mm × 240 mm × 190 mm were used as the bond characteristic test piece, as presented in Figure 2. The volume percentages of the vertical holes for whole and half concrete hollow blocks were 33.8% and 33.9%. The RAC block masonry was prepared in accordance with standard test methods for basic mechanical properties of masonry GB/T 5129-2011 [26]. The thickness of mortar joint was 10 mm. Compressive prisms were manufactured by two full blocks and two half blocks with the dimensions of 390 mm × 240 mm × 590 mm, as presented in Figure 3. The design parameters of specimens are listed in Table 3. Three test pieces were produced for each code, and a total of 27 specimens were tested.

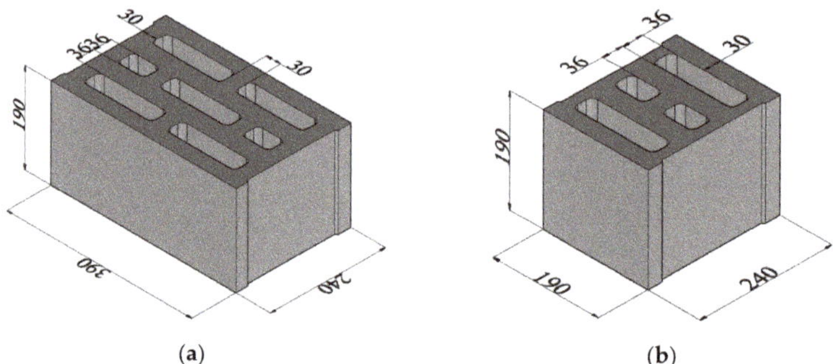

Figure 2. Diagrams of blocks with dimensions of (**a**) 390 mm × 240 mm × 190 mm; (**b**) 190 mm × 240 mm × 190 mm.

Figure 3. Specimens employed to test the compressive strength: (**a**) left for curing; (**b**) dimensions (mm).

Table 3. Design parameters of specimens.

Code	r (%)	f_1 (MPa)	f_2 (MPa)
B-0-1	0	10.5	14.8
B-0-2		10.5	12.0
B-0-3		10.5	8.0
B-50-1	50	7.53	12.0
B-50-2		7.53	8.0
B-50-3		7.53	6.2
B-100-1	100	5.31	8.0
B-100-2		5.31	6.2
B-100-3		5.31	3.1

Note: r: replacement ratio of RA, f_1: compressive strength of concrete blocks, f_2: compressive strength of mortar.

2.3. Experimental Procedures

The Chinese standard test methods for the concrete block and brick GB/T 4111-2013 [27] were employed to guide the compressive strength test. The rough surfaces of specimen were coated with high-strength gypsum in order to ensure the uniform load distribution during compression test.

The servo-hydraulic actuator (YE-200A) with force-controlled 2000 kN capacity was used in the study. The transverse and vertical deformations of specimen were measured by the dial indicator. The load and test scenario are shown in Figure 4.

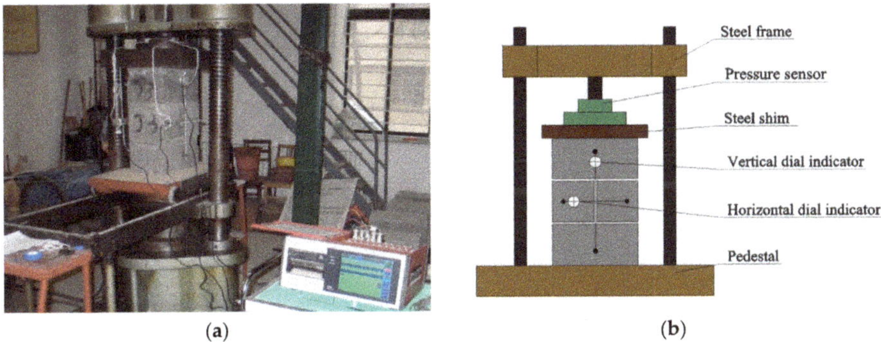

Figure 4. Test setup for the compressive strength: (**a**) experimental picture; (**b**) schematic diagram.

Firstly, 5% of the estimated failure load of specimen was applied to check the performance of test apparatus. Three to five repeated loads were carried out within the range of 5% to 20% of the estimated failure load. The relative error of axial deformation value of the two opposite surfaces should not exceed 10%, otherwise the position of the test piece should be adjusted. The 10% of estimated failure load of specimen was served as the increment of each stage with a loading rate of 1 kN/s. The deformation value of each load grade should be recorded.

3. Experimental Results and Discussions

3.1. Description of Experimental Phenomena

The failure characteristics of RAC hollow block masonry are presented in Figure 5. It can be noticed that the specimens with a different replacement ratio of RAs have the similar failure characteristics. The failure process can be summarized as the following phases:

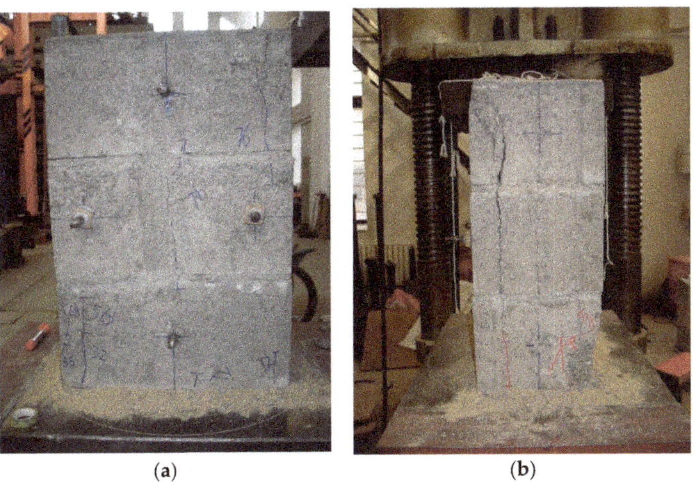

Figure 5. Failure modes of compressive strength: (**a**) front view; (**b**) side view.

Phase I: Before the test piece cracks, the value of the compression machine and dial indicator increased steadily with the increase of the load, and there were no abnormal phenomena on the surface of the specimen. The end of the first stage was accompanied by the appearance of the first micro-crack. Meanwhile, the micro-crack will not extend when the load stops increasing, and the specimen remains stable at this stage.

Phase II: The specimen from the initial crack to the limit state was regarded as the second stage. With the emergence of slight vertical cracks on the narrow side of the recycled concrete hollow masonry, the bearing capacity of the specimens began to increase tardily. The vertical or oblique crack was generated on the wide side of the specimen, accompanied by the persistent augment in load. The dial indicator increased rapidly. With the further enhancement of the load, the penetrating cracks were generated on the narrow side of the specimen, which divided the test piece into several small columns. The RAC hollow block masonry reached ultimate bearing capacity when the load did not grow persistently. The reading of the hydraulic press and dial indicator fluctuates obviously. Then the second stage terminated.

Phase III: When the limit load was exceeded, the number of the hydraulic press was decreased fiercely, and the deformation of the specimen increased sharply. The specimen was crushed and destroyed eventually.

It can be seen from the failure status of RAC hollow block masonry that the top face and the connection of the two ribs were weak links, where the penetration cracks will appear firstly, and the whole specimen was destroyed finally. This was mainly attributed to by the fact that there were more holes and ribs in the recycled concrete block with three rows of holes, and there were still process grooves in the end of the holes at the junction of the two ribs on the top surface, which was effortless in forming a failure surface. The compression test values of the RAC hollow block masonry are listed in Table 4.

Table 4. Compression strength test results of RAC block masonry.

Code	P_c (kN)	P_u (kN)	P_c/P_u
B-0-1	623.1	802.3	0.78
B-0-2	513.2	771.3	0.66
B-0-3	480.3	712.3	0.67
B-50-1	401.1	581.2	0.69
B-50-2	373.8	479.2	0.78
B-50-3	299.7	422.1	0.71
B-100-1	255.6	387.1	0.66
B-100-2	250.7	347.9	0.72
B-100-3	231.5	330.1	0.70

Note: P_c: cracking load of concrete masonry, P_u: ultimate load of concrete masonry.

3.2. *Analysis of Test Results*

3.2.1. The Stress–Strain Curve of RAC Masonry

The stress–strain diagrams of the RAC hollow block masonry with different replacement ratios of RA are shown in Figure 6a. The effects of mortar strength for stress–strain curves are also presented in Figure 6b–d. The ascending section of the stress–strain curve was only discussed in this research due to the difficult measurement on the descent section by the servo-hydraulic actuator.

It can be noticed from Figure 6a that the variation trend of stress–strain diagram for different specimen was similar. The strength and deformation of specimen with NAs were higher than that of the specimen with RAs, and these values decreased with the replacement ratio of RA enhanced. This result may be attributed to there being more ITZs in the RAC block. As can be seen from Figure 6b–d, the deformation of specimens enhanced with the increase of mortar strength no matter what the replacement ratio of RA was.

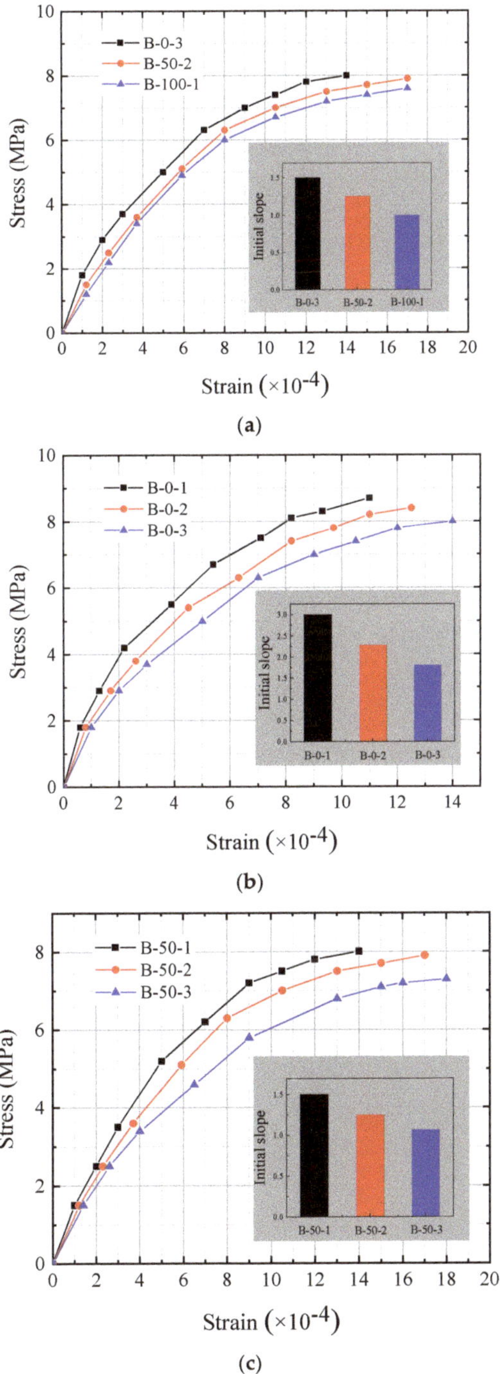

Figure 6. *Cont.*

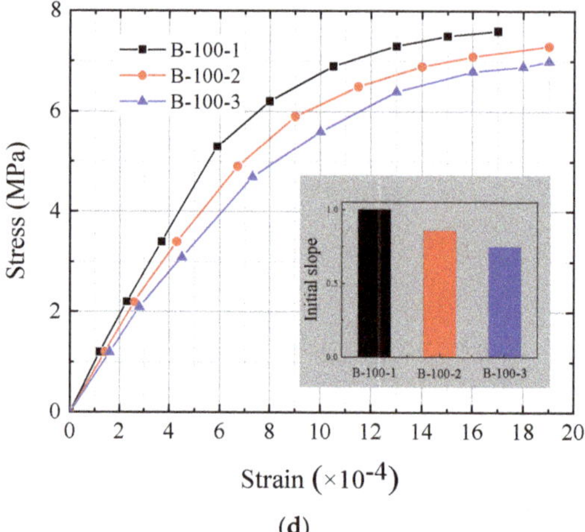

(d)

Figure 6. Stress–strain curves of RAC block masonry under compression: (**a**) effect of RA replacement ratio; (**b**–**d**) effect of mortar strength.

The elastic modulus of the specimen can be described by the slope of the stress–strain curve. The bar diagram shows that the increase in the replacement ratio of RA played an inhibitory influence for the elastic modulus of the test piece, while the effect of the mortar strength was reversed.

3.2.2. Analysis of Influencing Factors

The effects of the block and mortar strength on the compressive strength of RAC block masonry structure are shown in Figures 7 and 8, respectively.

It can be observed from Figure 7 that the compressive strength of RAC block masonry was enhanced with the increase of block strength. The difference between NAC and RAC blocks masonry may be attributed to more ITZs present in RAC blocks. The weak ITZs in concrete were more likely to generate micro-cracks under the pressure, which led to the destruction of the concrete masonry.

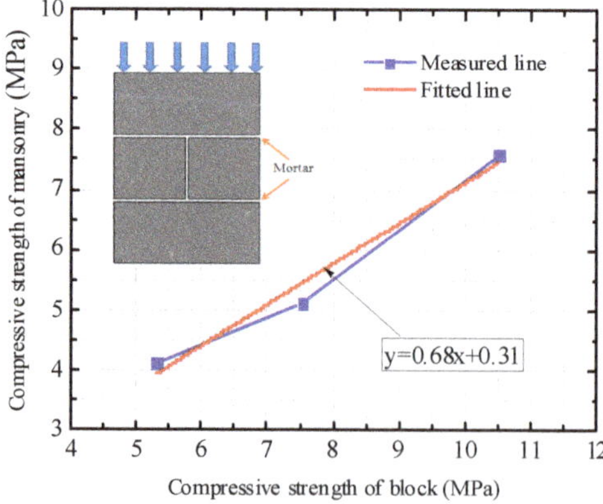

Figure 7. Effect of RAC block strength on the compressive strength of RAC blocks masonry.

According to Figure 8, it can be found that the mortar strength also had a significant effect on the strength of concrete block masonry. The increase of mortar strength was encouraging to improve the strength of concrete masonry. Moreover, the lower the concrete block strength was, the more sensitive the concrete block masonry strength was to mortar strength. The results may be due to the characteristic of recycled concrete block being unstable, which weakened the contribution of block strength to the compressive strength of concrete masonry and reinforced the effect of mortar strength relatively.

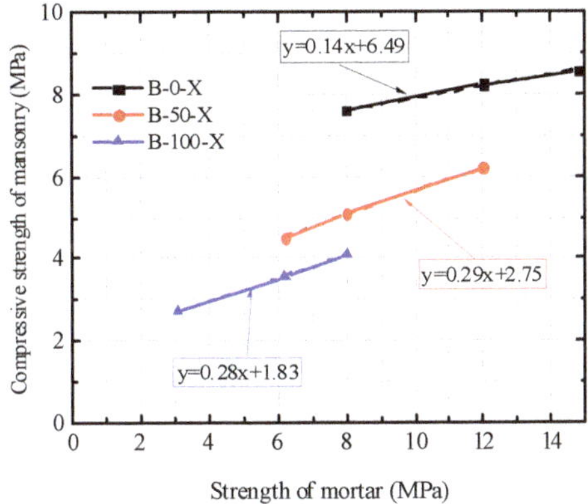

Figure 8. Effect of mortar strength on the compressive strength of RAC blocks masonry.

This paper compared the effects of mortar strength and block strength on masonry structure performance. A 40% increase in block strength will result in a 39% increase in compressive strength of concrete block masonry, but a 40% increase in mortar strength will only increase the compressive strength of concrete block masonry by 8%. It can be found that the strength of block is more sensitive to the compressive strength value of concrete block masonry than that of mortar.

3.3. The Calculation Formula of Recycled Concrete Hollow Block Masonry Strength

It was significant to research the formula for calculating the compressive strength of RAC hollow block masonry for its application. It can be found from the above results that the evolution law and internal mechanism of compressive strength were similar between NAC and RAC hollow block masonry.

The calculation formula of NAC block masonry applies to this study [28], as follows:

$$f_{\mathrm{m}} = k_1 f_1^{\alpha}(1 + 0.07 f_2)k_2 \tag{1}$$

where f_{m} was the calculated value of masonry compressive strength; f_1 and f_2 were the average compressive strength of block and masonry respectively; α and k_1 were the effect coefficients of block shape, size, masonry method, and other factors of different types of masonry; α and k_1 were equal to 0.9 and 0.46 for concrete block; k_2 was the influence coefficient of mortar strength on the compressive strength of masonry, and k_2 was 1.0 in this study. α was still 0.9 in the study due to there being the same size between NAC and RAC blocks. Regression analysis was carried out on the strength test value of RAC block masonry, and $k_1 = 0.538$ was obtained, which was higher than 0.46 in the specification. Guo et al. [29] reported similar results. For safety reasons, the parameter values in the specification should be adopted because the strength of RAC blocks was discrete relatively. The results were calculated according to formula 1 and are listed in Table 5.

It can be found that the measured value of RAC block masonry compressive strength was higher than the calculated value by formula in specification. The ratio of the measured value to the calculated value was concentrated between 1.1 and 1.2, which can be indicated by the formula for the compressive strength of NAC block masonry being suitable for RAC block masonry.

The standard value and design value of material strength were significant in the design of masonry structure. The probability distribution of material strength should conform to the normal distribution, and the standard value of material strength could be determined by the 0.05 quantile of probability distribution. According to the literature [28,30], the standard value f_k of its compressive strength can be calculated as follows:

$$f_k = f_m - 1.645\alpha = f_m(1 - 1.645\delta) \quad (2)$$

where α was the standard deviation of compressive strength, δ was the variation coefficient of compressive strength, and the value of δ was 0.21 (the average variation coefficient of specimen with 50% recycled aggregate substitution rate) in this paper.

The ratio of the standard value to the component coefficient was defined as the design value of material performance.

The strength design value f of RAC block masonry structure was the strength representative value adopted by masonry structural members when they were designed according to the ultimate state of bearing capacity. Considering the influence of geometric parameter variation, calculation mode uncertainty, and other factors on reliability, the calculation formula of f is as follows:

$$f = \frac{f_k}{\gamma_f} \quad (3)$$

where γ_f was the partial factor for material performance of RAC block masonry structure, and γ_f was taken as 1.612 when the construction quality was considered to be B.

Table 5. The measured and calculated values of compressive strength for RAC block masonry.

Code	f'_m (MPa)	f_m (MPa)	f'_m/f_m
B-0-1	8.57	7.78	1.10
B-0-2	8.24	7.03	1.17
B-0-3	7.61	5.96	1.28
B-50-1	6.21	5.76	1.08
B-50-2	5.12	4.42	1.16
B-50-3	4.51	4.06	1.11
B-100-1	4.12	4.94	1.20
B-100-2	3.55	4.08	1.15
B-100-3	2.72	3.05	1.12

Note: f'_m: experiment values of compressive strength, f_m: calculated values of compressive strength.

The construction quality was difficult to guarantee due to the RAC block masonry being prepared by RAs with high water absorption and porosity. It was necessary to increase the γ_f to 1.7 in order to ensure a reliable strength design value. The standard and design values of compressive strength for RAC block masonry are listed in Table 6.

Table 6. Standard values and design values of compressive strength for RAC block masonry.

Code	f_1 (MPa)	f_2 (MPa)	f_m (MPa)	
			Standard Values	Design Values
B-0-1	10.50	14.8	5.61	3.30
B-0-2	10.50	12.0	5.39	3.17
B-0-3	10.50	8.0	4.98	2.93
B-50-1	7.53	12.0	4.06	2.39
B-50-2	7.53	8.0	3.35	1.97
B-50-3	7.53	6.2	2.95	1.73
B-100-1	5.31	8.0	2.56	1.75
B-100-2	5.31	6.2	1.96	1.36
B-100-3	5.31	3.1	1.35	0.95

4. The Carbon Emissions Assessment on the Life Cycle of Recycled Concrete Block

4.1. Description of Life-Cycle Assessment (LCA) Methodology

The methodology employed in this study was in accordance with the ISO 14040 and 14044 standards [31,32], which defined the environmental impact assessment principles for concrete and its mixtures from "cradle to gate". The functional unit of this research was per cubic meter of concrete with different replacement ratio of RAs (0%, 50%, 100%). The cradle-to-gate system boundaries of the NAC and RAC production are illustrated in Figure 9. Carbon emission calculation can be divided into the following three stages: the raw material production process (C1); transporting raw materials to batching plant (C2); the preparation of concrete blocks (C3).

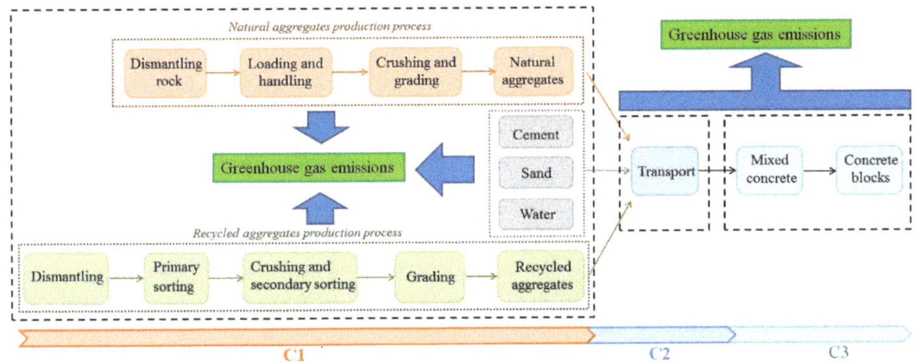

Figure 9. The cradle-to-gate concrete blocks production processes.

4.2. Assumptions and Description of Life-Cycle Inventory (LCI)

Five different concrete mixtures were specified to determine the life-cycle inventory. The assumptions and restrictions associated with the life-cycle inventory for each concrete component will be presented in detail in the following paragraphs.

The carbon emission generated per ton of cement produced was 0.80 kg, and the transportation distance of cement was 150 km. The electricity and fuel consumption required to produce per ton of natural crushed stone were 1.17 kW/t and 0.723 L/t, respectively, and converted into carbon emissions of 3.12 kg. The carbon emission per ton of sand produced was 3.66 kg. The transportation distance of natural gravel and sand was 200 km [33].

The typical preparation process of RA was as follows: The waste concrete was crushed by a forklift into the crusher, and then the broken concrete was screened by the sieving machine to obtain the recycled coarse aggregate with different particle sizes, and this process was assisted by some iron and impurity removal equipment. Statistical data on mass production of recycled aggregate were relatively lacking, as it was still in the stage of popularization in China. The data in reference [16] was referred to in this paper. The carbon emission generated per ton of RA produced was 1.61 kg, and the transportation distance of recycled aggregate was 30 km. The direct carbon emission coefficient of diesel truck transportation was 89.84 g/(km·t). These data apply only to Shaanxi province in China.

4.3. Discussion and Interpretation of Global Warming Potential (GWP)

The calculated GWP (kg CO_2-eq) related to the production of per unit volume of concrete blocks are listed in Table 7. The total index consists of the carbon emissions of each mixed material during quarrying, transportation, and production processes that take place within the system boundaries.

Table 7. The GWP (kg CO_2-eq) for each 1 m^3 of concrete blocks.

Periods	Details	B-0	B-50	B-100
C1	Cement	280	280	280
	NA	3.8	1.9	-
	RA	-	2.55	5.1
	Sand	2.31	2.31	2.31
C2	Transportation	35.6	23.5	12.5
C3	Block production	2.13	2.13	2.13
	Total	323.84	312.39	302.04

It can be understood from Table 7 that the production process of aggregates and the transportation of materials were main factors leading to differences in carbon emissions between NAC and RAC blocks. The total GWP for RAC block (with recycled aggregate replacement ratio was 100%) was 7.2% lower than that of NAC block. This opinion was inspiring for the promotion and application of RAC blocks. In addition, the calculated total GWP and the allocation of GWP by the major concrete ingredients and production processes are illustrated in detail in Figures 10 and 11.

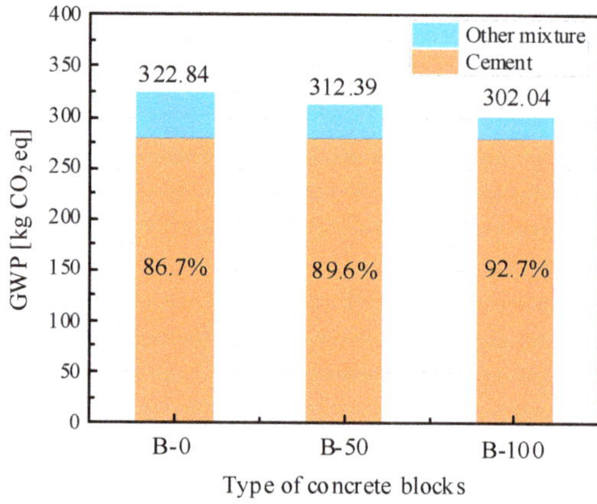

Figure 10. Comparison of total GWP for concrete blocks with different RA content.

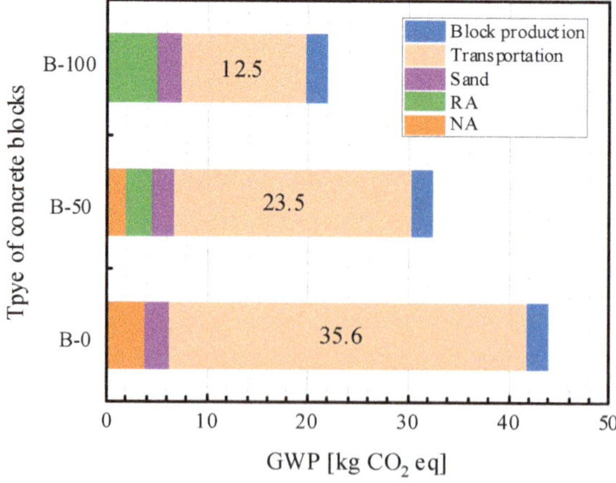

Figure 11. The GWP results for 1 m^3 of concrete blocks (excluding cement).

It is well known that cement production makes the greatest contribution to GWP, and it accounts for about 89.7% of total carbon emissions. Moreover, the transportation of raw materials to the concrete block plant was the second highest source of emission, which was about 7.5% of the total carbon emissions. The percentage of other concrete mixtures in total GWP was as follows: 1.2% for NA, 1.7% for RA, and 0.73% for sand.

According to the data of each stage, the transport of raw materials was the decisive reason that the carbon emission of concrete decreased with the increase of the replacement ratio of RA. The shorter transport distance was mainly due to the raw material of the RA being waste concrete, the source was less restricted by the region, and the processing plant can be located close to the mixing plant. However, from the life-cycle inventory, the carbon emission of RAC block will exceed that of the NAC block when the transported distance of RA exceeds a certain range. In addition, the carbonization of RA to absorb CO_2 was also of great significance for the total GWP. However, the measured mechanical performance of the concrete block decreased with the incorporation of RAs. It is necessary to limit the proportion of RAs to 50%. Therefore, it can be concluded that the RAC blocks prepared by RAs (replacement ratio was 50%) do have great significance towards improving the environmental quality.

5. Conclusions

In this study, the compressive properties and carbon emission assessment of RAC hollow block masonry were investigated by several methods, and the following conclusions can be obtained.

1. The compressive strength of RAC block decreased with the increases of RA replacement ratio. The compressive strength of RAC block masonry increased regularly with the strength of RAC block being enhanced, and the effect law of mortar performance was similar.
2. The calculation formula of compressive strength for the NAC block masonry was still applicable to the RAC block masonry. However, the partial coefficient for material performance should be enhancive appropriately for ensuring the construction quality of RAC block masonry.
3. The design values and standard values of the RAC block masonry strength with different replacement ratio of RA were calculated based on the experimental data in this study.
4. In the case of rational allocation of resources, such as controlling the distance between the RA production plant and the ready-mix plant, the environmental benefits of the promotion and application for RAC block masonry were very inspiring, attributed to the total GWP of the RAC block being lower than that of the NAC block.

Author Contributions: C.L. and C.Z. were responsible for the writing of this article. G.B. and Z.Q. were responsible for the evaluation and comments of the paper content, and J.W. was responsible for the organization of the article data.

Funding: This research was funded by the Natural Science Foundation of China [grant number 51878546], the Natural Science Basic Research Program of Shaanxi Province [grant number 2019JM-597], the Innovative Talent Promotion Plan of Shaanxi Province [grant number 2018KJXX-056], the Key Research and Development Program of Shaanxi Provincial [grant number 2018ZDCXL-SF-03-03-02], the Open Fund Project of Shaanxi Province [grant number XJKFJJ201904] and the Science and Technology Innovation Base of Shaanxi Province [grant number 2017KTPT-19].

Acknowledgments: The authors would like to acknowledge the National Natural Science Foundation of China (Grant NO. 51878546), the Natural Science Basic Research Program of Shaanxi Province (Grant NO. 2019JM-597), the Innovative Talent Promotion Plan of Shaanxi Province (Grant NO. 2018KJXX-056), the Key Research and Development Projects of Shaanxi Province (Grant NO. 2018ZDCXL-SF-03-03-02), the Open Fund Project of Shaanxi Province (Grant NO. XJKFJJ201904) and the Science and Technology Innovation Base of Shaanxi Province (Grant NO. 2017KTPT-19) for financial support.

Conflicts of Interest: All authors declare that this article has no conflict of interest, and the data (including figures, tables) and other relevant information used in the article are approved by all authors. The correspondent authors exercise power over the article on behalf of all authors.

References

1. Xiao, J.Z.; Li, W.; Fan, Y.H.; Huang, X. An overview of study on recycled aggregate concrete in China (1996–2011). *Constr. Build. Mater.* **2012**, *31*, 364–383. [CrossRef]
2. Liu, C.; Liu, H.W.; Zhu, C.; Bai, G.L. On the mechanism of internal temperature and humidity response of recycled aggregate concrete based on the recycled aggregate porous interface. *Cem. Concr. Compos.* **2019**, *103*, 22–35. [CrossRef]
3. García-Sanz-Calcedo, J.; Al-Kassir, A.; Yusaf, T. Economic and environmental impact of energy saving in healthcare buildings. *Appl. Sci.* **2018**, *8*, 440. [CrossRef]
4. He, S.Y.; Lai, J.X.; Wang, L.X.; Wang, K. A literature review on properties and applications of grouts for shield tunnel. *Constr. Build. Mater.* **2020**, *231*, 468–482.
5. Li, P.L.; Lu, Y.Q.; Lai, J.X.; Liu, H.Q. A comparative study of protective schemes for shield tunneling adjacent to pile groups. *Adv. Civ. Eng.* **2020**, *12*, 1874137.
6. GB/T 25177-2010. *Recycled Coarse Aggregate for Concrete*; Ministry of Housing and Urban-Rural Development of the People's Republic of China: Beijing, China, 2010.
7. Nie, X.X.; Feng, S.S.; Zhang, S.D.; Gan, Q. Simulation study on the dynamic ventilation control of single head roadway in high-altitude mine based on thermal comfort. *Adv. Civ. Eng.* **2019**, *12*, 2973504. [CrossRef]
8. Liu, C.; Fan, Z.Y.; Chen, X.N.; Zhu, C.; Wang, H.Y.; Bai, G.L. Experimental study on bond behavior between section steel and RAC under full replacement ratio. *KSCE J. Civ. Eng.* **2018**, *23*, 1159–1170. [CrossRef]
9. Suryawanshi, S.R.; Singh, B.; Bhargava, P. Characterization of recycled aggregate concrete. In *Advances in Structural Engineering*; Springer: Berlin/Heidelberg, Germany, 2015; pp. 1813–1822. [CrossRef]
10. Debieb, F.; Courard, L.; Kenai, S.; Degeimbre, R. Mechanical and durability properties of concrete using contaminated RAs. *Cem. Concr. Compos.* **2010**, *32*, 421–426. [CrossRef]
11. Leite, F.C.; Motta, R.S.; Vasconcelos, K.L.; Bernucci, L. Laboratory evaluation of recycled construction and demolition waste for pavements. *Constr. Build. Mater.* **2011**, *25*, 2972–2979. [CrossRef]
12. Rao, M.C.; Bhattacharyya, S.K.; Barai, S.V. Influence of field recycled coarse aggregate on properties of concrete. *Mater. Struct.* **2011**, *44*, 205–220. [CrossRef]
13. Wang, Y.; Zhang, S.H.; Niu, D.T.; Su, L.; Luo, D.M. Strength and chloride ion distribution brought by aggregate of basalt fiber reinforced coral aggregate concrete. *Constr. Build. Mater.* **2020**, *234*, 117390. [CrossRef]
14. Bai, G.L.; Zhu, C.; Liu, C.; Liu, H.W. Chloride ion invasive behavior of recycled aggregate concrete under coupling flexural loading and wetting-drying cycles. *KSCE J. Civ. Eng.* **2019**, *23*, 4454–4462. [CrossRef]
15. JGJ/T 240-2011. *Technical Specification for Application of Recycled Aggregate*; Ministry of Housing and Urban-Rural Development of the People's Republic of China: Beijing, China, 2011.
16. Marinković, S.; Radonjanin, V.; Malešev, M.; Ignjatović, I. Comparative environmental assessment of natural and recycled aggregate concrete. *Waste Manag.* **2010**, *30*, 2255–2264. [CrossRef] [PubMed]
17. Liu, C.; Lv, Z.Y.; Zhu, C.; Bai, G.L.; Zhang, Y. Study on calculation method of long term deformation of RAC beam based on creep adjustment coefficient. *KSCE J. Civ. Eng.* **2019**, *23*, 260–267. [CrossRef]
18. Marrocchino, E.; Fried, A.N.; Koulouris, A.; Vaccaro, C. Micro-chemical/structural characterization of thin layer masonry: A correlation with engineering performance. *Constr. Build. Mater.* **2009**, *23*, 582–594. [CrossRef]
19. Matar, P.; Dalati, R.E. Using recycled concrete aggregates in precast concrete hollow blocks. *Mater. Werkst.* **2012**, *43*, 388–391. [CrossRef]
20. Poon, C.S.; Chan, D. Paving blocks made with recycled concrete aggregate and crushed clay brick. *Constr. Build. Mater.* **2006**, *20*, 569–577. [CrossRef]
21. Poon, C.S.; Kou, S.C.; Lam, L. Use of recycled aggregates in molded concrete bricks and blocks. *Constr. Build. Mater.* **2002**, *16*, 281–289. [CrossRef]
22. Poon, C.S.; Lam, C.S. The effect of aggregate-to-cement ratio and types of aggregates on the properties of pre-cast concrete blocks. *Cem. Concr. Compos.* **2008**, *30*, 283–289. [CrossRef]
23. Corinaldesi, V. Mechanical behavior of masonry assemblages manufactured with recycled-aggregate mortars. *Cem. Concr. Compos.* **2009**, *31*, 505–510. [CrossRef]
24. JGJ 52-2006. *Standard for Technical Requirements and Test Method of Sand and Crushed Stone (or Gravel) for Ordinary Concrete*; Chinese Building Press: Beijing, China, 2006.

25. GB/T 14685-2011. *National Pebble and Crushed Stone Code for Buildings*; Chinese Building Press: Beijing, China, 2011.
26. GB/T 5129-2011. *Standards for Test Method of Basic Mechanical Properties of Masonry*; Chinese Building Press: Beijing, China, 2011.
27. GB/T 4111-2013. *Test Methods for the Concrete Block and Brick*; Chinese Building Press: Beijing, China, 2013.
28. GB 50003-2011. *Code for Design of Masonry Structures*; China Building Industry Press: Beijing, China, 2011.
29. Guo, Z.G.; Sun, W.M.; Peng, Y.; Liu, J.L. Experimental study on the compression behavior of recycled aggregates concrete small hollow block masonry. *Build. Struct.* **2011**, *41*, 127–128. [CrossRef]
30. Xu, S.F.; Xiong, Z.M.; Wei, J. *Masonry Structure*; China Science Press: Beijing, China, 2004.
31. ISO 14040. *Environmental Management Life Cycle Assessment Principles and Framework*; International Organization for Standardization: Geneva, Switzerland, 2006.
32. ISO 14044. *Environmental Management Life Cycle Assessment Requirements and Guidelines*; International Organization for Standardization: Geneva, Switzerland, 2006.
33. Gao, Y.X.; Wang, J.; Xu, F.L.; Lin, X.H.; Chen, J. Carbon emissions assessment of green production for ready-mix concrete. *Concrete* **2011**, *1*, 110–112. [CrossRef]

Article

Influence of Multiple Openings on Reinforced Concrete Outrigger Walls in a Tall Building

Han-Soo Kim *, Yi-Tao Huang and Hui-Jing Jin

Department of Architecture, Konkuk University, Seoul 05029, Korea; icyhyt@konkuk.ac.kr (Y.-T.H.); hg1827@daum.net (H.-J.J.)
* Correspondence: hskim@konkuk.ac.kr; Tel.: +82-2-2049-6110

Received: 1 October 2019; Accepted: 13 November 2019; Published: 15 November 2019

Featured Application: The reinforced concrete outrigger wall with multiple openings can be used to replace the conventional steel outrigger trusses in tall building structures.

Abstract: Outrigger systems have been used to control the lateral displacement of tall buildings. Reinforced concrete (R.C.) outrigger walls with openings can be used to replace conventional steel outrigger trusses. In this paper, a structural model for an R.C. outrigger wall with multiple openings was proposed, and the effects of the multiple openings on the stiffness and strength of the outrigger walls were evaluated. The equivalent bending stiffness of the outrigger wall was derived to predict the lateral displacement at the top of tall buildings and internal shear force developed in the wall. The openings for the passageway in the wall were designed by the strut-and-tie model. The stiffness and strength of the outrigger wall with multiple openings was analyzed by the nonlinear finite element analysis. Taking into consideration the degradation in stiffness and strength, the ratio of the opening area to the outrigger wall area is recommended to be less than 20%. The degradation of stiffness due to openings does not affect the structural performance of the outrigger system when the outrigger has already large stiffness as the case of reinforced concrete outrigger walls.

Keywords: outrigger wall; multiple openings; deep beam; stiffness; shear strength; tall building

1. Introduction

With the development of tall building structures, the outrigger system has become one of the most popular structural systems that control the lateral displacement of such buildings. Several researchers and engineers have been studying and developing the outrigger system because it performs well in controlling the lateral displacement at the top of a tall building by reducing the overturning moment [1]. For example, the Shanghai Tower, Hong Kong IFC2, and Taipei 101 are successful applications of the system in tall buildings [2]. Most of the previous research studies focused on the optimum location of outriggers and lateral stiffness of the whole structure. Taranath [3] assumed a rigid outrigger beam for single-outrigger structures, and proposed that the optimum location of an outrigger is 0.455 of the total height from the top. McNabb and Muvdi [4] proposed that the optimum locations for two outriggers are 0.312 and 0.685 of the total height from the top based on Taranath's research. Smith and Nwaka [5] presented generalized results for optimum locations in multi-outrigger structures by assuming rigid outriggers in flexure. Smith and Salim [6] proposed equations for the optimum locations of outriggers by considering their flexibility. Hoenderkamper and Bakker [7] considered the bending and racking shear stiffness of the outrigger truss to determine the optimum location of the outrigger. The research studies on outrigger-braced structures have been summarized and some further studies have been attempted by Wu and Li [8]. Recently, Kim et al. [9] proposed a dual-purpose outrigger system to reduce the lateral displacement and differential column shortening.

Even though some researchers assumed equivalent beams for the outriggers to derive equations for the optimum locations, most of the outriggers built were made of steel trusses. Although there are a few applications of reinforced concrete outrigger walls, such as Chong Qing Raffles City [10], discussing the walls is not enough. Furthermore, too little attention has been paid to the reinforced concrete outrigger walls with openings.

In theory, the deeper the outrigger, the stiffer the structure [11]. It means that a deep outrigger, such as an outrigger wall, can provide more effective lateral load-resistance for high-rise structures. At the same time, reinforced concrete outrigger walls can be cost-effective systems because of the same construction procedure as reinforced concrete core walls. However, as large concrete structures, the reinforced concrete outrigger walls are heavy and space consuming. As a result, it is necessary to solve the problems associated with making better use of the space occupied by outrigger walls. In this study, a model for arranging multiple openings on reinforced concrete outrigger walls, as shown in Figure 1, is proposed. Moreover, the study investigates the influence of multiple openings on the stiffness and strength of the outrigger walls and whole tall building structure.

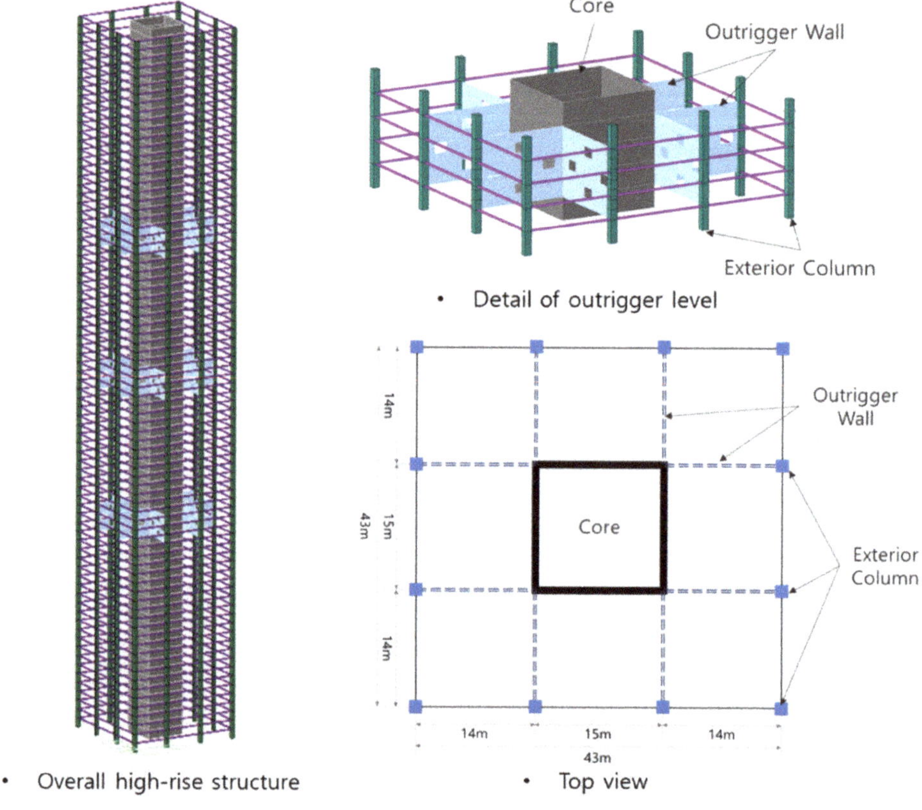

Figure 1. A typical tall building structure with reinforced concrete outriggers walls with multiple openings.

Given the aspect ratio of the outrigger walls, they can be designed as deep beams with openings. Kong and Sharpe [12] proposed a shear strength equation for deep reinforced concrete beams with web openings. Tan et al. [13] have predicted the capacity of deep beams with openings using the strut-and-tie model. Tang and Tan [14] paid more attention in the evaluation of the ultimate shear strength of deep beams, but there was only an opening in every shear region while the opening is placed in the critical load path. Although several studies have been performed on deep beams with web openings, there are still difficulties in predicting the structural behavior of deep beams with multiple openings. In this study, analytic equations used to predict the internal shear forces on outrigger walls due to the lateral loads are derived and the results are verified through numerical examples. After that, an outrigger wall with four openings was proposed through the strut-and-tie model. Subsequently, the linear and nonlinear finite element method and strut-and-tie model were used to evaluate the influence of the size of multiple openings on the stiffness and shear strength of reinforced concrete outrigger walls. The study ends up investigating the influence of multiple openings on the lateral displacement of whole tall building structures by using the proposed analytic equations.

2. Design of Reinforced Concrete Outrigger Wall with Multiple Openings

In this section, analytic equations used to predict the lateral displacement at the top of the building and shear forces developed in each outrigger wall are derived and verified with the result of finite element analysis. Subsequently, a structural model for the outrigger wall with four openings is proposed.

2.1. Effect and Demand of Outrigger Walls

The internal shear force of the outrigger walls due to lateral loads is derived based on the simplified model of the core wall and outrigger structural system [15]. A simplified model of tall building structures with multiple outriggers is shown in Figure 2. The core wall is connected to the perimeter columns through the outriggers. When lateral loads are applied at the central core wall, the outrigger develops axial forces in the perimeter columns. The limitation of the simplified model is that only flexural deformation of the outrigger is considered, and the length of the outrigger is assumed as the distance from the center of the core to that of the perimeter column. These assumptions are different from the actual behavior of outrigger walls. To enhance the accuracy of this simplified model, an equivalent bending stiffness $(EI)_O$ was proposed by considering the shear deformation of beams and the clear span of outrigger walls.

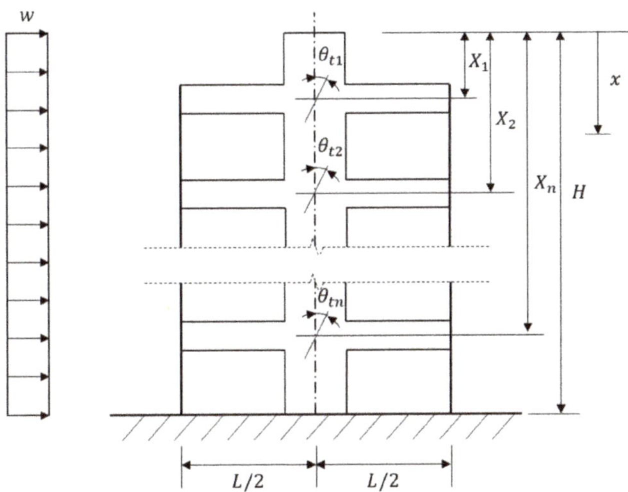

Figure 2. Simplified analysis model with multiple outriggers. The core wall is connected to the perimeter columns through the outriggers.

Because the outrigger walls belong to the category of deep beams [16], the total deformation is the sum of flexural and shear deformations, and the actual flexural distance is the clear span l_o, as shown in Figure 3. According to the principle of virtual works, for the outrigger wall with a moment of inertia I_o and cross-sectional area A_o, the equivalent bending stiffness $(EI)_o$ can be obtained by the following equation.

$$(EI)_o = \left(1 + \frac{c}{2l_o}\right)^3 EI_o / \left(1 + \frac{12EI_o}{G\kappa_s A_o L^2}\right) \tag{1}$$

where E and G are the modulus of elasticity and shear modulus, respectively. κ_s is the shear coefficient to account for the shear deformation, which is 0.83 for a rectangular section. c is the width of the core wall, and L is the distance from the center of the core wall to the center of perimeter column, as shown in Figure 3.

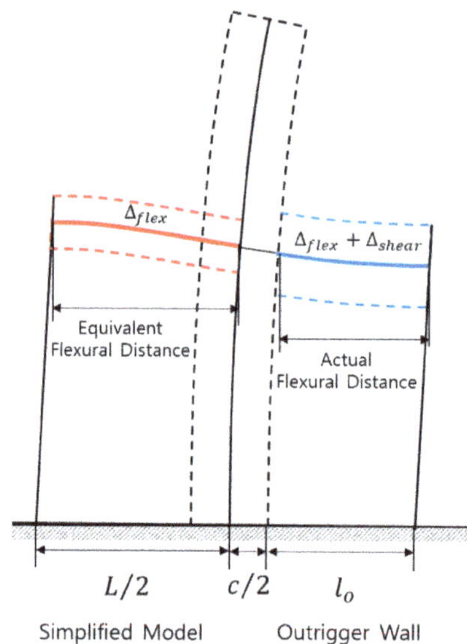

Figure 3. Shear and flexural deformations of the outrigger wall are considered and the clear span rather than center-to-center span is used in the equivalent bending stiffness of the outrigger wall.

For the core, the rotation θ_{ti} at the ith outrigger level due to the lateral load w and the restraining moment M_{ri} can be given by the following equation.

$$\theta_{ti} = \frac{1}{(EI)_t}\left[\int_{x_1}^{x_2}\left(\frac{wx^2}{2} - M_{r1}\right)dx + \int_{x_2}^{x_3}\left(\frac{wx^2}{2} - M_{r1} - M_{r2}\right)dx + \ldots + \int_{x_n}^{H}\left(\frac{wx^2}{2} - M_{r1} - \ldots - M_{rn}\right)dx\right] \tag{2}$$

where $(EI)_t$ represents the bending stiffness of the core wall. At the same outrigger level, the rotation of the inboard end of the outrigger θ_{oti}, where it attaches to the core, is due to the axial forces of perimeter columns and overturning moment M_{ri}. It was noted that the restraining moment for the core and overturning moment for the outrigger walls have the same value and opposite directions. Hence, M_{ri} is used to represent both. The rotation θ_{oti} can be obtained by the following equation.

$$\theta_{oti} = \frac{2M_{r1}(H - X_1)}{L^2(EA)_c} + \frac{2M_{r2}(H - X_2)}{L^2(EA)_c} + \ldots + \frac{2M_{ri}(H - X_i)}{L^2(EA)_c} + \frac{M_{ri}L}{12(EI)_{oi}} \tag{3}$$

where $(EA)_c$ represents the axial stiffness of perimeter columns.

Because of $\theta_{ti} = \theta_{oti}$, the restraining moment M_{ri} can be obtained in the following matrix form.

$$\begin{bmatrix} M_{r1} \\ M_{r2} \\ \vdots \\ M_{ri} \\ \vdots \\ M_{rn} \end{bmatrix} = \left[\frac{w}{6(EI)_t}\right] \times \begin{bmatrix} B_1 + C(H - X_1) & C(H - X_2) & \cdots & C(H - X_i) & \cdots & C(H - X_n) \\ C(H - X_2) & B_2 + C(H - X_2) & \cdots & C(H - X_i) & \cdots & C(H - X_n) \\ \vdots & \vdots & \vdots & \vdots & \vdots & \vdots \\ C(H - X_i) & C(H - X_i) & \cdots & B_i + C(H - X_i) & \cdots & \vdots \\ \vdots & \vdots & \vdots & \vdots & \vdots & C(H - X_n) \\ C(H - X_n) & C(H - X_n) & \cdots & C(H - X_n) & \cdots & B_n + C(H - X_n) \end{bmatrix}^{-1} \begin{bmatrix} H^3 - X_1^3 \\ H^3 - X_2^3 \\ \vdots \\ H^3 - X_i^3 \\ \vdots \\ H^3 - X_n^3 \end{bmatrix} \quad (4)$$

where the parameter B_i and C are defined in the following equations.

$$B_i = \frac{L}{12(EI)_{oi}} \qquad C = \frac{1}{(EI)_t} + \frac{2}{L^2(EA)_c} \quad (5)$$

where $(EI)_{oi}$ represents the equivalent bending stiffness of the ith outrigger wall, as given in Equation (1).

Applying the restraining moment M_{ri} on the core, the displacement at the top of the building Δ_{top} can be obtained by using the following equation.

$$\Delta_{top} = \frac{wH^4}{8(EI)_t} - \frac{1}{2(EI)_t}\sum_{i=1}^{n} M_{ri}\left(H^2 - X_i^{\,2}\right) \quad (6)$$

The outriggers are subjected to the overturning moment M_{ri} and shear forces at the far edge due to the axial forces of the perimeter columns. According to the resulting rotation θ_{oti}, the internal shear forces of the outrigger walls can be calculated by dividing M_{ri} by L as follows.

$$\begin{bmatrix} V_1 \\ V_2 \\ \vdots \\ V_i \\ \vdots \\ V_n \end{bmatrix} = \left[\frac{w}{6(EI)_t}\right] \times \begin{bmatrix} B_1 + C(H - X_1) & C(H - X_2) & \cdots & C(H - X_i) & \cdots & C(H - X_n) \\ C(H - X_2) & B_2 + C(H - X_2) & \cdots & C(H - X_i) & \cdots & C(H - X_n) \\ \vdots & \vdots & \vdots & \vdots & \vdots & \vdots \\ C(H - X_i) & C(H - X_i) & \cdots & B_i + C(H - X_i) & \cdots & \vdots \\ \vdots & \vdots & \vdots & \vdots & \vdots & C(H - X_n) \\ C(H - X_n) & C(H - X_n) & \cdots & C(H - X_n) & \cdots & B_n + C(H - X_n) \end{bmatrix}^{-1} \begin{bmatrix} H^3 - X_1^3 \\ H^3 - X_2^3 \\ \vdots \\ H^3 - X_i^3 \\ \vdots \\ H^3 - X_n^3 \end{bmatrix} \quad (7)$$

2.2. Verification by Numerical Examples

The accuracy of the proposed equations for lateral displacement at the top of tall buildings and shear forces developed at the outrigger walls are verified by comparing with the results from linear finite element analysis. MIDAS-Gen [17] was used for the linear finite element analysis because it is widely used in the structural design practice. The outrigger walls and cores were modeled with the quadrilateral plane stress elements with incompatible modes [17].

Six models were analyzed to verify the proposed equations. Models A1~A3 were the single-outrigger structures. Model A4 and A5 were the two-outrigger structures. Model A6 included the three-outrigger structures. The overall height of the structure is $H = 280$ m, with story height being equal to 3.5 m with 80 stories. The additional dimensional conditions are shown in Figure 4. The uniform lateral load w is applied at the center of the core wall. The floor diaphragm was neglected to assign the same boundary conditions as assumed in Equations (6) and (7). Except for the locations, all outriggers have the same dimensions and elastic modulus of 36.4 GPa. The lateral displacement at the top of the structure Δ_{top} and internal shear forces of the outrigger walls V from two analysis methods are presented in Table 1. The difference shown in Table 1 is the percentage of the absolute value of the difference between the two results, which is compared to the results from the finite element method. As shown in Table 1, the differences in the top drift and shear forces between two analysis results were less than 5%. A possible explanation for the discrepancy might be that the effective bending stiffness given in Equation (1), in which the clear distance and shear deformation are included, does not represent the actual deformation of the outrigger walls. However, these differences can be regarded as permitted in a structural design practice. Taken together, these results can hold the view that the proposed analytical equations for the outrigger walls that introduce the equivalent bending stiffness are valid.

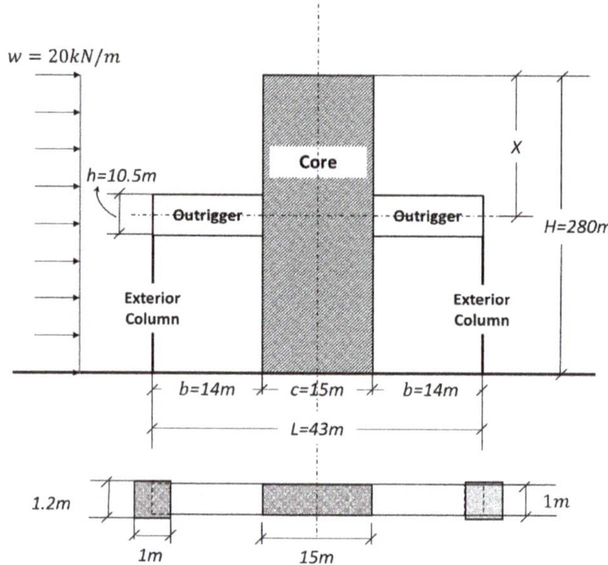

Figure 4. Dimensions of the analysis model to verify the proposed analytical equations.

Table 1. Results from Equations (6) and (7) and finite element analysis (FEA).

Model	Outrigger	X (m)	Δ_{top} (m)		Diff. (%)	V (kN)		Diff. (%)
			Equation (6)	FEA		Equation (7)	FEA	
A1	O11	89.25	0.4964	0.5134	3.30	6790	6876	1.25
A2	O21	138.25	0.4715	0.4905	3.87	8267	8331	0.77
A3	O31	183.75	0.5782	0.5985	3.39	9842	9840	0.02
A4	O41	89.25	0.3701	0.3789	2.34	3669	3808	3.65
	O42	183.75				6273	6225	0.77
A5	O51	68.25	0.3831	0.3908	1.96	3908	4028	2.97
	O52	208.25				6967	6890	1.11
A6	O61	68.25	0.3412	0.3429	0.42	2111	2220	4.92
	O62	138.25				3597	3528	1.95
	O63	208.25				5232	5155	1.51

2.3. Arrangement of Multiple Openings

Outrigger walls behave as cantilever deep beams, which have nonlinear distribution of stresses. It is not easy to analyze and design outrigger walls through design methods and formulas developed for slender beams. In this study, the strut-and-tie model was used to perform a preliminary design of the reinforced concrete outrigger walls. The strut-and-tie model was proven to be a desirable approach for designing reinforced concrete members with discontinuous regions (D-regions). The method is based on a truss analogy, which was first presented to explain the contribution of transverse reinforcement to the shear strength of a beam by Wilhelm Ritter in 1899. Following this, Schlaich et al. [18] developed a strut-and-tie model as a design method for the D-region, and later discussed some designing details [19]. Furthermore, the strut-and-tie model method has been adopted by some design provisions, such as American Concrete Institute (ACI) 318-14 [16] as a design method for structures with D-regions.

By using the strut-and-tie model, which assumes that all the stresses are condensed in struts and ties, horizontal and diagonal reinforcements were placed in the region with high tensile stresses. Meanwhile, in a practical situation, lateral loads coming from different directions lead to the fact that every half-side outrigger should be able to support upward and downward forces. This means that the tension zones could likely become compressive zones in an outrigger wall, and compressive zones should also be able to resist tension forces. Thus, every half-side of an outrigger wall with four reinforcements can be obtained by a symmetric superposition of the side supported upward forces and side supported downward forces, as shown in Figure 5 in which the blue lines represent steel reinforcements.

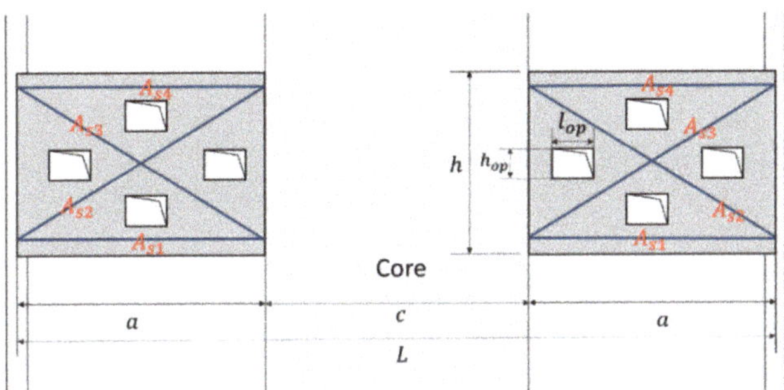

Figure 5. Outrigger walls with multiple openings. The blue lines represent the steel reinforcements to resist tension forces.

According to American Concrete Institute (ACI) 318–14, the distributed transverse reinforcement should be positioned for compressive strength of concrete f_c' not exceeding 42 MPa (6000 psi). However, there is some difficulty in the placement of the distributed transverse reinforcement in an outrigger wall with multiple openings, which is shown in Figure 5. In this study, the compressive strength of concrete that was used in the outrigger walls was greater than 42 MPa in order to evade the requirement for web reinforcement. Therefore, the outrigger wall was reinforced with four main reinforcements along the tension zone. The sectional area of each reinforcement was given by A_{s1}, A_{s2}, A_{s3}, and A_{s4}. These reinforcements divide the outrigger wall into four parts, as shown in Figure 6, in which the yellow colored regions were subjected to low stresses. It is essential to install openings in the regions subjected to low stresses.

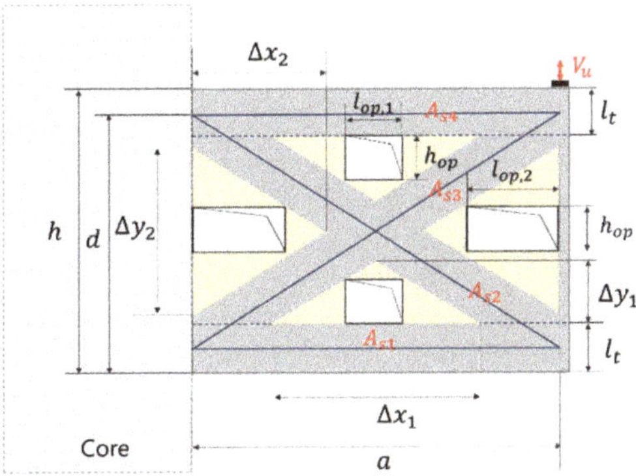

Figure 6. Arrangement of multiple openings in a reinforced concrete outrigger wall. The height of openings was assumed to be 2 m for it to be used as a passageway.

Through geometrical symmetry, the right triangular region was the same as the left one with the length, Δx_2, and height, Δy_2. On the other hand, the bottom triangular region was the same as the top one with length, Δx_1, and height, Δy_1. A half-side outrigger wall has a shear span, a, an effective depth, d, and overall depth of outrigger wall, h, as shown in Figure 6. The effective width of the tie, l_t, is equal to $2(h-d)$ and is also a limitation to the location of the top and bottom openings.

In order to get the maximum rectangular openings, rectangular openings should be inscribed in the triangular zones. In a practical view, the height of rectangular openings is assumed to be 2 m since it used to be the passageway. With the specified opening height, h_{op}, the maximum length of the top and bottom openings, $l_{op,1}$, as well as the maximum length of the left and right openings, $l_{op,2}$, can be obtained by the following geometrical calculation.

$$l_{op,1} = \Delta x_1 \frac{(\Delta y_1 - 2)}{\Delta y_1} \qquad l_{op,2} = \Delta x_2 \frac{(\Delta y_2 - 2)}{\Delta y_2} \tag{8}$$

where Δx_1, Δy_1, Δx_2, and Δy_2 are defined in Figure 6.

If the four openings are assumed to be the same size, among the length of the four openings, l_{op} is the smaller one compared to $l_{op,1}$ and $l_{op,2}$.

$$l_{op} = min\left(l_{op,1},\ l_{op,2}\right) \tag{9}$$

The total area of openings A_{op} can also be obtained using the equation below.

$$A_{op} = 4 l_{op} h_{op} \tag{10}$$

3. Evaluation of the Influence of Multiple Openings

In order to study the influence of the size of openings on the outrigger wall and whole structure, structural performances associated with the stiffness and strength are investigated by using both linear and nonlinear analyses. Since the strength of the reinforced concrete outrigger wall cannot be evaluated by linear analysis, the strength of the outrigger wall is obtained by the nonlinear finite element analysis and compared with the results from the strut-and-tie models presented in ACI 318-14. On the other hand, the initial stiffness of the outrigger wall can be estimated by linear analysis. Therefore, the stiffness of the wall was evaluated by linear and nonlinear finite element analyses, and the results from both analyses were compared.

A single-outrigger-braced tall building with a 3-story-high outrigger system located in 39th, 40th, and 41st floor of the tall building structure shown in Figure 1 was studied as a numerical example. The overall depth, h, of the outrigger wall was 10.5 m. As a symmetrical structure, only the half-side outrigger wall subjected to downward forces was analyzed. Through the proposed design procedure, the cross-sectional areas of the diagonal reinforcements (A_{s2} and A_{s3}) and horizontal reinforcements (A_{s1} and A_{s4}) were assumed to be 0.057 m^2 and 0.099 m^2, respectively. The outrigger wall had a shear span, a, of 14 m and an effective depth, d, of 10 m. The compressive strength of concrete f_c' was 60 MPa.

The size of the openings was changed through 12 analyses models. The models were named by assigning a capital M followed by a number i ($i = 0, 1, 2, \ldots, 11$). Among them, model M0 refers to the outrigger wall without opening a reference model. All the openings have the specified height ($h_{op} = 2$ m). Since the 250 mm × 250 mm size of the rectangular element was used in the finite element mesh, the length of the opening starting from 0 increased by 0.5 m steps and reached their limitations in model M11. The opening ratios (A_{op}/A_{out}) are defined as the area ratios of the four openings and outrigger walls. This study aimed to find out whether and how the stiffness and the shear strength of outrigger walls change by increasing the opening ratio. In order to illustrate the influence upon the stiffness and shear strength, parameters expressing the variation are introduced. The parameter β_K is the ratio of the stiffness K_i of each outrigger wall to the stiffness K_0 of the outrigger wall without openings. The parameter β_V is the ratio of the shear strength V_i to the shear strength V_0 of the outrigger wall without openings.

3.1. Finite Element Modelling

In this study, Abaqus [20], which is a commercial nonlinear finite element analysis program, was used. Several researchers have used Abaqus in analyzing the nonlinear behavior of reinforced concrete structures [21–24]. The program provides three material models for concrete. These are the smeared crack concrete, brittle crack concrete, and concrete damaged plasticity models. In this paper, the concrete damaged plasticity (CDP) model was used because it can represent the complete inelastic behavior of concrete, both in compression and tension [25,26]. The CDP is a continuum, plasticity-based, damage model for concrete. It assumes that the main two failure mechanisms of concrete material are tensile cracking and compressive crushing, which are represented by the uniaxial tension and compression behavior. Under uniaxial tension, the stress-strain response follows a linear elastic relationship until the value of the failure stress is reached. The failure stress corresponds to the onset of micro-cracking in concrete material. Beyond the failure stress, the formation of micro-cracks is represented macroscopically with a softening stress-strain response, which induces strain localization in concrete structures. Under uniaxial compression, the response is linear up to the value of the initial yield. In the plastic regime, the response is typically characterized by stress hardening, which is followed by strain softening beyond ultimate stress.

The CDP model in Abaqus is a plasticity material model that requires the definition of the yield surface and requires hardening and flow rules. The model makes use of the yield function proposed by Lee and Fenves [27] to account for varying degrees of evolution of concrete strength under tension and compression. The CDP model assumes a non-associated potential plastic flow. The flow potential is the Drucker-Prager hyperbolic function, which requires the dilation angle (ψ), flow potential eccentricity (ϵ), ratio of the initial biaxial compressive yield stress to the initial uniaxial compressive yield stress (σ_{b0}/σ_{c0}), and the ratio of the second stress invariant on the tensile meridian to that on the compressive meridian (K_c). The viscosity parameter should be entered to overcome some of the convergence difficulties caused by material models exhibiting softening behavior and stiffness degradation. The values of the parameters for the plasticity model used in this study are shown in Table 2 below.

Table 2. Plasticity parameter of concrete damaged plasticity (CDP) specified for this study.

Parameter	Value
ψ	38
ϵ	0.1
σ_{b0}/σ_{c0}	1.16
K_c	0.667
Viscosity parameter	0.0001

CDPs require a stress-strain relationship of concrete in uniaxial compression and tension. The relationship suggested by Carreira and Chu [28], as shown in Figure 7, was used for the compression model. For the tension model, the stress-strain relationship proposed by Wahalathantri et al. [29] was used. The tension model as shown in Figure 8 includes the tension stiffening to simplify the post cracking behavior for concrete and the interaction between reinforcement and concrete. The CDP model was used in predicting the shear behavior of reinforced and prestressed concrete deep beams and confirmed that the shear strengths predicted by the finite element analysis agreed well with those obtained by experiments [30].

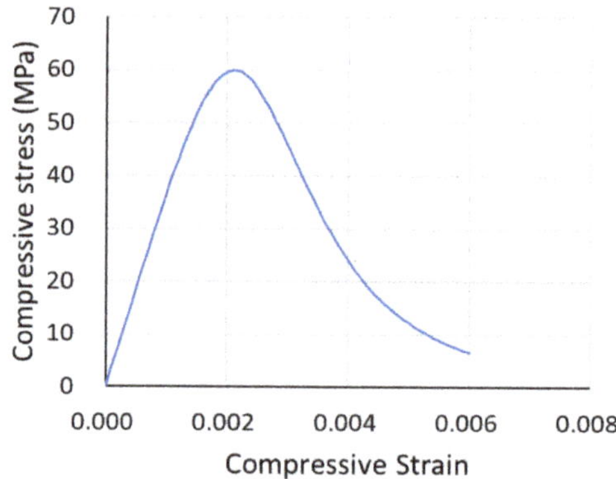

Figure 7. Compressive stress-strain relationship of concrete.

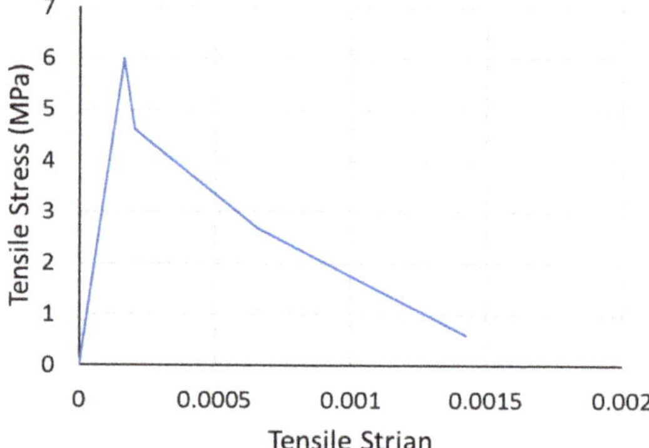

Figure 8. Compressive stress-strain relationship of concrete.

When applying downward force on the loading plate for the analysis model used in this study, the reinforcements with cross-sectional areas A_{s1} and A_{S3} were subjected to compression. On the other hand, the reinforcements with cross-sectional areas A_{s2} and A_{s4} were subjected to tensile forces. To model the behavior of stress reduction caused by buckling in a reinforcement subjected to compression, the material model was defined differently from the material model of the reinforcement subjected to tension. For the reinforcement with cross-sectional areas A_{s2} and A_{s4}, which were subjected to tension, only the yield strength of 414 MPa and modulus of elasticity of 200 GPa were specified by assuming perfect plasticity, as shown in Figure 9. The compressive reinforcements were assumed to be buckled at a compressive strain of 0.003 and reduced the strength to 10% of the yield strength, as shown in Figure 9.

The structures used in this study were symmetrical structures. For the convenience of analysis, half of the entire outrigger wall was taken and modeled two-dimensionally. The half-side outrigger wall was modeled with half of the core wall and the loading plate at which the vertical displacement was applied. The loading plate replaced the perimeter column and was assumed to be perfectly bonded to the concrete and modeled to share the nodes with concrete. Both concrete and loading plates were modeled by using CPS4R, which is a quadrilateral plane stress element with reduced integration. The reinforcements were represented with a T2D2 truss element and were modeled by embedding them in concrete. It was assumed that the reinforcement and concrete were perfectly bonded and the interaction between two materials was indirectly represented by the tension stiffening, as shown in Figure 8. Vertical displacement was applied to the loading plate in the form of displacement control. The left, top, and bottom sides of the core wall were constrained as pinned boundary conditions.

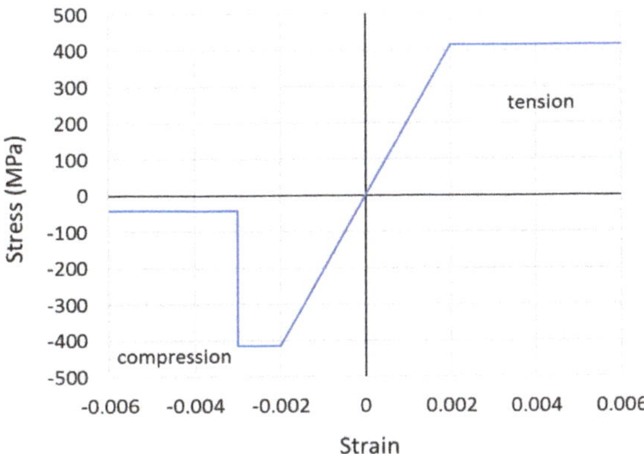

Figure 9. Material model of reinforcement subjected to tension and compression.

3.2. Results of Finite Element Analysis

The absolute maximum principal stress distribution calculated by the nonlinear finite element analysis is shown in Figure 10. In the outrigger wall without web openings (M0), the results show that the compressive strut was formed from the loading plate, and the bottom of the outrigger wall and tensile zone was formed at the top of the outrigger wall. In the outrigger wall with web openings (M8), compressive strut was more clearly formed at the diagonal line. Additionally, the bottom strut was formed separately from the diagonal strut. It was also observed that the cross-section of the diagonal strut was slightly defected by the openings.

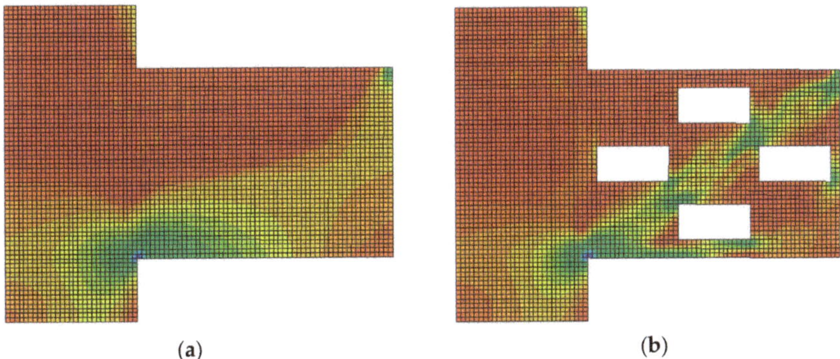

Figure 10. Absolute maximum principal stress distribution of (**a**) M0 and (**b**) M8.

The maximum plastic strain distribution at the ultimate state is shown in Figure 11. The maximum plastic strain represents the tensile strain in concrete, and the distribution shows the crack pattern of the analysis models. In the M0 model, the widest crack developed in the flexural critical section at which the outrigger wall meets the core wall. Several flexural cracks developed along the top reinforcement. The diagonal crack initiated at the loading plate and developed along the diagonal reinforcement, which was subjected to compression, was also noticeable. In model M1, many cracks developed along the top and diagonal reinforcements, which were subjected to tension. It was also noticed that wide local cracks were developed at the bottom left corner of the openings. In M8, the crack initiated at the loading plate and propagated along the top reinforcement, which formed a large arc along the corners of openings. In M11, in which the compressive struts interfered by the openings, the ultimate state developed by the local failure at the corners of the top and right openings.

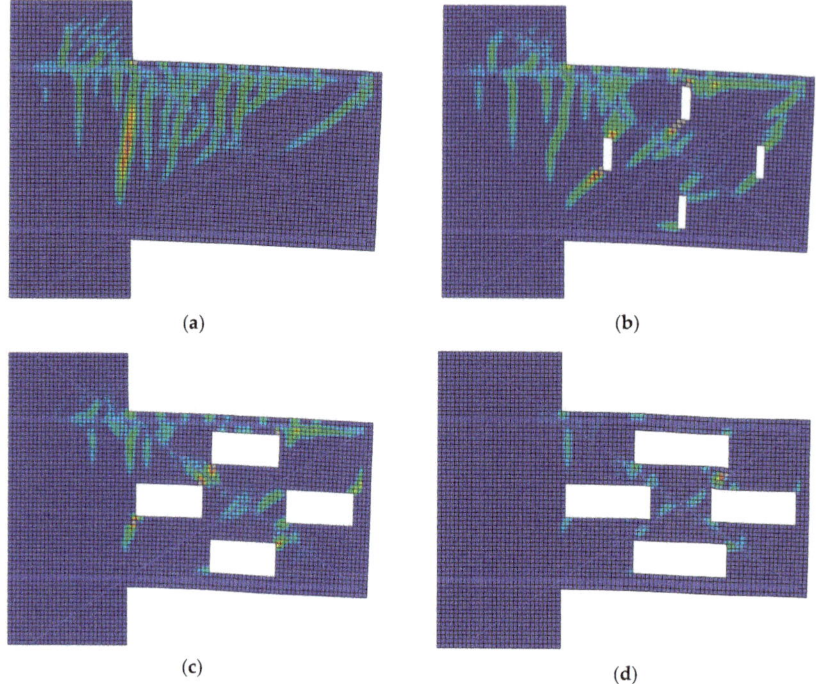

Figure 11. Plastic strain distribution at the ultimate state of (**a**) M0, (**b**) M1, (**c**) M8, and (**d**) M11.

Figure 12 shows the relation between the applied vertical displacement and total vertical reactions measured at the pinned nodes from the nonlinear analysis. It can be clearly observed that the outrigger walls experienced a progressive decrease in both stiffness and strength as the size of the opening increased. However, the ultimate displacement of all the models, except M11, were around 60 mm and the initial yielding was developed at about 10 mm.

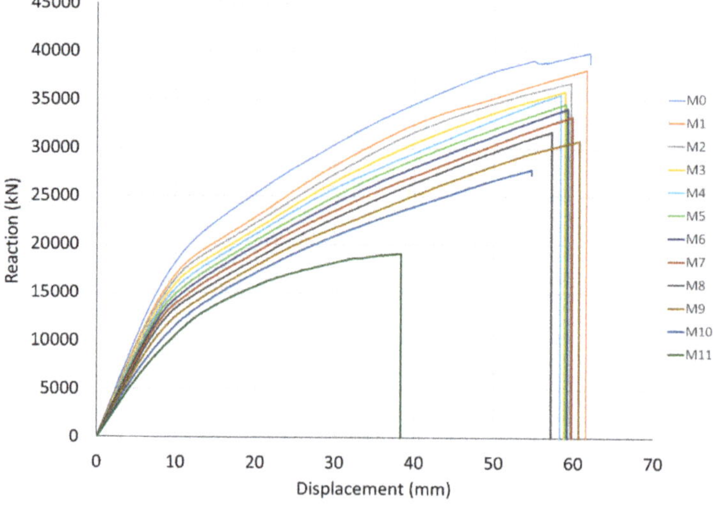

Figure 12. Displacement-reaction curves of nonlinear analysis.

In order to quantify the degradation in stiffness due to the size of openings, the initial tangential slope, K_{ti}, calculated from the load-displacement curves shown in Figure 12 and compared it with the slope K_i from the linear analysis. In order to evaluate the stiffness of outrigger walls, all the analysis models were constructed as the same as the nonlinear analysis, with an exception of the plastic modeling removed in the linear analysis. The slope of the linear analysis K_i is defined as the vertical force developing a unit deflection at the loading point of the i-th model in the linear analysis. The term K can be defined as the tangential stiffness of the outrigger walls. The initial tangential slopes K_{ti} from the nonlinear analysis and stiffness K_i from the linear analysis were identical.

To include the effect of initial cracking in the nonlinear analysis, the secant stiffness, K_{si}, were defined as the slope from the origin to the point with vertical displacement of 10 mm and are shown in Table 3 with the stiffness K_i from the linear analysis. As mentioned before, β_K is defined as the ratio of K_i to K_0 or K_{si} to K_{s0} of the outrigger wall without openings. β_K is regarded as an index to show the degradation of stiffness of the outrigger walls. It can be observed that, even though the secant stiffness was slightly less than the tangential stiffness, K_i and the degradation ratio β_K were almost identical.

Table 3. Evaluation of stiffness of the outrigger wall.

Model	A_{op}/A_{out}	Linear Analysis		Nonlinear Analysis	
		K_i (MN)	β_K	K_{si} (MN)	β_K
M0	0.00	2124	1.00	1817	1.00
M1	0.03	1973	0.93	1690	0.93
M2	0.05	1914	0.90	1642	0.90
M3	0.08	1858	0.87	1593	0.88
M4	0.11	1801	0.85	1542	0.85
M5	0.14	1742	0.82	1487	0.82
M6	0.16	1685	0.79	1435	0.79
M7	0.19	1625	0.76	1383	0.76
M8	0.22	1557	0.73	1332	0.73
M9	0.24	1468	0.69	1248	0.69
M10	0.27	1365	0.64	1156	0.64
M11	0.29	1234	0.58	1056	0.58

Figure 13 shows the degradation ratios in stiffness as the size of the opening increases. The ratios of the stiffness decline almost linearly with a gentle slope from M0 to M8. When the opening area exceeds 22% (M8), the decreasing trends are also regarded as being linear, but they become steeper. From the similar trends in the results of linear and nonlinear analyses, it can be concluded that the loss of area due to multiple openings is a more influential factor on the stiffness of outrigger walls than the material degradation.

The maximum vertical reaction, R_i, in the nonlinear analysis can be defined as the strength of the outrigger walls. The ratio β_R refers to the ratio of the vertical reaction at the ultimate state of the i-th models R_i to R_0 of the M0 model. The maximum vertical reactions and degradation ratio in strength from the nonlinear analysis are shown in Table 4 and Figure 14. The reduction in strength is a function of the opening ratio, which is similar to the results of stiffness. However, the outrigger wall began to experience larger losses in strength when the opening area reached 24% (M9) in which the shear strength reduced to 77%. Comparing the reduction of stiffness and strength in model M0 to model M8, where the sizes of openings were less than 22% of the overall area of the outrigger wall, the gentle reductions in stiffness and strength were similar, while the sharp reduction occurred earlier in the stiffness (M8) than in strength (M9).

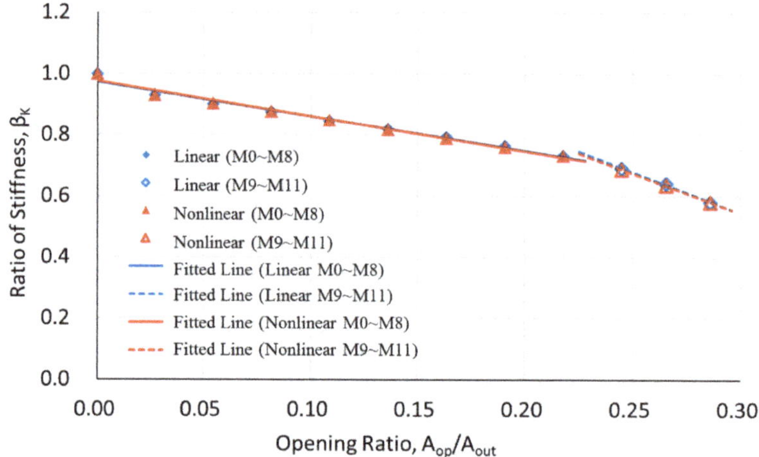

Figure 13. Degradation of stiffness of the outrigger wall according to the size of openings.

Table 4. Evaluation of strength of the outrigger wall.

Model	A_{op}/A_{out}	Nonlinear Analysis	
		R_i (kN)	β_R
M0	0.00	39,932	1.00
M1	0.03	38,193	0.96
M2	0.05	36,798	0.92
M3	0.08	35,938	0.90
M4	0.11	35,604	0.89
M5	0.14	34,635	0.87
M6	0.16	34,150	0.86
M7	0.19	33,314	0.83
M8	0.22	31,729	0.79
M9	0.24	30,815	0.77
M10	0.27	27,853	0.70
M11	0.29	19,083	0.48

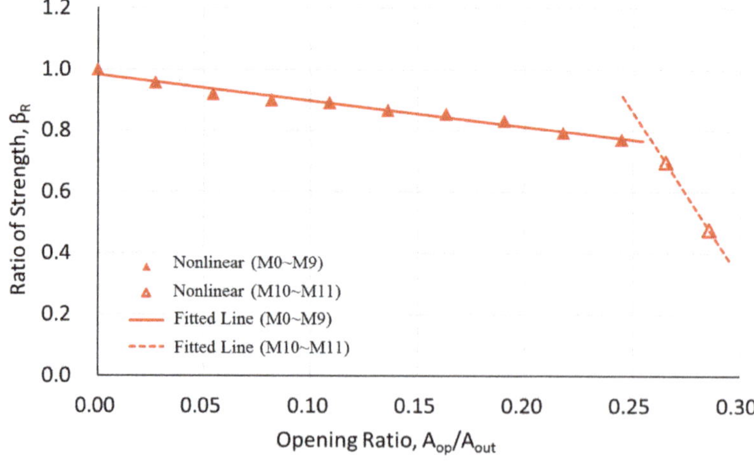

Figure 14. Degradation of strength of the outrigger wall according to the size of the openings.

From the analysis results, it can be concluded that the reinforced concrete outrigger walls can accommodate four openings without significantly decreasing the stiffness and strength of the outrigger walls if the openings do not severely interfere with the critical load path.

3.3. Strength Predicted by Strut-and-Tie Models

According to Chapter 23 of ACI318-14, the nominal shear strength of the outrigger walls can be calculated by using the strut-and-tie model. In this study, the nodal zones are assumed to be stiff enough. Therefore, only the strengths of strut and tie were considered. The diagonal strut in the central zone surrounded by four openings has an angle θ with the horizontal tie, as shown in Figure 15.

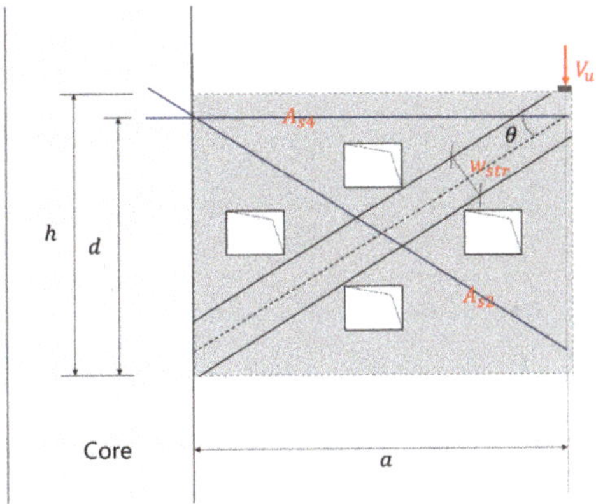

Figure 15. The strut-and-tie model for the outrigger wall with multiple openings.

The shear strength of the outrigger wall by the strut-and-tie model V_{stm} was the smaller one of the shear strengths provided by the strut and the tie, as shown by the following equations.

$$V_n = F_{ns} sin\theta \leq 0.85\beta_s f_c' b_w w_{str} \quad (11)$$

$$V_n = F_{nt2} sin\theta + F_{nt4} tan\theta \leq A_{s2} f_y sin\theta + A_{s4} f_y tan\theta \quad (12)$$

where F_{ns} is the nominal compressive strength of a strut. β_s is the reduction factor to account for the bottle-shaped strut. For a strut of a uniform cross-sectional area, β_s = 1.0 and w_{str} is the width of strut. F_{nt2} and F_{nt4} are the nominal tensile strengths of the top and diagonal reinforcements, respectively.

The indices representing the relative strength degradation β_V as increasing the size of the openings are presented in Figure 16. It can be observed that the β_V remained 1.0 before the opening ratio reached 19% (M0–M7). This means that the steel reinforcements yielded before the failure of concrete until the size of the openings reached 19%. When the size of the openings increased beyond 19%, the openings reduced the width of the strut and, consequently, reduced the strength of the strut and strength of the outrigger walls. The strength ratio β_V began decreasing sharply from model M8 and dropped to 39% in model M11. These results indicate that the failure of concrete preceded the yielding of reinforcements in model M11. When comparing Figure 14 from the nonlinear finite element analysis and Figure 16 from the strut-and-tie model, it can be noticed that the strut-and-tie model is not conservative in predicting the shear strength of outrigger walls with low opening ratios.

3.4. Influence on Lateral Stiffness of Tall Buildings

The influence of openings on the stiffness and strength of reinforced concrete outrigger walls was investigated in the previous sections. In this section, the influence of openings on the lateral stiffness of the whole structure was investigated by applying the degraded stiffness of outrigger walls due to openings given in Table 3 to the proposed, analytical Equations (6) and (7). The equivalent bending stiffness of the outrigger wall without openings given in Equation (1) can be easily converted to the equivalent bending stiffness of the outrigger wall with openings, as shown by the following equation.

$$(EI)_{oi} = \beta_K\,(EI)_o \tag{13}$$

where $(EI)_{oi}$ is the equivalent bending stiffness of the i-th analysis models.

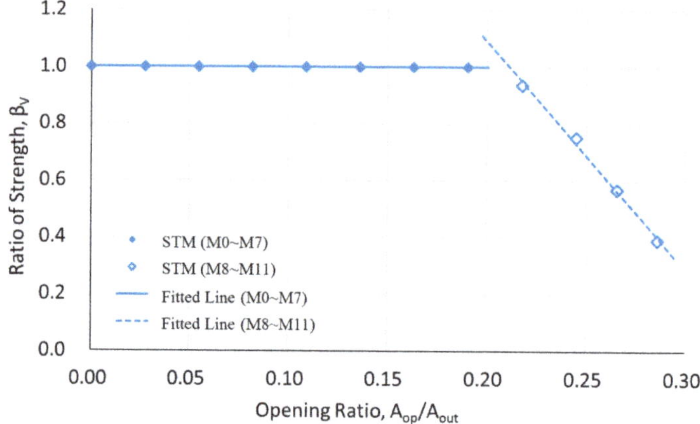

Figure 16. Strength of outrigger walls predicted by the strut-and-tie model.

The lateral displacement at the top of the building and shear forces developed in each outrigger are summarized in Table 5. The lateral displacement and shear forces were almost the same, even though the minimum value of $\beta_K = 0.58$ (M11) was applied. According to a study by Kim [31], the relation between the lateral displacement at the top and stiffness of the outrigger is nonlinear. Moreover, the lateral displacement is not sensitive to the stiffness of the outrigger when the outrigger has sufficiently high stiffness, such as the outrigger walls in this study. Even though the internal shear force developed in each outrigger remained almost the same as seen in Table 5, the strength of the outrigger wall with openings was reduced significantly, as seen in Table 4, since the size of the openings increased. Therefore, the structural safety in terms of strength should be carefully checked when designing the reinforced concrete outrigger walls.

Table 5. The lateral displacement at the top of the building and shear forces developed in the outrigger walls predicted by Equations (6) and (7).

Model	Outrigger	Δ_{top} (m)		Diff. (%)	V (kN)		Diff. (%)
		M0	M11		M0	M11	
A1	O11	0.4964	0.5050	1.73	6790	6732	0.85
A2	O21	0.4715	0.4833	2.50	8267	8173	1.14
A3	O31	0.5782	0.5935	2.65	9842	9679	1.66
A4	O41	0.3701	0.3777	2.01	3669	3707	1.04
	O42				6273	6133	2.23
A5	O51	0.3831	0.3918	2.22	3908	3938	0.77
	O52				6967	6786	2.60
A6	O61	0.3412	0.3470	1.70	2111	2144	1.56
	O62				3597	3594	0.08
	O63				5232	5089	2.73

4. Conclusions

After studying outrigger systems, designing multiple openings on outrigger walls and evaluating the influence of the size of openings on outrigger walls, several conclusions can be summarized as follows.

The introduction of equivalent bending stiffness allows for the derivation of internal shear forces on the outriggers. The proposed equations include the effect of clear span and shear deformation of outrigger walls. In comparison with the results from the numerical analysis, the errors between the results of developed equations and results of finite element analysis were less than 5%. This proves that the shear reaction on outrigger walls due to lateral loads derived from a simplified model can be used for the preliminary design of outrigger walls.

To use the space occupied by the outrigger walls more efficiently, a development of an outrigger wall with four reinforcements and four openings was proposed through the strut-and-tie model. The stiffness and strength of outrigger walls experience reductions in different levels since the size of openings increases. The decrease in trends can be represented by two linear functions of the opening ratios, which are the gentle and steep linear functions. The linear finite element analysis can replace the time-consuming nonlinear finite element analysis when evaluating the stiffness degradation of outrigger walls with multiple openings. Taking into consideration the degradation in stiffness and strength, the ratio of the opening area is recommended to be less than 20%. The degradation of stiffness due to openings does not significantly affect the global behavior of the whole structure when the outrigger already has a large stiffness, as the case of reinforced concrete outrigger walls.

Author Contributions: Conceptualization, methodology, validation, writing—review and editing, supervision, H.-S.K. Formal analysis, data curation, writing—original draft preparation, Y.-T.H. Formal analysis, investigation, H.-J.J.

Funding: This work was supported by Konkuk University in 2017.

Conflicts of Interest: The authors declared no conflict of interest.

References

1. Choi, H.S.; Joseph, L. Outrigger system design considerations. *Int. J. High-Rise Build.* **2012**, *1*, 237–246.
2. Ali, M.M.; Moon, K.S. Structural developments in tall buildings: Current trends and future prospects. *Archit. Sci. Rev.* **2007**, *50*, 205–223. [CrossRef]
3. Taranath, B.S. Optimum belt truss location for high-rise structures. *Struct. Eng.* **1975**, *53*, 18–21.
4. McNabb, J.W.; Muvdi, B.B. Drift reduction factors for belt high-rise structures. *Eng. J. AISC* **1975**, *12*, 88–91.
5. Smith, B.S.; Nwaka, I.O. The behaviour of multi-outrigger braced tall buildings. *ACI SP-63* **1980**, 515–541.
6. Smith, B.S.; Salim, I. Parameter study of outrigger-braced tall building structures. *J. Str. Div. ASCE* **1981**, *107*, 2001–2014.
7. Hoenderkamp, J.C.D.; Bakker, M.C.M. Analysis of high-rise braced frames with outriggers. *Struct. Des. Tall Spec. Build.* **2003**, *12*, 335–350. [CrossRef]
8. Wu, J.R.; Li, Q.S. Structural performance of multi-outrigger-braced tall buildings. *Struct. Des. Tall Spec. Build.* **2003**, *12*, 155–176. [CrossRef]
9. Kim, H.S.; Lee, H.L.; Lim, Y.J. Multi-objective optimization of dual-purpose outriggers in tall buildings to reduce lateral displacement and differential axial shortening. *Eng. Struct.* **2019**, *189*, 296–308. [CrossRef]
10. Wang, A. Multi-dimensional hybrid design and construction of skyscraper cluster-innovative engineering of Raffles City. *Int. J. High-Rise Build.* **2017**, *6*, 261–269. [CrossRef]
11. Ho, G.W.M. The evolution of outrigger system in tall buildings. *Int. J. High-Rise Build.* **2016**, *5*, 21–30. [CrossRef]
12. Kong, F.K.; Sharp, G.R. Structural idealization for deep beams with web openings. *Mag. Concr. Res.* **1977**, *29*, 81–91. [CrossRef]
13. Tan, K.H.; Tong, K.; Tang, C.Y. Consistent strut-and-tie modeling of deep beams with web openings. *Mag. Concr. Res.* **2003**, *55*, 65–75. [CrossRef]

14. Tang, C.Y.; Tan, K.H. Interactive mechanical model for shear strength of deep beams. *J. Struct. Eng.* **2004**, *130*, 1534–1544. [CrossRef]
15. Smith, B.S.; Coull, A. *Tall Building Structures: Analysis and Design*; John Wiley & Sons: New York, NY, USA, 1991.
16. ACI Committee 318. *Building Code Requirements for Structural Concrete (ACI 318-14) and Commentary (318R-14)*; American Concrete Institute: Farmington Hills, MI, USA, 2014.
17. *Midas-Gen On-Line Manual*; MIDAS Information Technology Co. Ltd. Available online: http://manual.midasuser.com/EN_Common/Gen/855/whnjs.htm. (accessed on 10 September 2019).
18. Schlaich, J.; Schäfer, K.; Jennewein, M. Towards a consistent design of structural concrete. *PCI J.* **1987**, *32*, 74–150. [CrossRef]
19. Schlaich, J.; Schäfer, K. Design and detailing of structural concrete using strut-and-tie models. *Struct. Eng.* **1991**, *69*, 113–125.
20. Abaqus. *Abaqus Analysis User's Guide*; Dassault Systems Simulia Corp. Available online: http://130.149.89.49:2080/v2016/books/usb/default.htm, (accessed on 10 September 2019).
21. Wosatko, A.; Pamin, J.; Polak, M.A. Application of damage-plasticity models in finite element analysis of punching shear. *Comput. Struct.* **2015**, *151*, 73–85. [CrossRef]
22. Tao, Y.; Chen, J.F. Concrete damage plasticity model for modeling FRP-to-concrete bond behavior. *J. Compos. Const.* **2015**, *19*. [CrossRef]
23. Mohamed, K.; Farghaly, A.S.; Benmokrane, B.; Neale, K.W. Nonlinear finite-element analysis for the behavior prediction and strut efficiency factor of GFRP-reinforced concrete deep beams. *Eng. Struct.* **2017**, *137*, 145–161. [CrossRef]
24. Genikomsou, A.S.; Polak, M.A. Finite element analysis of punching shear of concrete slabs using damaged plasticity model in ABAQUS. *Eng. Struct.* **2015**, *98*, 38–48. [CrossRef]
25. Ombres, L.; Verre, S. Flexural strengthening of RC beams with steel-reinforced grout: Experimental and numerical investigation. *J. Compos. Const.* **2019**, *23*. [CrossRef]
26. Fortunato, G.; Funari, M.F.; Lonetti, P. Survey and seismic vulnerability assessment of the Baptistery of San Giovanni in Tumba (Italy). *J. Cult. Herit.* **2017**, *26*, 64–78. [CrossRef]
27. Lee, J.; Fenves, G.L. Plastic-damage model for cyclic loading of concrete structures. *J. Eng. Mech.* **1998**, *124*, 892–900. [CrossRef]
28. Carreira, D.J.; Chu, K.H. Stress-strain relationship for plain concrete in compression. *ACI Struct. J.* **1985**, *82*, 797–804.
29. Wahalathantri, B.L.; Thambiratnam, D.P.; Chan, T.H.T.; Fawzia, S. A material model for flexural crack simulation in reinforced concrete elements using Abaqus. In Proceedings of the First International Conference on Engineering, Designing and Developing the Built Environment for Sustainable Wellbeing, Brisbane, Australia, 28 April 2011; pp. 260–264.
30. Kim, H.J.; Kim, H.S. Finite element analysis to determine shear behavior of prestressed concrete deep beams. *J. Comput. Struct. Eng. Inst. Korea* **2019**, *32*, 165–172. [CrossRef]
31. Kim, H.S. Optimum design of outriggers in a tall building by alternating nonlinear programming. *Eng. Struct.* **2017**, *150*, 91–97. [CrossRef]

Article

Evaluation of Self-Compacting Concrete Strength with Non-Destructive Tests for Concrete Structures

Miguel C. S. Nepomuceno * and Luís F. A. Bernardo

C-MADE—Centre of Materials and Building Technologies, Department of Civil Engineering and Architecture, University of Beira Interior, 6201-001 Covilhã, Portugal; lfb@ubi.pt

* Correspondence: mcsn@ubi.pt

Received: 30 October 2019; Accepted: 22 November 2019; Published: 26 November 2019

Abstract: Self-compacting concrete (SCC) shows to have some specificities when compared to normal vibrated concrete (NVC), namely higher cement paste dosage and smaller volume of coarse aggregates. In addition, the maximum size of coarse aggregates is also reduced in SCC to prevent blocking effect. Such specificities are likely to affect the results of non-destructive tests when compared to those obtained in NVC with similar compressive strength and materials. This study evaluates the applicability of some non-destructive tests to estimate the compressive strength of SCC. Selected tests included the ultrasonic pulse velocity test (PUNDIT), the surface hardness test (Schmidt rebound hammer type N), the pull-out test (Lok-test), and the concrete maturity test (COMA-meter). Seven sets of SCC specimens were produced in the laboratory from a single mixture and subjected to standard curing. The tests were applied at different ages, namely: 1, 2, 3, 7, 14, 28, and 94 days. The concrete compressive strength ranged from 45 MPa (at 24 h) to 97 MPa (at 94 days). Correlations were established between the non-destructive test results and the concrete compressive strength. A test variability analysis was performed and the 95% confidence limits for the obtained correlations were computed. The obtained results for SCC showed good correlations between the concrete compressive strength and the non-destructive tests results, although some differences exist when compared to the correlations obtained for NVC.

Keywords: self-compacting concrete; non-destructive test methods; compressive strength; ultrasonic pulse velocity test; surface hardness test; pull-out test; maturity test; within-test variability; normal vibrated concrete

1. Introduction

Since the middle of last century, concrete has undergone several developments and continued to show to be a remarkably versatile material for many applications in civil constructions. Nowadays, concrete still continues to be the preferential construction material to integrate many structural members for different kinds of constructions and infrastructures, such as buildings, bridges, dams, among others. In particular, many architectural structures have been and still are built using concrete as the key building material to achieve often complex spatial geometries. In this particular case, self-compacting concrete (SCC) has been widely used because it can be placed easily in complicated formwork and with high degree of reinforcement without the need of vibration. In addition, SCC produces a smooth and well-finished surface at the end of concreting, which is an important factor in many architectural structures with exposed concrete.

In the last years, maintenance and rehabilitation concerns have also covered concrete as a building material. Its preservation is essential to ensure the stability of the structure, so as not to impair its use nor jeopardize the safety of the users during the intended lifetime.

The need to ensure the quality of concrete structures, the verification of aging of older structures and the premature degradation of recent ones justify, because of the associated high costs, special attention to the study of maintenance and rehabilitation issues. Nowadays, if existent concrete structures show unacceptable deterioration, a correct assessment of the actual state of concrete is required to quantify the available safety margin. The effective concrete compressive strength is usually a key parameter that is required to perform this evaluation. Other situations that require an evaluation of concrete strength include the quality control of precast or in-situ concrete application, namely to decide when handling and transport precast units, to evaluate the concrete compressive strength for application of prestress, or to support the decision to remove formwork or temporary supports for structural elements.

The use of non-destructive tests (NDT) to evaluate concrete quality and to estimate its in-situ compressive strength has been well-known for some decades [1–5]. Many of these tests and devices were initially developed for normal vibrated concrete (NVC) of normal strength range. However, in the nineties of the last century, some of these tests and devices were adapted for high-strength NVC [6–9]. Specific studies concerning the application of NDT in SCC are still scarce [10]. The procedures to apply the most firmly established NDT methods can be found in the normative documents from different countries. However, it is important to mention that such procedures can present small differences between them and the selection of the most appropriate method should be decided previous to test, to avoid divergences when interpreting the results [4,5].

The range of available tests vary from the most economical, simple, and easy to use (e.g., surface hardness test, ultrasonic pulse velocity test, and pull-out test using Lok-test system), to the most complicated (e.g., pull-off test using Bond-test system with partial coring and pull-out test using CAPO-test system) and expensive ones (e.g., Windsor Probe Test System and Maturity meter). Careful selection of the types of tests to be combined in each situation is critical to achieve both accuracy of results and cost savings [1,3,4]. When such tests are applied, it is necessary to evaluate the variables which can affect the test results and the correlations. Some tests are more sensitive and/or reliable than others, but all of them can differ in terms of the within-test variability and repeatability of test results [10–13].

In general, the interpretation and validation of NDT results should involve three distinct phases [10]: processing of collected data, analysis of within-test variability, and quantitative evaluation of the property under analysis. Relevant information can be obtained by the analysis of within-test variability, by comparing the obtained results in a location with the typical one for the NDT method in use, either to provide a measure of the quality control or to detect abnormal circumstances in NDT application [10]. A good planning of a research when inspecting a concrete structure should also include the procedures for data treatment and interpretation of in-situ test results prior to the inspection. When monitoring concrete compressive strength during construction, it is usually sufficient to compare test results with the limits established by trials made at the start of the contract, but in other complex situations, like in old structures, the prediction of the actual concrete compressive strength could be required for calculation design. Depending on the purpose of the research, either for estimation of the in-situ concrete compressive strength for conformity checking, either for design calculations, many questions concerning the conversion between the mean value of compressive strength and the characteristic value or the minimum in-situ design value, or either about the safety factor coefficient to apply, may lead to complex discussions, because of the basic differences between in-situ concrete and the standard test specimens upon which most specifications are based [1,4].

Most of the NDT give a measure of a property of concrete on the surface or near to the surface that can be related to concrete strength (surface hardness, resistance to penetration of a probe, pull-out force of a 25 mm ring placed at 25 mm depth, pull-off force to extract a cylindrical disk glued to the surface or near the surface to measure direct tensile strength, internal fracture test, among others). The obtained readings of NDT may be correlated with compressive strength experimentally. However, placing, compacting, and curing may turn the concrete in the surface zone unrepresentative of the concrete at deeper levels, and care should be taken to ensure that the correlations adopted are relevant to the circumstances of use. One of the NDT methods that does not cause any damages on concrete surface and can be used to evaluate the interior mass of concrete elements is the ultrasonic pulse velocity test. However, the ultrasonic pulse velocity in reinforced concrete is significantly affected by the presence of steel reinforcement, and this can impair significantly the results [1,3,4]. It is worth mentioning that, because of its versatility, the ultrasonic pulse velocity test can also be used in rocks to evaluate its mechanical properties, e.g., for structural diagnosis of old rock masonry of historical heritage building [14]. Likewise, the surface hardness test (type-L) can also be used to estimate the compressive strength in rocks for the same purpose.

The number and type of variables affecting the correlations with the concrete compressive strength may differ for each NDT method. Some correlations can include correction parameters to attend for some of these variables in order to broad the range of application of the test, while others are missing with respect to this. As examples, some of the referred variables that can affect the correlations are the following ones [1,3,4]: differences in concrete mix proportions (quantity, nature, shape, and texture of aggregates; type and amount of cement; paste to aggregate ratio, water to cement ratio, among others); differences in moisture conditions (saturated or dry); type and size of the test specimens used to establish the correlation; surface carbonation (which changes the relationship between the superficial and inner concrete); the age of concrete and used curing type (in some methods the correlations are different for concrete with short ages compared to concrete over 28 days); differences in concrete surface finish (metal formwork may lead to differences in the surface layer when compared with wood formwork); the used equipment (similar equipment with the same technical reference may have different correlation); the used procedure (different procedures may lead to different results); different stress states in the tested element may affect the readings; the mass of the test specimen, among others.

When selecting the most appropriate method some factors are crucial, such as: the purpose of the testing, practical factor related to the nature and position of the concrete under evaluation, the availability and reliability of surface damage, size of member to be tested, the complexity and preparation of the operation, access requirements and test positions. In some circumstances, the selection of a NDT which is quicker to carry out and less damaging can be more useful to mapping areas of different quality in a structural member (without the need to use correlations with compressive strength) and to locate appropriate areas for testing by other methods, more destructive and usually more expensive, but more precise, including the extraction of a small number of cores.

For the majority of the NDT, it is recommended that a specific correlation is obtained for the type of concrete under investigation to achieve higher accuracy. However, there are some NDT, such as the pull-out test, for which the use of general correlations is allowed to estimate concrete compressive strength with reasonable accuracy for a wide range of concrete mixes of NVC [15]. According to BS 1881-207:1992 [15], even for the pull-out test, special correlations are required for lightweight concretes or other mixes with less common constituents. It could be the case of SCC, in which the mix proportions differ from NVC in order to achieve the required fresh properties.

From the aforementioned, it can be stated that the specificities of SCC when compared with NVC, namely higher cement paste dosage and smaller volume of coarse aggregates, smaller coarse aggregates, the absence of vibration, among others, are susceptible to affect the correlations with the concrete compressive strength. In this sense, the present study evaluates the applicability of some NDT to SCC in order to estimate the concrete compressive strength. Selected tests included the ultrasonic pulse velocity test (PUNDIT), the surface hardness test (Schmidt rebound hammer type N), the pull-out test (Lok-test), and the concrete maturity test (COMA-meter).

The tests used in this research work were selected based on its user-friendly characteristics. The ultrasonic pulse velocity test and the surface hardness test can be used either in new or old concrete, being easy to operate, do not produce damage on the concrete surface, results are immediately available, are of low cost and require only the maintenance of the equipment. These allow a more extensive analysis of structures covering a larger extension. The pull-out test (Lok-test) can be used only in new concrete, since the insert has to be placed in formwork prior to casting. However, *Germann Instruments A/S* have developed the CAPO-test system (Cut And Pull-Out) which allows to perform the pull-out test in old concrete. In this system, the capo-insert (25-mm diameter ring) is placed in hardened concrete at a depth of 25 mm by drilling a 18-mm central hole with a drill unit, using a diamond recess router to open an inside hole at 25 mm depth and an expansion unit to fully expand a 25-mm diameter ring inserted in the hole. The geometry and mechanism of fracture in CAPO-test system is similar to Lok-test system, allowing the use of the same correlations to compressive strength. Both systems measure the force by which a 25-mm disc or ring placed in a depth of 25 mm is pulled out of the concrete through a 55 mm inner diameter counterpressure placed on the testing surface. The concrete maturity test (COMA-meter) is the one that is exclusively used in fresh concrete because the capillary tube has to be placed just after casting.

The 95% confidence limits for the estimation of the concrete compressive strength will vary significantly according to the type of selected NDT and the reproducibility of the used correlations. The data actually available concerns only the NVC. In this context, it has been referred that even using correlations specifically developed for a given concrete and under well reproduced in-situ conditions, it is unlikely that the 95% confidence limits for the estimation of the concrete compressive strength are better than ± 20%, ± 25%, and ± 10% of the mean value, when using the ultrasonic pulse velocity test, surface hardness test, and pull-out test, respectively [1,15–17]. When using specifically developed correlations and under ideal laboratory conditions it is probable that this difference would be reduced to ± 10% and ± 15% from the estimated mean value for ultrasonic pulse velocity test and surface hardness test, respectively [1,16,17]. Without specific correlations this difference could arise ± 50% for the ultrasonic pulse velocity test [1,16]. Even for the pull-out test, when using general correlations, such as those suggested by Lok-test and CAPO-test manufacturers, such interval would probably be widened to ± 20% of the mean value [1,15].

2. Experimental Program

The experimental program was developed in three stages. In the first stage, the mix proportions of a SCC with average compressive strength at 28 days of 90 MPa was studied and characterized. In the second stage, seven sets of concrete test specimens were produced: P1, P2, P3, P7, P14, P28, and P94. For each set, the number corresponds to the concrete age (days). In the third stage, the selected NDT were applied, namely: the ultrasonic pulse velocity test (PUNDIT), the surface hardness test (Schmidt rebound hammer type N), the pull-out test (Lok-test), and the concrete maturity test (COMA-meter).

2.1. Study and Characterization of the SCC

The design of the mix proportions for the SCC was performed according to the methodology proposed by Nepomuceno et al. [18–20]. The characterization of the fresh and hardened concrete properties was performed according to NP EN 206-9: 2010 [21].

2.1.1. Material

To produce the SCC, the following materials were selected: Portland cement (CEM I 42.5R) with density 3140 kg/m^3; fly ash with density 2380 kg/m^3; modified carboxylate-based superplasticizer supplied by Sika Portugal, SA with the commercial name Sika ViscoCrete 3005 having a density 1050 kg/m^3; fine-rolled natural sand (Sand 0/2) with density 2600 kg/m^3 and fineness modulus 2.104; rolled natural sand from river with medium grain size (Sand 0/4) with density 2640 kg/m^3 and fineness modulus 3.035; crushed granite aggregate (Gravel 3/6) with density 2710 kg/m^3 and fineness modulus 5.311; and crushed granite aggregate (Gravel 6/15) with density 2700 kg/m^3, fineness modulus 6.692 and maximum size 19.1 mm.

The optimum proportions of fine aggregates to fit with the fine aggregate reference curve was obtained by combining, in absolute volume ratio, 50% of Sand 0/2 and 50% of Sand 0/4, resulting in a mixture with fineness modulus 2.569. The coarse aggregates were combined in absolute volume ratio of 65% Gravel 3/6 and 35% Gravel 6/15 to fit with the coarse aggregate reference curve, resulting in a mixture with fineness modulus 5.794.

2.1.2. Mix Proportions of the SCC

The mix design of the mortar phase of SCC was performed based on the methodology proposed by Nepomuceno et al. [18], which considers the volumetric ratio of each fine aggregate (s1, s2,.., sn) in the total volume of fine aggregates (Vs), the powder mixture proportions (cement replacement by the addition), the ratio between the volume of powder and fine aggregates (Vp/Vs), the ratio between the volume of water and powder (Vw/Vp) and the percentage mass ratio between the superplasticizer and the powder (Sp/p%). Thus, considering the selected cement type and the intended average compressive strength, a water to cement ratio W/C (in mass) of 0.35 was estimated. Next, parameter Vp/Vs was set to be 0.80 and, based on the W/C ratio, cement type, and addition selected, the percentage of cement replacement by the addition was estimated as 30%. Parameters Vw/Vp and Sp/p% were obtained experimentally using the procedure described by Nepomuceno et al. [18]. The following values were obtained: Vw/Vp = 0.77 and Sp/p% = 0.70. The volumetric ratio of each fine aggregate, defined in Section 2.1.1, is 0.5 of Sand 0/2 and 0.5 of Sand 0/4.

According to the methodology proposed by Nepomuceno et al. [19,20], to complete the mix design of SCC, the following parameters are needed: the volumetric ratio of each coarse aggregate (g1, g2, ... , gn) in the total volume of coarse aggregates (Vg), the volume of voids in concrete (Vv) and finally, the ratio between the volume of mortar and coarse aggregates (Vm/Vg). The following parameters were defined: Vv = 0.03 m^3, Vm/Vg was estimated to be 2.279 considering the required fresh properties. The volumetric ratio of each coarse aggregate, defined in Section 2.1.1, is 0.65 for Gravel 3/6 and 0.35 for Gravel 6/15. The SCC mix proportions are presented in Table 1.

Table 1. Mix proportions of SCC (contents per cubic meter).

Constituent Materials	Dosage
Portland cement CEM I 42.5R (kg)	487.5
Fly ash (kg)	158.4
Superplasticizer (liters)	4.3
Water (liters)	170.8
Sand 0/2 (kg)	360.4
Sand 0/4 (kg)	366.0
Gravel 3/6 (kg)	521.1
Gravel 6/15 (kg)	279.6

2.1.3. Fresh Properties of SCC

The evaluation of the SCC fresh properties was performed by measuring the spread in the slump-flow test (Figure 1), the fluidity in V-funnel test (Figure 2), and the passing ability in L-box test (Figure 3). The obtained results are presented in Table 2 (where Dm is the average diameter in slump-flow test, t is the V-funnel time and H2/H1 is the concrete heights ratio in L-box test) and fit the defined objectives. These tests were further complemented by visual observation of the fresh concrete to evaluate the segregation resistance. As shown in Figure 1, the concrete has spread uniformly with a very homogeneous distribution of aggregates and without any visible segregation or bleeding.

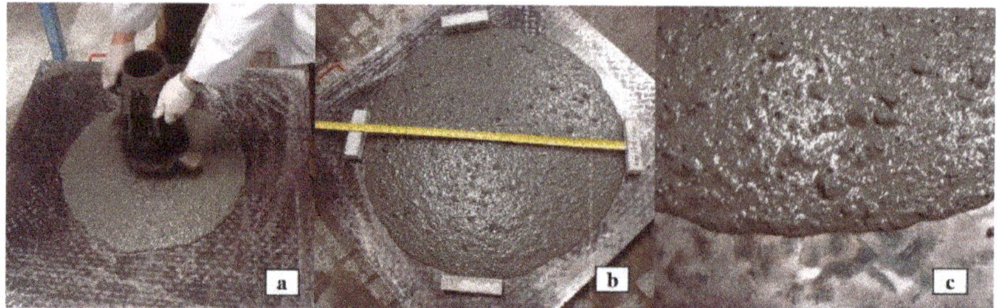

Figure 1. Slump-flow test: (**a**) start test, (**b**) measuring the diameter, (**c**) absence of segregation or bleeding.

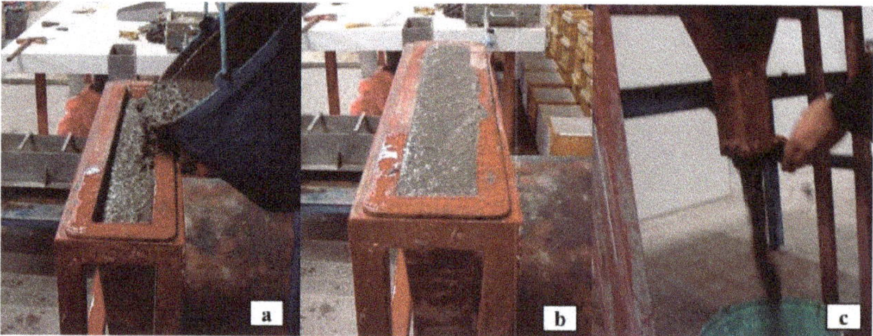

Figure 2. V-funnel test: (**a**) fill of V-funnel, (**b**) ready to test, (**c**) flowing of concrete.

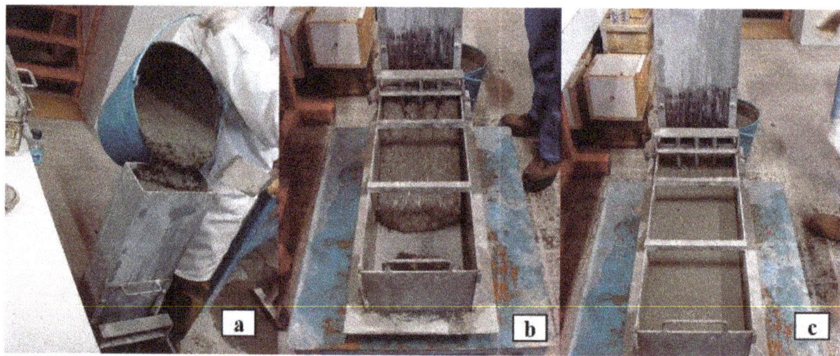

Figure 3. L-box test: (**a**) fill the L-box, (**b**) concrete flow, (**c**) prepared for readings after stop flow.

Table 2. Fresh properties of SCC.

Slump-Flow Dm (mm)	V-Funnel t (s)	L-Box H2/H1
780	15.6	0.92

2.1.4. Production of Specimens for Testing

Seven series of SCC test specimens were produced, all cast on the same day and from a single concrete mixture. Fresh SCC was placed in the formworks without any kind of vibration (Figure 4). Each series consisted of a 200 mm cubic specimen to accommodate the five pull-out probes (one per face, see Figure 4b) and the maturity meter (Figure 5), and four 150 mm cubic specimens for the remaining tests (ultrasonic pulse velocity test, surface hardness test and compressive strength).

After molding, all test specimens were protected with plastic sheet to prevent the premature loss of moisture and stored in the laboratory for 24 h (Figure 6a). After 24 h, the test specimens were demolded (Figure 6b) and then placed in a curing chamber. The curing of concrete test specimens follows the EN 12390-2:2000 [22]—testing hardened concrete—Part 2: Making and curing specimens for strength tests. The automatic curing chamber was programmed to keep a temperature of 20 °C and a relative humidity (RH) of 95%. However, a small fluctuation occurred and the temperature varied between 18 and 20 °C, while the RH varied between 90 and 95%.

Figure 4. Concrete molding: (**a**) cubes of 150 mm side, (**b**) cube of 200 mm side, (**c**) concrete placing.

Figure 5. Concrete maturity test: (**a**) COMA meter, (**b**) and (**c**) placing the closed capillary tube.

Figure 6. Concrete test specimens: (**a**) protection, (**b**) demolding.

2.2. Hardened Properties of SCC

The evaluated hardened state properties were the concrete compressive strength and the density. The average concrete compressive strength results (f_{cm}) are presented in Table 3 for each series corresponding to different ages of the same concrete. The density at 28 days was found to be around 2300 kg/m^3. In Table 3, S_d is the standard deviation and C_v is the coefficient of variation. Figure 7 shows graphically the evolution of the compressive strength (f_{cm}) with the age of concrete (in days) when submitted to standard curing conditions.

The characteristic value of the concrete compressive strength at 28 days (f_{ck}) is 88.5 MPa, considering the standard deviation (S_d) of 1.33 MPa, the mean value (f_{cm}) of 90.70 MPa, and a margin parameter for the probability distribution of strength of 1.64 (assuming a normal distribution). According to NP EN 206-9:2010 [21], the SCC concrete class can be classified as C70/85, which corresponds to a high strength SCC.

Table 3. Hardened properties of SCC.

Series	Age (days)	f_{cm} (MPa)	S_d (MPa)	C_v (%)
P1	1	45.31	1.48	3.26
P2	2	58.16	2.21	3.81
P3	3	64.06	2.09	3.26
P7	7	71.92	3.00	4.16
P14	14	81.47	6.16	7.56
P28	28	90.70	1.33	1.46
P94	94	97.00	1.79	1.84

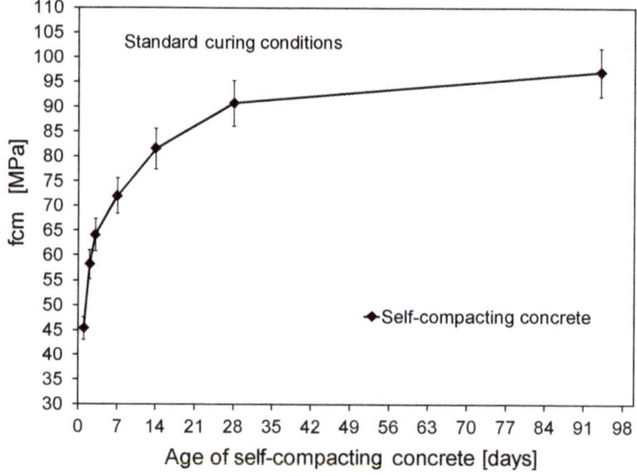

Figure 7. Variation of the concrete compressive strength with the age under standard cure.

2.3. Non-Destructive Tests

All test procedures and correlations for the selected NDT followed the recommendations from BS 1881-201:1986 [23].

2.3.1. Ultrasonic Pulse Velocity Test

The ultrasonic pulse velocity test was performed in accordance with BS 1881-203:1986 [16] using an apparatus (PUNDIT) connected to electro-acoustic transducers with frequency 54 kHz, produced by ELE International. Prior to the tests on each series, the calibration of the apparatus was checked (Figure 8). Then, four measurements were made in the 150 mm cubic test specimens, between two parallel faces, perpendicular to cast direction. The recorded readings are shown in Table 4, where V is the average value of four individual readings of the recorded velocity.

Figure 8. Calibration of the PUNDIT apparatus.

Table 4. Ultrasonic pulse velocity test results.

Series	Age (days)	V (km/s)	S_d (km/s)	C_v (%)
P1	1	4.31	0.035	0.82
P2	2	4.44	0.011	0.24
P3	3	4.57	0.025	0.56
P7	7	4.64	0.030	0.64
P14	14	4.80	0.063	1.31
P28	28	4.84	0.022	0.45
P94	94	4.83	0.015	0.31

2.3.2. Surface Hardness Test

Surface hardness tests were performed in accordance with BS 1881-202:1986 [17] by applying a Schmidt rebound hammer type N with an impact energy of 2.207 Nm, produced by ELE International. Prior to this test, and after performing the ultrasonic pulse velocity test, the average concrete compressive strength until failure was measured by using three of the four 150 mm cubic test specimens produced in each set. The remaining four 150 mm cubic test specimen of each set was loaded with a compressive stress state equivalent to 1/10 of the average concrete compressive strength previously measured, in order to confine the test specimen between the steel plates of the compressive testing machine (Figure 9b). The main purpose was to subject the specimen to a certain load to prevent bouncing during test and to simulate the concrete under loading, as in real situation. By testing a free specimen, it will bounce and the result will not represent correctly the concrete surface hardness. The Schmidt rebound hammer was used horizontally and nine readings were recorded in the test specimen, in a molded face perpendicular to the concrete cast direction (Figure 9c). Prior to the surface hardness tests, the calibration of the apparatus was checked (Figure 9a). The recorded readings are shown in Table 5, where R represents the average value of nine individual readings of rebound number.

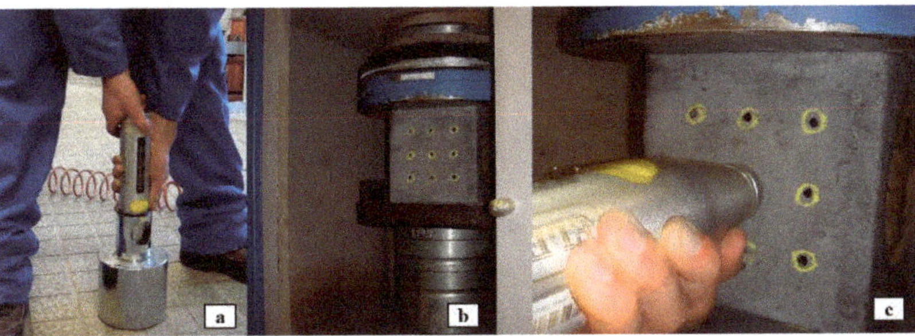

Figure 9. Surface hardness test: (**a**) calibration of apparatus, (**b**) confinement of specimen, (**c**) testing.

Table 5. Surface hardness test results.

Series	Age (days)	R	S_d	C_v (%)
P1	1	37.44	0.63	1.69
P2	2	41.00	1.20	2.92
P3	3	43.06	1.07	2.49
P7	7	45.39	1.39	3.06
P14	14	45.44	2.24	4.93
P28	28	47.56	1.26	2.65
P94	94	49.61	0.65	1.31

2.3.3. Pull-Out Test

Pull-out tests were performed in accordance with BS 1881-207:1992 [15] and using an apparatus with maximum loading capacity of 150 kN, from *Germann Instruments A/S* and based on the *Lok-test* system. Prior to the tests, the calibration provided by the manufacturer to convert the value of the pull-force recorded with the apparatus to the actual pull-force P (in kN) was checked. This checking was performed by using a load cell (Figure 10b) connected to a data logger (Figure 10a). The calibration correlation provided by the manufacturer was validated through several consecutive loading-unloading cycles between 10 to 60 kN, as shown in Figure 11.

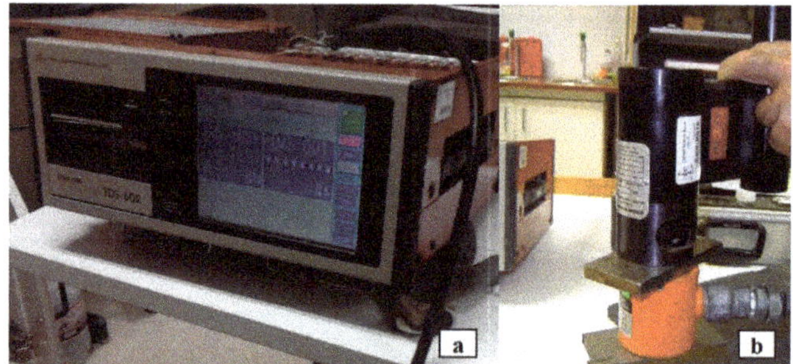

Figure 10. Pull-out test: (**a**) data logger, (**b**) calibration checking.

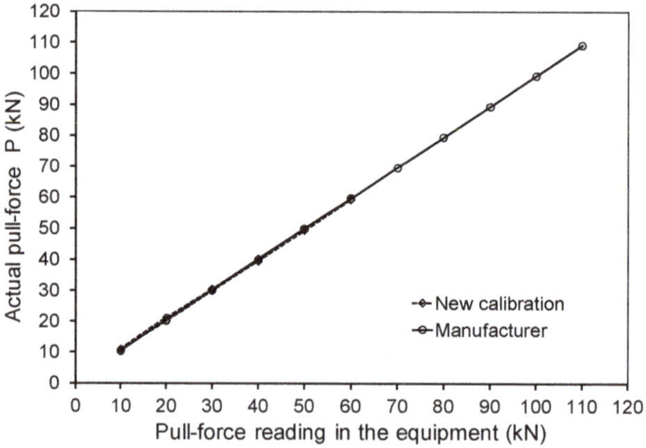

Figure 11. Comparison between new calibration and that provided by the manufacturer.

Pull-out tests were applied on five faces of the 200 mm cubic test specimens according to the arrangement of the probes illustrated in Figure 4b. The probe geometry of the Lok-test system is characterized by a disc with 25 mm diameter located 25 mm in depth (Figure 4b). The test procedure consists of the following steps: Removal of the insert stem that is screwed to the cast-in disk (Figure 12a); screwing a pull-bolt flange to the cast-in disk (Figure 12b); screw of the coupling in the head of the pull bolt flange (Figure 12c); connection of the hydraulic jack to the coupling (Figure 13a); loading the instrument by turning slowly the telescoping handle clockwise about two seconds per each full lap in order to keep a loading rate of 0.5 ± 0.2 kN/s (Figure 13b) and finally, register the peak load (Figure 13c). In the present research, a video of the screen of the hydraulic jack was made to register the progress of the gauge pointer (Figure 13c), since when it reaches the peak load it quickly jumps to zero scale and the measurement could be lost.

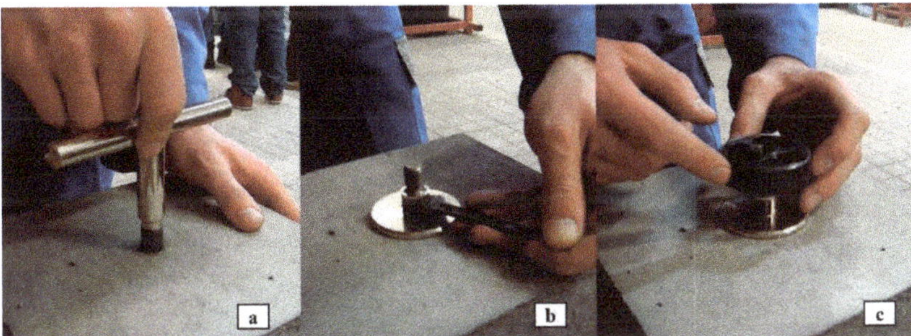

Figure 12. Pull-out test: (**a**) removal of the stem, (**b**) screw of pull-bolt flange (**c**) screw of coupling.

Figure 13. Pull-out test: (**a**) connection of the hydraulic jack, (**b**) loading, (**c**) reading.

After the pull-force load measured in the equipment is reached and recorded, the concrete cone trunk fragment was removed (Figure 14a) in order to visualize the geometry of the failure surface and to deduce the validity of the result (Figure 14b). If the insert is not to be reused, it is not necessary to fully extract the insert after achieving the maximum load, reducing the surface damage in the concrete surface. The pull-force load recorded in the apparatus was then converted into the actual pull-force load P (in kN) and the average value from the five readings obtained for each series was calculated. The recorded readings are shown in Table 6, where P represents the average value of the five individual readings of pull-force load, in kN.

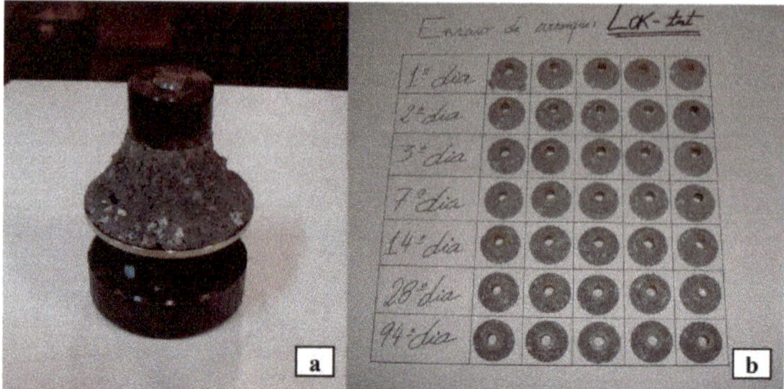

Figure 14. Pull-out test: (**a**) and (**b**) concrete cone trunk fragments.

Table 6. Pull-out test results.

Series	Age (days)	P (kN)	S_d (kN)	C_v (%)
P1	1	26.81	1.50	5.58
P2	2	34.51	1.50	4.34
P3	3	35.69	1.13	3.15
P7	7	43.20	1.29	2.98
P14	14	47.74	3.23	6.76
P28	28	53.07	4.27	8.04
P94	94	60.96	2.25	3.69

2.3.4. Maturity Test

For the concrete maturity test, meters of range 0 to 14 M_{20} days and type COMA-Meter (COncrete MAturity-Meter) were selected. They consist of a closed capillary tube containing a special liquid (Figure 5a). Right prior to starting the test, the capillary tube was broken at its upper ending, and was immediately inserted into the threaded protective casing and then placed into the fresh concrete (Figure 5b,c). From that time, the liquid inside the capillary tube begins to evaporate because of the temperature of the concrete. Fixed to the tube, exists a blade that shows an equivalent maturity scale in days (M_{20}). Right after the concrete casting, maturity meters were placed in the 200 mm cubic test specimens of series P1, P2, P3, P7, and P14. The maturity results M_{20} after 1, 2, 3, 7, and 14 days are shown in Table 7, where M_{20} (days) is the average value of the five individual readings.

Table 7. Maturity test results.

Series	Age (days)	M_{20} (days)	S_d (days)	C_v (%)
P1	1	1.26	0.089	7.10
P2	2	2.29	0.114	4.98
P3	3	3.26	0.119	3.66
P7	7	6.64	0.263	3.96
P14	14	11.92	0.698	5.85

3. Presentation and Discussion of the Results

3.1. Ultrasonic Pulse Velocity Test

Figure 15 illustrates the graph with the correlation obtained between the ultrasonic pulse velocity (V) and the average SCC compressive strength (f_{cm}), as well as the corresponding exponential curve (continuous curve) used to fit the results with a correlation coefficient of about 0.97. Figure 16 compares the obtained correlation for SCC from Figure 15 to that obtained by Nepomuceno and Lopes [4,9] for

NVC, with identical materials but with maximum coarse aggregate size of 25 mm. From Figure 16, it can be stated that a slight difference exists between the correlations. The small observed deviations are probably due to the differences of properties of the elastic medium because of the different mortar/coarse aggregate ratios. However, the results seem to show that for NVC the ultrasonic pulse velocity test loses sensitivity for the estimate of the concrete compressive strength for speeds above 4.6 km/s, while for SCC the same is observed to occur for higher speeds, from about 4.8 km/s. By analyzing the results presented in Table 4, it can be observed that S_d varied from 0.011 to 0.063 km/s with the average being 0.029 km/s. These results are similar to those obtained in NVC of compressive strength up to 82 MPa [4,9], the correlation of which is shown in Figure 16, namely S_d shows a variation from 0.006 to 0.086 km/s and a mean value of 0.028 km/s. Likewise, for SCC, the C_v varied from 0.2 to 1.3% with the average being 0.6% (Table 4). These results are similar to those for NVC [4,9], where C_v shows a variation between 0.1 to 1.9% with the average being 0.6%. In a previous analysis on repeatability undertaken by Nepomuceno and Lopes [10], bringing together NVC and SCC results, no evidence was found to conclude about the statistical parameter (S_d or C_v) that better represents the repeatability of ultrasonic pulse velocity test.

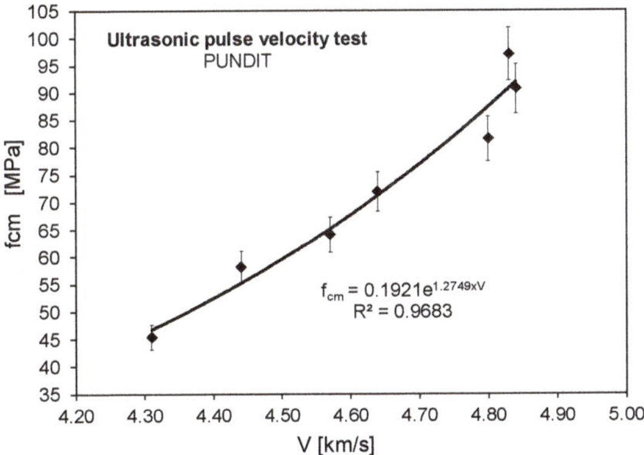

Figure 15. Ultrasonic pulse velocity versus average concrete compressive strength.

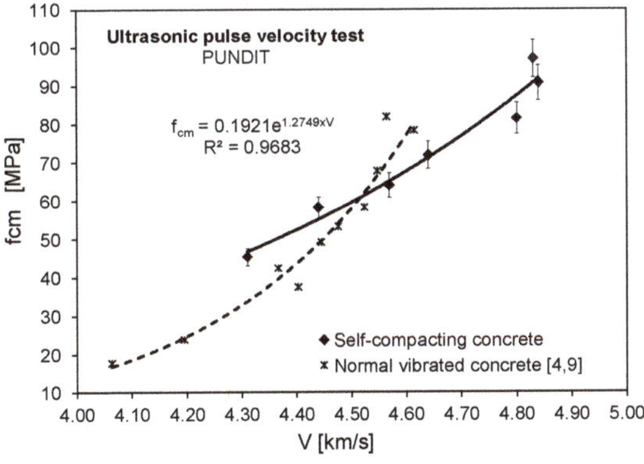

Figure 16. Comparison between correlations for SCC and NVC using ultrasonic pulse velocity test.

3.2. Surface Hardness Test

Figure 17 illustrates the graph with the correlation obtained between the rebound number (R) and the mean concrete compressive strength (f_{cm}) for the studied SCC as well as the corresponding equation for the straight line (continuous line) used to fit the results with a correlation coefficient of about 0.96. In Figure 18 the obtained correlation for SCC from Figure 17 is compared to that obtained by Nepomuceno and Lopes [4,9], as previously referred in Section 3.1. From Figure 18 some differences are visualized between SCC and NVC, which are probably due to the different mortar/coarse aggregate ratios and different maximum size of the coarse aggregate. Also, the wall effect and the external vibration used to compact NVC certainly contributed to densify the near surface area of concrete. In fact, from Figure 18 it seems that, for concretes with equal compressive strength, NVC presents higher surface hardness when compared to SCC.

The results of Table 5 show that S_d varied from 0.63 to 2.24, the average being 1.21. Again, there is no significant difference when testing NVC of compressive strength up to 82 MPa [4,9], which correlation is shown in Figure 18, since the S_d varied from 0.66 to 1.93, the average being 1.11. Similar analysis was done for the C_v, showing that for SCC the C_v varied from 1.3 to 4.9%, the average being 2.7% (Table 5), while for NVC [4,9] the C_v varies between 1.3 to 5.0% with the average being 3.0%. These results are according to those reported by Bungey [24], which indicates as typical a C_v of 4% when testing different locations of the same element. The values of S_d and C_v are quite similar for NVC and SCC and no abnormal circumstances were detected. Previous analysis on the repeatability of surface hardness test, undertaken by Nepomuceno and Lopes [10], bringing together NVC and SCC results here reported, have revealed that C_v tends to be slightly lower as concrete compressive strengths increases, while S_d remains almost constant, which indicates that S_d is the statistical parameter which better represents the repeatability.

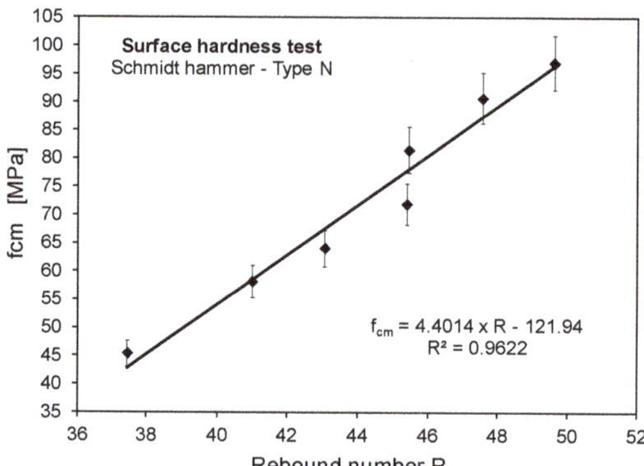

Figure 17. Rebound number versus average concrete compressive strength.

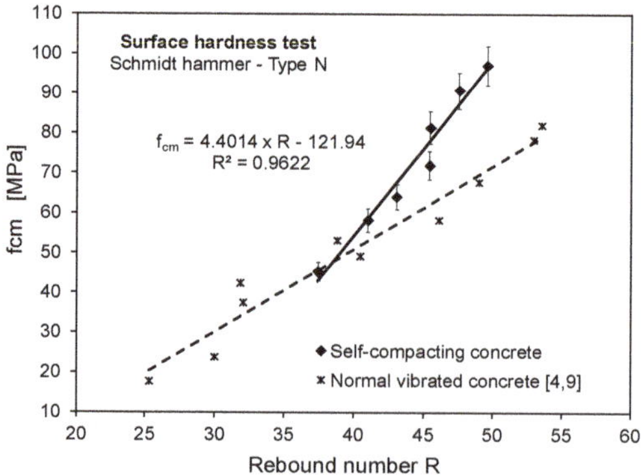

Figure 18. Comparison between correlations for SCC and NVC using surface hardness test.

3.3. Pull-Out Test

Figure 19 illustrates the graph with the correlation obtained between the pull-out force (P) and the average concrete compressive strength (f_{cm}) for the studied SCC, as well as the corresponding equation for the straight line (continuous line) used to fit the results, with a correlation coefficient of about 0.98. Figure 20 illustrates the comparison between the correlation obtained in the present research work for SCC (Figure 19) and those obtained for NVC by Nepomuceno and Lopes [4,7] and by Krenchel and Peterson [2]. For both SCC and NVC, the correlations in Figure 20 show the same tendency and high correlation coefficients. However, for the same concrete compressive strength, the pull-out force is higher for NVC when compared to SCC. This observation can be explained because of the higher amount of mortar and the existence of smaller aggregates in the surrounding area of the probe for SCC. For NVC, the existence of larger aggregates in the surrounding area provides higher resistance to failure in the zone of the compressive arm between the probe and the counterpressure ring.

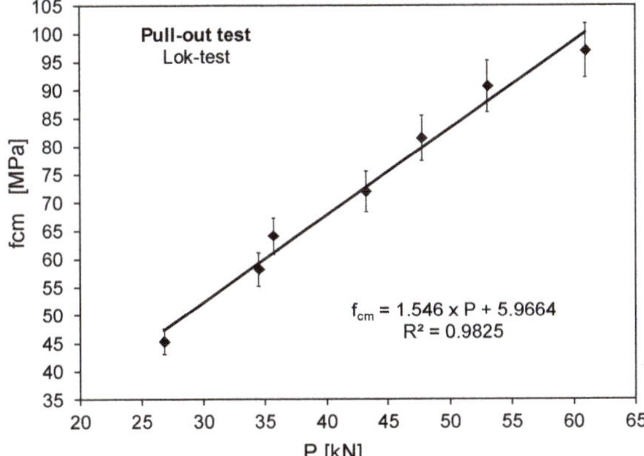

Figure 19. Pull-out force versus average concrete compressive strength.

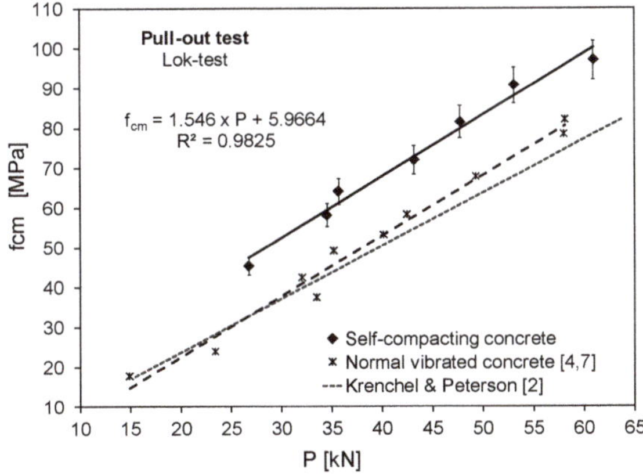

Figure 20. Comparison between correlations for SCC and NVC using pull-out test.

At the ending of the last century, many independent analytical and experimental studies have been developed to understand how failure mechanism works during the pull-out test. A consensus has been achieved regarding the existence of a triaxial state of stress highly non uniform on the concrete involving the insert during extraction [1,3]. In spite of some divergence as far as the basic failure mechanism is concerned, a consensus exists regarding the fact that the last pull-out load is influenced by the same properties which influence the concrete compressive strength [1,3]. As previously mentioned, the pull-out test is assuming by the BS 1881-207:1992 [15] as being a very reliable test method, for which the use of general correlations is allowed with reasonable accuracy for a wide range of NVC mixes. However, the same standard points out that special correlations are required for lightweight concretes or other mixes with less common constituents. The different mix proportions of an SCC compared to NVC and the absence of vibration can justify the differences in the concrete hardness in the surface or near to the surface. Such differences were also detected when using the surface hardness test.

Table 6 shows that for SCC the S_d varied from 1.13 to 4.27 kN with the average being 2.17 kN. These values are similar to those obtained in NVC of compressive strength up to 82 MPa [4,7], the correlation of which is shown in Figure 20, presenting an S_d from 1.24 to 5.19 kN with the average being 2.53 kN. Krenchel and Petersen [2] have reported as typical a S_d from 1.9 kN to 2.5 kN when using 150 mm cubes and nearly 2.8 kN for larger specimens. For SCC, the C_v varied from 3.0 to 8.0% with the average being 4.9% (Table 6), while for NVC [4,7] of high strength the C_v varied from 2.9 to 7.0% with the average being 5.1%. Krenchel and Petersen [2] reported as typical a C_v between 6.8% and 7.5% when using 150 mm cube specimens and nearly 9.9% for larger specimens. The BS 1881-207:1992 [15] indicates a typical C_v of 7%. Previous analysis on repeatability of pull-out test, undertaken by Nepomuceno and Lopes [10], bringing together NVC and SCC reported results, have revealed that S_d is the statistical parameter which better represents the repeatability, which contradicts the Carino's report [3].

3.4. *Maturity Test*

Figure 21 shows the graph with the correlation between the maturity days (M_{20}) and the average SCC compressive strength (f_{cm}) as well as the logarithmic curve (continuous curve) used to fit the results with a correlation coefficient of about 0.98. Based on the values presented in columns 2 and 3 of Table 7, a small lag can be observed between the maturity days M_{20} and the effective curing days after 7 days. This can be explained because the temperature of the curing chamber was, on average, slightly below 20 °C. Anyway, Figure 21 shows that the maturity test is effective to estimate the SCC compressive strength. Table 7 shows that S_d of five individual readings of M_{20} varied from 0.09 to 0.70 days with the average being 0.26 days, while the C_v varied from 3.7 to 7.1% with the average being 5.1%. From Table 7 it can be observed that, except for the first reading (one day), the C_v remains almost constant as SCC compressive strength increases, while the S_d tends to increase. These results can lead to the conclusion that C_v will better represents the repeatability of the maturity meter test.

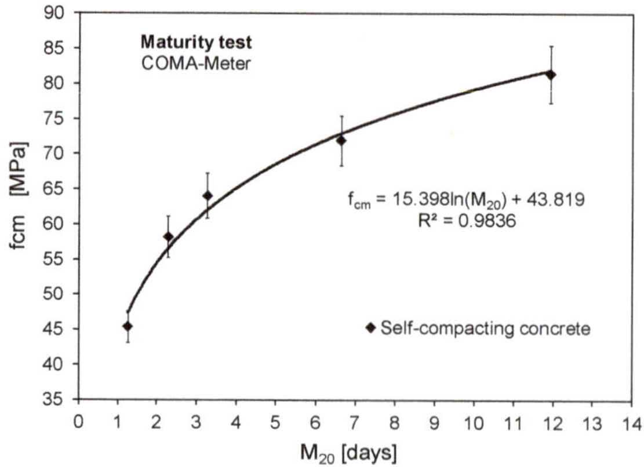

Figure 21. Maturity days versus average concrete compressive strength.

4. Conclusions

Concerning the main achievement in this article the following conclusions can be drawn:

(1) The obtained results showed good correlations between the SCC compressive strength and the NDT test readings. However, some differences were observed, when comparing with the correlations obtained for NVC, being more evident for the surface hardness test and the pull-out test. Thus, when testing SCC with NDT, general correlation should be used with precaution.
(2) Surface hardness of SCC seems to be lower than that measured in NVC for the same level of concrete compressive strength, and this can be attributed to differences in mixture proportions and the used method to densify the concrete. The absence of vibration in SCC (it compacts by its self-weigh), the higher cement paste dosage, lower volume of coarse aggregates, and lower maximum aggregate size, together with the wall effect, can introduce differences near to the concrete surface. External vibration used to compact NVC specimens certainly have contributed to densify the near surface area of concrete.
(3) Pull-out force measured in SCC was lower than that measured in NVC for the same level of concrete compressive strength, and this can be attributed to the same causes reported for surface hardness. In fact, both methods evaluate the concrete in the surface or near to the surface.
(4) The analysis of the within-test variability allows to conclude that the standard deviation (S_d) and the coefficient of variation (C_v) which could be expected in a location of SCC with compressive strength up to 97 MPa, when using surface hardness test (9 readings), ultrasonic pulse velocity test (4 readings), and pull-out test (5 readings) are quite similar to that obtained in a location of NVC of compressive strengths up to 82 MPa, when using the same type of equipment and number of individual readings, respectively.

Author Contributions: M.C.S.N. supervised the experimental tests, processed the recorded data, and performed the comparative analyzes. M.C.S.N. and L.F.A.B. wrote the paper.

Funding: This work was financed by Portuguese national funds through FCT—Foundation for Science and Technology, IP, within the research unit C-MADE, Centre of Materials and Building Technologies (UID/ECI/04082/2019 project), University of Beira Interior, Portugal.

Acknowledgments: The authors would like to thank João J.B. Costa for his support in laboratory tests.

Conflicts of Interest: The authors declare no conflict of interest.

References

1. Bungey, J.H.; Millard, S.G. *Testing of Concrete in Structures*, 3rd ed.; Chapman & Hall: London, UK, 1996.
2. Krenchel, H.; Petersen, C.G. In-situ Pullout Testing with Lok-test. Ten Years' Experience. In Proceedings of the International Conference on In Situ/Non-Destructive Testing of Concrete, Ottawa, ON, Canada, 2–5 October 1984.
3. Carino, J.N. Pullout Test. In *Handbook on Non-Destructive Testing of Concrete*; CRC Press Inc.: Boca Raton, FL, USA, 1991; pp. 39–82.
4. Nepomuceno, M.C.S. Ensaios Não Destrutivos em Betão (Non-destructive Tests on Concrete). In *Provas de Aptidão Pedagógica e Capacidade Científica*; UBI: Covilhã, Portugal, 1999. (In Portuguese)
5. Nepomuceno, M.C.S.; Lopes, S.M.R. Non-destructive Tests on Concrete. In *Journal of Concrete Technology Today Incorporating Structural Steel*; Trade Link Media: Singapore, 2002; pp. 14–20.
6. Lopes, S.M.R.; Nepomuceno, M.C.S. A Comparative Study of Penetration Resistance Apparatus on Concrete. In Proceedings of the Fourth International Conference on Composites Engineering, Kona, HI, USA, 6–12 July 1997; University of New Orleans: New Orleans, LA, USA, 1997; pp. 615–616.
7. Lopes, S.M.R.; Nepomuceno, M.C.S. Evaluation of In-place Concrete Strength by Near-to-surface Tests. In Proceedings of the 12th European Ready Mixed Concrete Congress (ERMCO1998), Lisbon, Portugal, 23–26 June 1998; pp. 338–347.

8. Lopes, S.M.R.; Nepomuceno, M.C.S. High Strength Concrete: Penetration Resistance Tests on High Strength Concrete. In Proceedings of the 1st International Conference on High Strength Concrete, Kona, HI, USA, 13–18 July 1997; ASCE: Reston, VA, USA, 1999; pp. 425–433.
9. Lopes, S.M.R.; Nepomuceno, M.C.S. Non-Destructive Tests on Normal and High Strength Concrete. In Proceedings of the 26th Conference on Our World in Concrete & Structures, Singapore, 27–28 August 2001; CI-Premier PTE LTD-Singapore: Singapore, 2001; Volume 20, pp. 53–67.
10. Nepomuceno, M.C.S.; Lopes, S.M.R. Analysis of Within-Test Variability of Non-Destructive Test Methods to Evaluate Compressive Strength of Normal Vibrated and Self-Compacting Concretes. *IOP Conf. Ser. Mater. Sci. Eng.* **2017**, *245*, 032025. [CrossRef]
11. Alwash, M.; Breysse, D.; Sbartaï, Z.M.; Szilágyi, K.; Borosnyói, A. Factors affecting the reliability of assessing the concrete strength by rebound hammer and cores. *Constr. Build. Mater.* **2017**, *140*, 354–363. [CrossRef]
12. Szilágyi, K.; Borosnyói, A.; Zsigovics, I. Extensive statistical analysis of the variability of concrete rebound hardness based on a large database of 60 years experience. *Constr. Build. Mater.* **2014**, *53*, 333–347. [CrossRef]
13. El Mir, A.; Salem, G. Nehme. Repeatability of the rebound surface hardness of concrete with alteration of concrete parameters. *Constr. Build. Mater.* **2017**, *131*, 317–326. [CrossRef]
14. Ramos, L.F.; Miranda, T.; Mishra, M.; Fernandes, F.M.; Manning, E. A Bayesian approach for NDT data fusion: The Saint Torcato church case Study. *Eng. Struct.* **2015**, *84*, 120–129. [CrossRef]
15. British Standard BS 1881-207:1992. *Testing Concrete. Recommendations for the Assessment of Concrete Strength by Near-to-surface Tests*; British Standards Institution: London, UK, 1992.
16. British Standard BS 1881-203:1986. *Testing Concrete: Recommendations for Measurement of Velocity of Ultrasonic Pulses in Concrete*; British Standards Institution: London, UK, 1986.
17. British Standard BS 1881-202:1986. *Testing Concrete: Recommendations for Surface Hardness Testing by Rebound Hammer*; British Standards Institution: London, UK, 1986.
18. Nepomuceno, M.C.S.; Pereira-de-Oliveira, L.A.; Lopes, S.M.R. Methodology for mix design of the mortar phase of self-compacting concrete using different mineral additions in binary blends of powders. *Constr. Build. Mater.* **2012**, *26*, 317–326. [CrossRef]
19. Nepomuceno, M.C.S.; Pereira-de-Oliveira, L.A.; Lopes, S.M.R. Methodology for the mix design of self-compacting concrete using different mineral additions in binary blends of powders. *Constr. Build. Mater.* **2014**, *64*, 82–94. [CrossRef]
20. Nepomuceno, M.C.S.; Pereira-de-Oliveira, L.A.; Lopes, S.M.R.; Franco, R.M.C. Maximum coarse aggregate's volume fraction in self-compacting concrete for different flow restrictions. *Constr. Build. Mater.* **2016**, *113*, 851–856. [CrossRef]
21. NP EN 206-9:2010. *Regras adicionais para o betão autocompactável (BAC)*; Instituto Português da Qualidade: Caparica, Portugal, 2010. (In Portuguese)
22. EN 12390-2:2000. *Testing Hardened Concrete—Part 2: Making and Curing Specimens for Strength Tests*; European Committee for Standardization: Brussels, Belgium, 2000.
23. British Standard BS 1881-201:1986. *Testing Concrete: Guide to the Use of Non-Destructive Methods of Test for Hardened Concrete*; British Standards Institution: London, UK, 1986.
24. Bungey, J.H. Concrete Strength Variations and In-place Testing. In *2nd Australian Conference on Engineering Materials*; University of New South Wales: Sydney, Australia, 1981; pp. 85–96.

Article

Real Cyclic Load-Bearing Test of a Ceramic-Reinforced Slab

Albert Albareda-Valls *, Alicia Rivera-Rogel, Ignacio Costales-Calvo and David García-Carrera

Department of Technology in Architecture, School of Architecture, Polytechnic University of Catalonia, Av Diagonal 649, 08034 Barcelona, Spain; alicia.rivera@upc.edu (A.R.-R.); nacho.costales@upc.edu (I.C.-C.); david.garcia.carrera@upc.edu (D.G.-C.)

* Correspondence: albert.albareda@upc.edu; Tel.: +34-639-523-624

Received: 21 January 2020; Accepted: 29 February 2020; Published: 4 March 2020

Featured Application: This paper contributes to understanding the behavior of ceramic slabs which were widely used during the 1950s and 1960s, especially for industrial buildings. Knowing the ultimate capacity of these slabs under uniform loading, especially under cyclic performance, is crucial to validating these material types in existing buildings. Analytical approaches to the capacities of these slabs are not enough to certify the ultimate loading performance, usually in industrial buildings.

Abstract: Ceramic-reinforced slabs were widely used in Spain during the second half of the 20th century, especially for industrial buildings. This solution was popular due to the lack of materials at that time, as it requires almost no concrete and low ratios of reinforcement. In this study, we present and discuss the results of a real load-bearing test of a real ceramic-reinforced slab, which was loaded and reloaded cyclically for a duration of one week in order to describe any damage under a high-demand loading series. Due to the design of these slabs, the structural response is based more on shear than on bending due to the low levels of concrete and the geometry and location of re-bars. The low ratio of concrete makes these slabs ideal for short-span structures, mainly combined with steel or RC frames. The slab which was analyzed in this study covers a span of 4.88 m between two steel I-beams (IPN400), and corresponds to a building from the mid-1960s in the city of Igualada (Barcelona, Spain). A load-bearing test was carried out up to 7.50 kN/m^2 by using two-story sacks full of sand. The supporting steel beams were propped up in order to avoid any interference in the results of the test; without the shoring of the steel structure, deflections would come from the combination of the ceramic slab together with the steel profiles. A process of loading and unloading was repeated for a duration of six days in order to describe the cyclic response of the slab under high levels of loading. Finally, vibration analysis of the slab was also done; the higher the load applied, the higher the fundamental frequency of the cross section, which is more comfortable in terms of serviceability.

Keywords: traditional slabs; ceramic-reinforced slabs; shear response; cyclic loading

1. Introduction

Ceramic slabs were intensively used in Spain during the 1950s and 1960s. The possibility of building rigid and resistant slabs with almost no concrete turned this material type into a very attractive option at that time due to the shortage of materials, especially concrete and steel. A wide range of ceramic cassettes were used for these slabs over more than three decades; some of them have not yet been catalogued or well identified. The idea consisted of building rigid horizontal slabs with the same requirements as precast reinforced concrete ones, but avoiding the beams, with only the minimum amount of in situ concrete needed (only enough to fill the interface between the ceramics and the rebar). Thus, steel reinforcements become crucial in the structural response of these slabs, as almost no concrete contributes; besides this, most of these slabs were built with smooth bars for reinforcing, which was quite typical at that time [1–3]. In 1955, the Spanish Ministry of Labor of that time approved the "Ordenanzas Técnicas y Normas Constructivas para viviendas de renta limitada", which was a document which established the Building Regulations for public housing. To be more specific, it was a document that authorized architects and engineers to prescribe any type of slab regardless of the amount of concrete and steel, always under their responsibility and after proper design [4]. From that moment, the use of ceramic slabs spread and became generalized around Spain, especially for housing, since these solutions meant a significant lower amount of concrete and even steel. The use of ceramic slabs was later extended to office buildings and industrial hangars, in combination sometimes with steel or RC frames [5,6]. Despite the extensive implementation of ceramic slabs, the literature on their real structural performance and durability is scarce. Although it is difficult to analytically justify their compliance by using only the current calculation criteria, it is surprising to see how most structures based on this material type remain impeccable, even with very surprising load capacities. In this study we analyze the structural behavior of these slabs by means of a real loading test performed in an industrial building made of metal frames and ceramic slabs (a typical case of industrial heritage of that time in Spain). As mentioned below, the loading test was performed by repeating several loading and unloading cycles in order to describe the damage evolution under cyclic loading.

1.1. Ceramic Slabs as a Material Type

As previously mentioned, ceramic slabs constitute a specific type of precast slab where the geometry of the lighteners (usually made of ceramics) allows them to contribute to the structural response; this fact obviously minimizes the amount of steel and, especially, concrete when compared with other similar material types. Many different geometries of ceramic lighteners existed in the past in Spain, all them specifically designed for ceramic-based solutions with higher or lower responsibility of these elements in the global response, as shown in Figure 1.

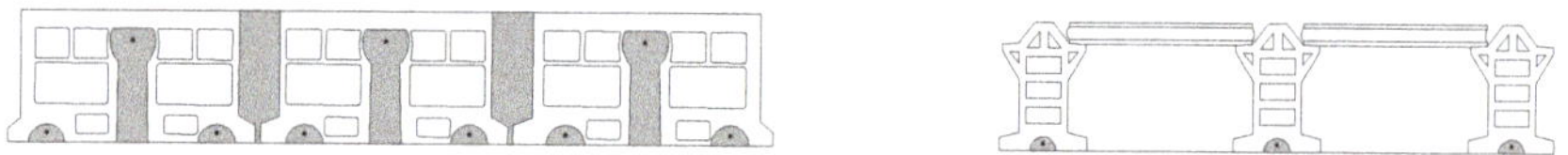

Figure 1. Different types of concrete solutions for typical Spanish "ceramic slabs".

Although these slabs were widely used during the 1950s and 1960s in Spain, they are in total disuse nowadays due to the lack of monolithic response and the evident limitations in terms of span. They are a very rigid solution with really low deformation ratios; thus, they do not work for covering mid- or long-span spaces. Besides this, this material type usually requires globally braced structures. The characteristic strengths which were usually used for the concrete in these slabs varied, starting from 17.5 MPa under compression. The quality of the steel was also variable, starting usually from 400 MPa according to the manufacturer. All the properties required by materials were specified in the document called "Normas para el Proyecto y Ejecución de Forjados de Ladrillo Armado" from 1941 [4], which could be translated as "Regulations oriented to the Project and the Execution of Reinforced Ceramic Slabs".

1.2. Determination of the Load-Bearing Capacity

In order to analytically obtain the load-bearing capacity of a ceramic-reinforced slab, the geometry and strength of components is obviously needed. Assuming that the analysis of these slabs is not easy, the load-bearing capacity may be approximated by using the concept of "homogenized section". This method allows for calculating the position of the neutral axis of the cross section by converting all components as if they were only one and thereby calculating the maximum moment under bending. Assuming that both ceramics and concrete only work under compression, the following distribution of forces can be established in the cross section by considering that deformation along the section always remains plain, as shown in Figure 2.

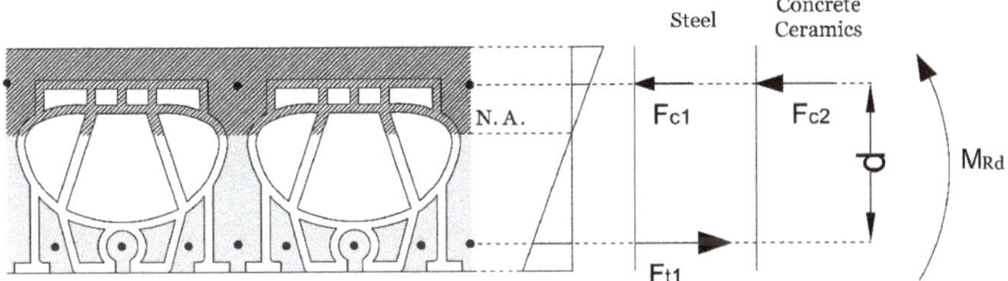

Figure 2. Scheme of forces in the equilibrium of the section subject to bending.

The position of the neutral axis (N.A., or axis corresponding to zero deformation, where components turn from being compressed to being tensioned) is determined by a simple balance of forces. F_{c1} and F_{c2} in Figure 2, corresponding to the capacity under compression of the upper steel bars and the capacity under compression of the upper layer made of concrete and ceramics, respectively, must be equal to F_{t1}, which is the capacity under tension of the lower steel bars only (since concrete is cracked in that area). Then, by knowing the position of the neutral axis (N.A. in Figure 2), it is possible to calculate the value of these forces and finally obtain the maximum resisted bending moment by multiplying one of these forces by the distance "*d*" between them (see Equation (1)).

$$M_{Rd} = F_{t1} \cdot d \tag{1}$$

Each of these forces may be obtained by multiplying the corresponding area of each component by its capacity under tension or compression, respectively.

In order to obtain the capacity under shear of these slabs, and assuming that they are not usually reinforced with transverse reinforcements, the shear capacity of concrete will be taken into consideration only according to the following expression:

$$V_{Rd} = \tau \cdot b_0 \cdot Z \tag{2}$$

where b_0 is the width of available concrete area in the cross section, adding all vertical ribs and subtracting the existing holes; Z is the distance between the compression and tension final resulting forces (as an approximation, it could be taken as 90% of the thickness of the slab), and τ is the shear strength of concrete, which is generally limited to 0.3 MPa.

It is worth pointing out that these slabs sometimes included inclined reinforcements along the nerves to absorb part of the shear stresses. Re-bars must be placed vertically or inclined in the cross section in order to be considered against shear. This allows an important increment of the shear capacity without need of extra material.

2. Description of the Analyzed Slab

The slab which was analyzed in this study is located in an industrial building built in the mid-1960s, with a steel-framed structure and reinforced ceramic slabs. Steel frames may be classified into two main categories, depending on their function: principal and secondary structure. The first one appears in upper floors, forming spans of 9.75 m × 7.0 m, while the second one only appears in basement floors, dividing the previous span in two; these are 4.88 m and covered by a reinforced ceramic slab as shown in Figure 3.

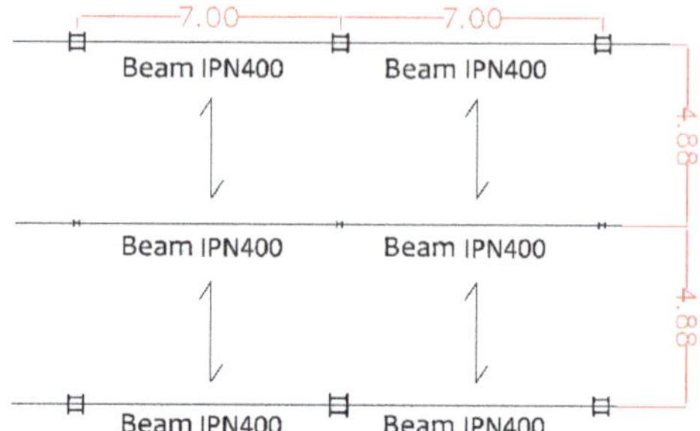

Figure 3. Dimensions of the steel-framed structure.

The principal columns are composite sections formed by two IPN400 (typical I-beam cross section of 400 mm height, widely used in Europe) linked through two 10 mm thick lateral plates, while secondary columns are formed by uncovered HEB160 profiles. Frames are completed with I-beams formed by IPN400 profiles and connected to the columns by means of screws. The total height of the basement floors is 3.20 m, including slab thickness, assuming that the finishing of these slabs is made of 3.0 cm terrazzo flooring over a base of 2.0 cm leveling mortar.

2.1. Description of the Slab

The slab which was analyzed here covers a span of 4.88 m, with a total thickness of 230 mm (30 mm of concrete upper layer, plus 200 mm of ceramic scaffolding). If we consider the width of the support on the metallic IPN400 profiles, the real span of the floor may be reduced to 4.72 m. According to the classification proposed by Seguí [4] in the manual "Recommendations for the Recognition, Diagnosis and the Therapy of Ceramics Slab", this particular slab would be classified into the AF3 category. This classification implies that the slab is reinforced with passive bars (A), that ceramics contribute to resist compression stresses, and that the concrete compression upper layer (F) was done completely in situ.

According to sample extractions and global analysis, the slab is reinforced by smoothed 5 mm bars of 414 MPa strength. Reinforcement was done at both sides of the slab, with upper and lower bars. Besides this, some of the main reinforcing bars were placed continuously, going from the upper face on support areas to the lower face in central areas, in order to optimize their performance. Overlap takes place in the lower face, once the bar is horizontal again, in order to coincide with positive reinforcement; as mentioned before, the inclined layout greatly improves the shear capacity of the section, as shown in Figure 4.

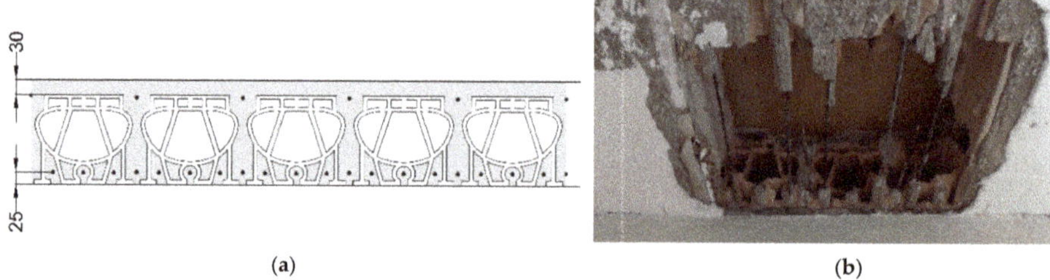

(a) (b)

Figure 4. (**a**) Typical section of the analyzed slab; (**b**) View of an inspection which was carried out on the analyzed slab to extract samples.

Steel bars that transfer stresses from the upper to the bottom face of the slab correspond to concrete ribs by simultaneously connecting the whole slab with the concrete upper layer, as shown in Figure 5.

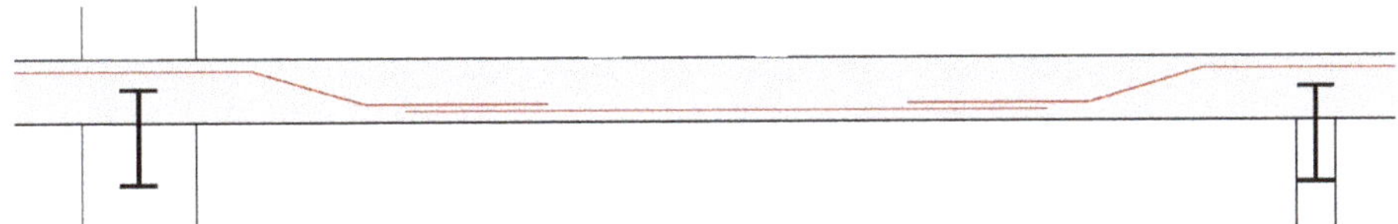

Figure 5. Reinforcements are inclined so as to maximize their performance against shear.

2.2. Properties of Re-Bars

One sample of an existing re-bar with a minimum length of 40 cm was extracted from the analyzed slab in order to carry out a pure tension test. The aim of this test was to obtain an estimated value of strength for the steel re-bars, a parameter which is necessary to carry out any other analysis or calibration. The existing bars were of 5 mm diameter and showed an elastic limit of about 410–450 MPa. Two cycles of pure tensile loading were applied to the sample using a HOYTOM machine, MEM 101/10M4 with serial number 2654 16A13034; the first cycle was up to 30% of the expected strength of the bar and the second one was up to failure in order to see the influence of loading and unloading on the steel.

The proportional limit was reached at 410 MPa, which implies that the elastic limit is similar or slightly beyond: a conservative value of 430 MPa was assumed for the analysis (see Figure 6 where results from the first and second cycles are shown). By comparing the curves between these first two cycles, a significant variation in the stiffness can be observed during the first stages of loading; this effect is mainly due to the lack of linearity of the extracted sample. The tested re-bar was not completely linear and regular from the beginning; it showed a permanent curvature derived from the demolition and extraction processes.

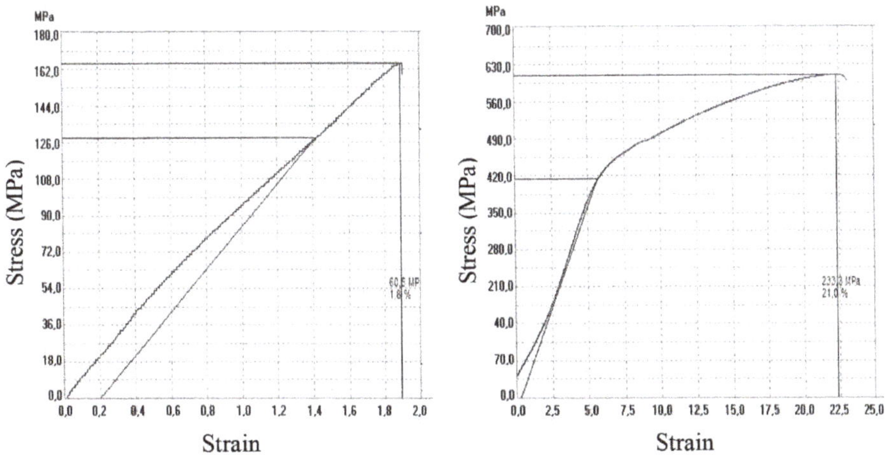

Figure 6. Stress–strain curves obtained from the tensile test on the extracted sample (first and second cycles).

2.3. Estimation of the Theoretical Load-Bearing Capacity

By considering a total of 4 ribs per meter according to the typical cross section in Figure 2, a minimum of 12 bars of 5 mm diameter per meter was considered in the lower section of the mid-span areas, while no bars were assumed within the upper part (only concrete); this corresponds to a typical distribution of reinforcements against positive bending moments (those appearing at mid-span areas). On the contrary, in supporting areas, there were 12 reinforcing bars in upper levels without any in the lower section. By using the approximation proposed by Seguí [4] in the ITEC manual to determine the load-bearing strength of ceramic-reinforced slabs, it is possible to estimate the ultimate load-bearing capacity.

In order to calculate the ultimate bending moment capacity, we firstly need the position of the neutral axis under positive (mid-span areas) and negative (supporting areas) bending moments. This may be obtained by starting with the total capacity of reinforcements:

$$F_{t1} = 12\ rebars{\cdot}19.62\ \text{mm}^2{\cdot}413\ \text{MPa} = 97.2\ \text{kN}. \tag{3}$$

The position of the neutral axis in the case of positive bending may be estimated by equating the total area of concrete which is needed in the upper layer with the maximum tensile capacity of reinforcing bars:

$$\begin{gathered} F_{c1} = F_{t1} \rightarrow h_{FN+}{\cdot}1000\ \text{mm}{\cdot}15\ \text{MPa} = 97.200\ \text{N}, \\ h_{FN+} = 6.48\text{mm}. \end{gathered} \tag{4}$$

Thus, the position of the neutral axis almost coincides with the upper layer of the slab, so the distance "d" is

$$d = 230\ \text{mm} - 6.48 - 25 = 198.52\ \text{mm} \tag{5}$$

and the final resisted positive bending moment is

$$M_{Rd+} = F_{t1}{\cdot}d = 97.2\ \text{kN}{\cdot}0.198\ \text{m} = 19.24\ \text{kNm}. \tag{6}$$

On the contrary, in the case of supporting areas, the tensile force appears in upper layers and it is the concrete of the ribs in the lower levels which resists compression stresses. These ribs sum to a total width of 60 cm (15 cm per ceramic piece, and four pieces per meter). Then, the neutral axis is

$$\begin{gathered} F_{c1} = F_{t1} \rightarrow h_{FN-}{\cdot}600\ \text{mm}{\cdot}15\ \text{MPa} = 97.200\ \text{N}, \\ h_{FN-} = 10.80\ \text{mm}. \end{gathered} \tag{7}$$

The distance "d" is 194.2 mm, and the ultimate resisted bending moment is slightly lower in this case:

$$M_{Rd-} = F_{t1}{\cdot}d = 97.2\ \text{kN}{\cdot}0.194\ \text{m} = 18.85\ \text{kNm}. \tag{8}$$

Regarding the load-bearing capacity, it is evident that the contribution of concrete is very low.

Thus, by assuming the span of the tested slab, and due to the hypothesis of continuity on both supporting edges thanks to the previous shoring of the steel structure, the maximum theoretical uniform load that could be resisted by this particular ceramic-reinforced slab (obviously considering only the bending moment) would be

$$M_{Rd-} = q{\cdot}\frac{L^2}{12} \rightarrow q = M_{Rd-}{\cdot}\frac{12}{L^2} = 18.85\ \text{kNm}{\cdot}\frac{12}{4.80^2} = 9.81\ \text{kN/m}^2. \tag{9}$$

The performance of a flat slab (especially with short span) is not only characterized by the bending strength; we obviously need to also consider the shear resistance. By again following the expressions proposed by Segui [4], the total capacity against shear of a ceramic-reinforced slab would be

$$V_{Rd} = 0.3\ \text{MPa}{\cdot}200\ \text{mm}{\cdot}194\ \text{mm} = 11.64\ \text{kN/mL}. \tag{10}$$

This value is relatively low compared with usual acting shear forces, especially compared with the bending capacity of the same slabs; this is the reason why there are bars inclined at 45° in the principal ribs of the section (see Figure 5), to enhance the shear response of the section. These inclined re-bars provide extra capacity against shear that may be quantified as

$$\Delta V = 4\ re - bars\ {\cdot}19.62\ \text{mm}^2{\cdot}413\ \text{MPa} = 32.41\text{kN}. \tag{11}$$

Finally, the maximum shear capacity of the slab by considering both contributions (concrete and steel) is about 44.05 kN/mL. Assuming a span of 4.88 m, as in case of the tested slab, the maximum uniformly distributed load that could be resisted according to shear would be

$$V_{Rd,\ total} = 44.05\frac{\text{kN}}{\text{mL}} = q * \frac{\text{L}}{2} \rightarrow q = V_{Rd,\ total} \cdot \frac{2}{\text{L}} = 44.05 * \frac{2}{4.88} = 18.35\ \text{kN}/\text{m}^2. \quad (12)$$

In conclusion, the theoretical failure loads (without safety factors) of the analyzed ceramic-reinforced slab, obtained by using the formulation proposed in the literature by Seguí [4], are shown in Table 1.

Table 1. Calculated failure loads of the slab (without safety factors). Adapted from [4].

Failure Mode	Maximum Theoretical Load
Positive bending (mid-span area)	19.62 kN/m^2
Negative bending (supporting areas)	9.81 kN/m^2
Shear	18.35 kN/m^2

Assuming the hypothesis of full continuity at both supporting edges, negative bending at the supports is the most restrictive mode of collapse, with a theoretical failure load of 9.81 kN/m^2.

3. Loading Test

3.1. Objective

The main objective of the test was to describe the real load-bearing capacity and behavior of the analyzed slab, as well as the cyclic loading–unloading response. Due to the industrial use of the building, the testing load was established as 7.5 kN/m^2 uniformly distributed over all the area, assuming that this load is already close to the predicted failure.

3.2. Methodology

After considering different options, the chosen method for the test was based on using 6.0 kN sacks, filled with sand, each with an area of 0.8 m × 0.8 m (this is 0.64 m^2). The sacks were placed one on top of the other, reaching a total load of 12 kN per pallet, thus achieving the desired load level distributed on the real area of the slab as shown in Figure 7.

Figure 7. General view of the loading test with sand sacks.

As previously mentioned, the main goal of the load-bearing test was to determinate the capacity of the ceramic-reinforced slab itself, independently of the capacity of the steel structure; this is the reason why the latter was completely shored up to avoid interference (see Figure 7). Otherwise, deflections of the steel beams would affect the results of the test, assuming that it was dealing strictly with the slab. This way, the load was only applied on the slab area by keeping the beams infinitely rigid; this makes the upper reinforcements start to work actively in terms of continuity of the negative bending moments. By restricting both steel beams against deformation, the tested slab may be considered continuous on the two supporting edges.

Considering the magnitude of the load, it was necessary to apply several safety measures in case of reaching collapse. On the one hand, a formwork panel was built a few centimeters below the slab itself in the second basement floor; in case of collapse, this formwork would avoid falling from a worrying height. On the other hand, a pair of reinforcement profiles were welded at half height between the pillars parallel to the slab in the first basement floor. This was done to guarantee that the pillars would not suffer excessive bending moments in case of collapse. These reinforcements guaranteed the integrity of the structure under any circumstance (see Figure 8).

Figure 8. General view of the shoring and formwork on basement level -2 for safety reasons.

3.3. Applied Load

An equivalent load of 7.50 kN/m^2 was applied by means of pallets full of sand sacks. These pallets were moved over the tested slab by using trans-pallets and following the shored area in order to place them uniformly. A total of 40 sacks with 6.0 kN of sand per sack were placed over an area of 33.04 m^2 of slab (a grid of 4 rows × 5 columns). A total load of 260 kN was applied on a limited area of ceramic-reinforced slab of 4.88 m span and supported by no more than five continuous negative re-bars per meter; that means 35 bars of 5 mm diameter per side. Since the pallets only occupy 0.64 m^2 in area, the load of two sacks (12.0 kN) turns into a uniform loading of 7.50 kN/m^2.

Taking into account that one of the objectives of the test was to determine the slab response under a cyclic process of loading and unloading, a set of seven cycles was carried out during a week (six days, one per day) in order to describe the structural performance; the loading and unloading processes were totally manual and were thus very slow and smooth. During the process of distributing the sacks, all possible health and safety measures were undertaken in order to minimize any possible risk for workers, since the testing load was similar to the ultimate one. Among other measures, the worker was fastened with a safety harness to the metallic structure of the upper floor (see Figure 9).

Figure 9. Process of loading and unloading (up to seven cycles in a week).

In order to monitor the load-bearing behavior of the slab, a real-time laser was installed in the lower level (basement level -2) together with a live webcam to observe cracks and movements from below. Real-time measurement in millimeters has a precision of three digits and has also been recorded to have the full sequence of deflections as shown in Figure 10. The laser was a BOSCH model 316514083004 with capacity up to 20 m and precision of ±3 mm, and the webcam had a resolution of 2560 × 1440 pixels.

Figure 10. Overview of the measurement system which was used.

4. Analysis of Results

4.1. Capacity Curve

The slab showed almost linear behavior in the first loading cycle, until a level of load of 5.00 kN/m^2 (see Figure 11). Although linear and elastic, significant remaining deformation (about 11 mm) was observed after the first cycle due to the relocation of the reinforcement bars. Note that the slab had not been subjected to these levels of load before, so the first time involved some plastic-like behaviors beyond approximately 5.00 kN/m^2.

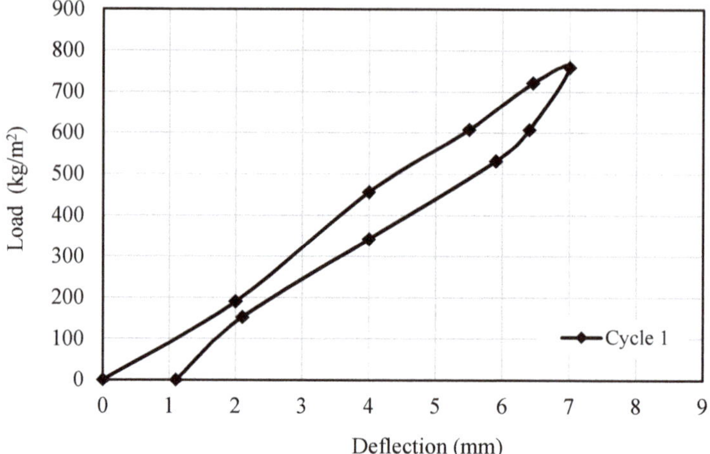

Figure 11. Obtained diagram of the loading and unloading process (Cycle 1).

Thus, the ultimate load which was previously calculated in Section 2.3 was resisted by the slab (9.80 kN/m^2, which results in a serviceability load of 6.5 kN/m^2 by assuming a safety factor of 1.50). Besides this, the measured deflection of the slab is small if we consider the height of the cross section and the global span (L/685).

The fact of having inclined reinforcement bars within the ribs makes the slab so rigid that the global response depends directly on the response of the steel bars. After a certain level of loading (5 kN/m^2) there is a change in the mechanical response; assuming that this is probably the first time that the slab was subjected to those levels of loading, beyond 5.00 kN/m^2, something changed to produce a plastic-like behavior, especially during the unloading process. This phenomenon could derive from the micro-crushing of ceramics and concrete by steel re-bars. Note that the unloading curve in Figure 11 shows changed stiffness at a certain level of load. This effect generated a remaining deflection of around 1 mm.

4.2. Cyclic Response

As mentioned before, six different cycles of loading–unloading were carried out on six consecutive days. The processes of loading, as well as unloading, were totally manual and smooth, placing the pallets slowly on the tested slab area.

From the curve obtained during the second cycle, we observed a significant change in the initial stiffness until 2.00 kN/m^2, as well as a smaller remaining deflection (almost nonexistent, less than 0.2 mm). This phenomenon may be explained by the relocation of the re-bars which occurred during the first cycle after reaching higher levels of load than usual. As has been mentioned, under a high level of loading, plastic phenomena take place. Micro-crushing of concrete and ceramics results in two different consequences: the first is the appearance of remaining strains after unloading; the second is typical plastic hardening due to geometrical relocation.

This hardening effect is clearly seen when reloading in the second and further cycles up to 2.00 kN/m^2, where the stiffness of the section is much higher than that of the original one. After a certain loading (2.00 kN/m^2), the slab behaves in exactly the same way as the original one until about 6.00 kN/m^2 of load, where micro-crushing tends to appear again. This hardening behavior informs of the proximity of collapse, since it is a clear signal of creep between reinforcements and concrete after micro-crushing (see Figure 12). However, the high rigidity of the material type (controlled by inclined steel re-bars) makes the solution really rigid for loading and unloading. This behavior is shear-based, with quite a short span and derived from having inclined bars.

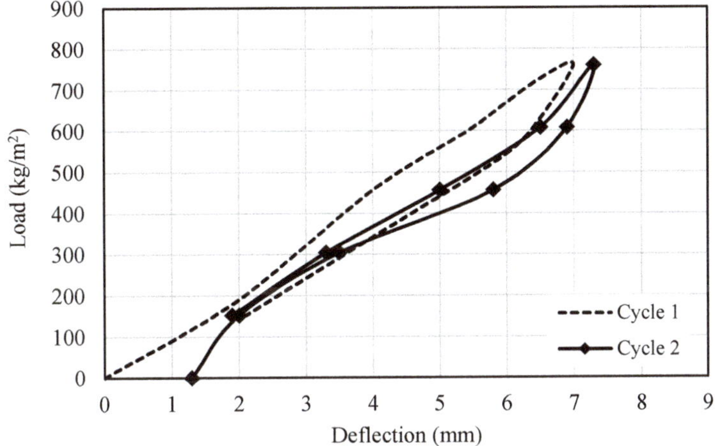

Figure 12. Overview of the loading and unloading process (Cycle 1 and Cycle 2).

From the curves which were obtained in the following cycles up to the sixth, no loss of ultimate loading capacity was observed. At each cycle, the load reaches a value of 7.50 kN/m^2, although each time there is a slight reduction of stiffness. After the second cycle, the pattern of loading and unloading is almost the same for all cycles, except for the first loading steps during the last two cycles, where plastic hardening plays a significant role (see Figure 13. This phenomenon goes hand in hand with a clear reduction of remaining strains; both effects are clear signals of a certain damage process within the slab when subjected repeatedly to high levels of loading and unloading in consecutive cycles.

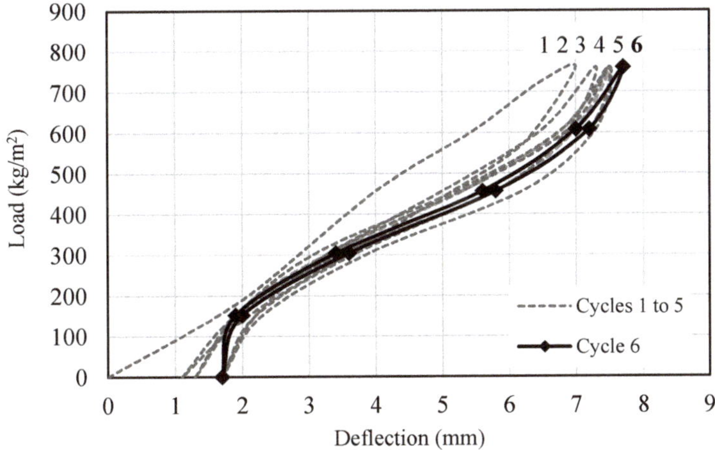

Figure 13. Overview of the loading and unloading process (several cycles).

All these results lead to the conclusion that this type of slab is very rigid due to its shear-based response and dependence on reinforcements. This rigidity does not decrease significantly when repeating cycles up to high rates of loading.

5. Vibrational Response

Vibrations which are induced by human walking or machines on slabs become seriously disturbing when the fundamental frequencies are below 7–8 Hz according to most Standards, including the

"Design of Floor Structures for Human Induced Vibrations" [7]. In slabs of relatively low self-weight compared with the load-bearing capacity, dynamic analysis is needed in order to avoid comfort problems during serviceability [8–13].

Estimation of the Elemental Frequency of the Slab Cross Section

In order to estimate the fundamental frequency of the structure, assuming full continuity of the slab and the beams at both supporting edges in the two cases, the following simplified expression according to [7] may be used:

$$f = 4/\pi \cdot \sqrt{(3 \cdot E \cdot I/(0.37 \cdot \mu \cdot L^4)} \tag{13}$$

where E is the estimated elastic modulus of the section, I is the moment of inertia of the cross section, μ is the total weight in a width of one meter of slab (including live load), and L is the span of the slab.

In the case of one-way slabs supported by steel beams, like in this case, the fundamental frequency of the whole structure (f_{ea}) may be calculated as the combination of both frequencies (the frequency f_1 of the slab and the frequency f_2 of the I-beam) by means of the Dunkerley [7] expression:

$$1/f_{ea} = 1/f_1^2 + 1/f_2^2 \tag{14}$$

We consider a moment of inertia of the equivalent section of 270,843,903 mm^4 (measured from section of Figure 4) and obtain an approximated elasticity modulus E of the whole slab by using the value of deflection obtained in the experimental test during the first cycle of loading:

$$\delta = 5 \cdot W \cdot \frac{L^4}{384 \cdot E \cdot I} \rightarrow E = 5 \cdot W \cdot L^4/(384 \cdot \delta \cdot I). \tag{15}$$

Then, by substituting the obtained value of deflection in the first cycle (7 mm) into the latter expression, generated by a uniform load of 7.50 kN/m^2, it is possible to estimate the secant elastic modulus:

$$E = 5 \cdot 350 \cdot \frac{4500^4}{384 \cdot 7 \cdot 270843903} = 97137 \text{ MPa}. \tag{16}$$

Regarding the steel I-beams, it is possible to calculate quite accurately the fundamental frequency by using the same expression [8]. The mentioned beams are IPN400 profiles (typical steel I-beam cross section of 400 mm height, widely used in Europe) with 7 m span, 104 mm^4 moment of inertia, and elastic modulus of 210,000 MPa, characteristic of steel.

Then, by considering all previous expressions and the real self-weight (slab + pavement = 2.2 + 1.0 = 3.2 kN/m^2), the fundamental frequencies of the whole structure were estimated for different loading levels.

It can be observed in Table 2 that the structure is not comfortable without live or service loads, with a fundamental frequency of almost 5 Hz, clearly below the recommended values; this may be directly demonstrated through simple walking on basement level -1, where vibration of the slab becomes notorious for the user. This is mainly due to the low self-weight ratio of the slab compared to its load-bearing capacity.

Table 2. Calculated fundamental frequencies.

Live Load	f_1 Slab	f_s I-Beam	f Total
0	8.41 Hz	0.45 Hz	4.84 Hz
5 kN/m^2	5.09 Hz	0.29 Hz	11.94 Hz
10 kN/m^2	4.02 Hz	0.23 Hz	19.05 Hz

Note that the higher the applied load, the higher the fundamental frequency of the structure, although in any case the frequencies are especially high. This is typical of low-weight structures with high load-bearing capacities. These are the reasons why this type of slab was widely used especially for industrial buildings, where heavy machines worked with high frequencies; slabs with high capabilities and low frequencies were needed.

6. Conclusions

Derived from the analysis and the real loading test in particular, the results show that the typical ceramic-reinforced slabs that were built during the 1960s and 1970s in Spain are rigid solutions with very high load-bearing capacities. The results from the test allow us to conclude that the analyzed slab, 230 mm thick and only 2.2 kN/m^2 in weight, resists up to 7.5 kN/m^2 extra live load.

Besides this, the structural response of the slab is mostly elastic up to advanced steps of loading, with all the advantages that this behavior implies. However, a slight plastic response was detected beyond approximately 5 kN/m^2, as this load level had never been reached before; the fact of going beyond the current load states implies a redistribution of the reinforcements within the section that results in plastic remaining strains (see the diagrams in Section 4). This effect is even enhanced when the load is cyclically repeated, although an evident hardening effect was also observed in the first early stages of loading. The hardening effect is directly derived from the relocation of re-bars, which takes place the first time and hardens during the following cycles.

The results from the loading test show that deflections on the analyzed slab were very low compared with those on other slab types. The maximum mid-span deflection of the analyzed slab was 7 mm under 7.50 kN/m^2, which corresponds to L/685. This is mainly due to the structural response of the solution, which is mainly based on shear behavior and steel reinforcements. Deflections of the slab are mostly those derived from steel re-bars, with little interference by concrete and its typical creep.

The slab response under vibration was better as the live load was increased. As a low self-weight solution with enhanced load-bearing capacity, problems derive more from serviceability than from ultimate strengths. This means that the analyzed slab, as a typical example of the ceramic-reinforced slabs which were used throughout Spain, works better when loaded in terms of dynamic response. When not loaded, the slab is uncomfortable for users due to the low fundamental frequency, which coincides with the vibration generated by humans walking. However, in the case of heavy machines with high frequencies, this is an ideal solution.

For all these reasons, the typical ceramic-reinforced slabs which were used during the 1960s, 1970s, and even later in Spain, especially for industrial buildings, are very powerful slab solutions from all points of view. It is curious to see how a specific solution for slabs which is currently outside of the Standards is so efficient from several points of view. Obviously, always limiting spans under 5.00 m, as in the analyzed case, means it works basically under shear. This study on ceramic-reinforced slabs verifies that these specific solutions with almost no concrete work efficiently for some specific cases, which is opposite to the current Standard's postulates.

Author Contributions: The four authors were involved in the conceptualization, methodology, formal analysis, investigation, and data curation of this study. Writing—original draft: A.R.-R.; A.A.-V. and I.C.-C.; Writing—review and editing: A.A.-V.; A.R.-R. and D.G.-C. All authors have read and agreed to the published version of the manuscript.

Funding: This research received no external funding.

Acknowledgments: We acknowledge the Serra Hunter Program for providing funds to carry out this research. Alicia Rivera-Rogel acknowledges the financial support received from the "Secretaría de Educación Superior, Ciencia, Tecnología e Innovación" (SENESCYT) for her doctoral studies at Polytechnic University of Catalonia (UPC).

Conflicts of Interest: The authors declare no conflict of interest.

References

1. Institut de Tecnologia de la Construcció – ITEC. *Recomendaciones para la Terapia de Forjados Unidireccionales de Viguetas Autoportantes de Hormigón ("Recommendations for the Therapy of Unidirectional Slabs of Self-Supporting Concrete Joists")*; Generalitat de Catalunya; Departamento de Política Territorial y Obras Públicas; Dirección General de Arquitectura y Urbanismo: Barcelona, Spain, 1992; Volume 1, pp. 1–132. (In Spanish)
2. Trias de Bes, J.; Casariego, P. The art of the technology: Construction of structures by domed ceramic with industrial systems. *Informes de la Construcción* **2016**, *68*, 169. [CrossRef]
3. Gemma, M.S. Las últimas construcciones de fábrica de ladrillo resistente: La generación de los años cincuenta a los setenta ("The last resistant brick factory constructions: The generation from the fifties to the seventies"). In *Actas del Séptimo Congreso Nacional de Historia de la Construcción, Santiago de Compostela*; Instituto Juan de Herrera: Santiago de Compostela, La Coruña, Spain, 26–29 October 2011. (In Spanish)
4. Víctor, S. Recomanacions per al reconeixement, la diagnosi i la teràpia de sostres ceramics. In *Recommendations for the Recognition, Diagnosis and the Therapy of Ceramics Slab*; Indian Technical and Economic Cooperation: Barcelona, Spain, 1995.
5. Laureà, M. Forjados cerámicos: Estudio de un colapso ("Ceramic slabs: Study of a collapse"). *Associació de Consultors d'Estructures* **2011**, *42*, 28–35. (In Spanish)
6. Ramón, B.J. Los forjados de cerámica armada antes y después de la primera instrucción de cálculo a rotura ("Reinforced ceramic floor slabs before and after the first breakage calculation instruction"). *Associació de Consultors d'Estructures* **2012**, *43*, 5–21.
7. Feldmann, M.; Heinemeyer, C.; Butz, C.; Caetano, E.; Cunha, A.; Galanti, A.; Goldack, O.; Hechler, O.; Hicks, S.; Keil, A.; et al. Design of Floor Structures for Human Induced Vibrations. In *JRC Scientific and Technical Reports*; Joint report, prepared under the JRC-ECCS cooperation agreement for the evolution of Eurocode 3 (programme of CEN TC 250); Publications Office of the European Union: Luxemburg, 2009; pp. 1–75.
8. Sedlacek, G.; Heinemeyer, C.; Butz, C.; VEILING, B.; Waarts, P.; Van Duin, F.; Hicks, S.; Devine, P.; Demarco, T. Generalisation of criteria for floor vibrations for industrial, office, residential and public building and gymnastic halls. In *European Commission—Technical Steel Research*; Report EUR 21972 EN; Office for Official Publications of the European Communities: Brussels, Belgium, 2006; pp. 1–343.
9. Wendell, V.; Ronaldo, B. Control of vibrations induced by people walking on large span composite floor decks. *Eng. Struct.* **2011**, *33*, 2485–2494.
10. Nguyen Huu, A.T. Walking Induced Floor Vibration Design and Control. Ph.D. Thesis, Swinburne University of Technology, Melbourne, VIC, Australia, 2013.
11. da Silva, J.G.; de Andrade, S.A.; Lopes, E.D. Parametric modelling of the dynamic behaviour of a steel–concrete composite floor. *Eng. Struct.* **2014**, *75*, 327–339. [CrossRef]
12. Mohammed, A.S.; Pavic, A.; Racic, V. Improved model for human induced vibrations of high-frequency floors. *Eng. Struct.* **2018**, *168*, 950–966. [CrossRef]
13. Machado, W.G.; da Silva, A.R.; das Neves, F.D.A. Dynamic analysis of composite beam and floors with deformable connection using plate, bar and interface elements. *Eng. Struct.* **2019**, *184*, 247–256. [CrossRef]

Article

FlexFlax Stool: Validation of Moldless Fabrication of Complex Spatial Forms of Natural Fiber-Reinforced Polymer (NFRP) Structures through an Integrative Approach of Tailored Fiber Placement and Coreless Filament Winding Techniques

Vanessa Costalonga Martins [1], Sacha Cutajar [1], Christo van der Hoven [1], Piotr Baszyński [2,*] and Hanaa Dahy [2,3]

1 ITECH Masters Program, University of Stuttgart, Keplerstr. 11, 70174 Stuttgart, Germany; costalonga.v@gmail.com (V.C.M.); cutajar.sacha@gmail.com (S.C.); christovanderhoven@gmail.com (C.v.d.H.)

2 BioMat Department: Bio-Based Materials and Materials Cycles in Architecture, Institute of Building Structures and Structural Design, University of Stuttgart, Keplerstr. 11, 70174 Stuttgart, Germany; hanaa.dahy@itke.uni-stuttgart.de

3 Faculty of Engineering, Department of Architecture (FEDA), Ain Shams University, Cairo 11517, Egypt

* Correspondence: piotr.baszynski@itke.uni-stuttgart.de

Received: 1 April 2020; Accepted: 29 April 2020; Published: 8 May 2020

Abstract: It has become clear over the last decade that the building industry must rapidly change to meet globally pressing requirements. The strong links between climate change and the environmental impact of architecture mean an urgent necessity for alternative design solutions. In order to propose them in this project, two emergent fabrication techniques were deployed with natural fiber-reinforced polymers (NFRPs), namely tailored fiber placement (TFP) and coreless filament winding (CFW). The approach is explored through the design and prototyping of a stool, as an analogue of the functional and structural performance requirements of an architectural system. TFP and CFW technologies are leveraged for their abilities of strategic material placement to create high-performance differentiated structure and geometry. Flax fibers, in this case, provide a renewable alternative for high-performance yarns, such as carbon, glass, or basalt. The novel contribution of this project is exploring the use of a TFP preform as an embedded fabrication frame for CFW. This eliminates the complex, expensive, and rigid molds that are traditionally associated with composites. Through a bottom-up iterative method, material and structure are explored in an integrative design process. This culminates in a lightweight FlexFlax Stool design (ca. 1 kg), which can carry approximately 80 times its weight, articulated in a new material-based design tectonic.

Keywords: natural fiber-reinforced polymers; NFRP; computational design; tailored fiber placement; coreless filament winding; rapid prototyping; industry 4.0; lightweight structure

1. Introduction

The environmental impact of commercial construction and fabrication methods has been well documented. The built environment, for instance, accounts for 36% of global energy use, with cement alone contributing to 8% of annual global carbon dioxide production [1–3].

The creation of complex forms, be it a wall or a chair, can become an especially resource-intensive exercise. This expense is not only in terms of the embodied energy of the artefact itself but also the process required to produce it. This often occurs when materials are deployed in a simple, homogenous, top-down manner. This is not a new problem, nor is it trivial, and there are a number of attitudes for using materials more responsibly. Some strategies come from incremental technological development, whilst others adopt techniques, technologies, and ideas from other industries, opening the door to new possibilities [4]. It is appropriate, then, to investigate how these alternative approaches may translate into more harmonious architectural design solutions.

Fiber-reinforced composites (FRCs) have been well used in the automotive, nautical, and aerospace industries over the last decades, but less so in architecture and product design. In contrast to some other processes, working with composites can allow for the strategic placement of material to create differentiated structure and geometry. This is typically done in an additive process and can result in products with a high specific strength. A superior strength-to-weight ratio can, in turn, translate into reduced material use, as well as reduced use of supporting materials, such as concrete or metals [5].

Two particularly novel composite fabrication methods for FRCs are coreless filament winding (CFW) and tailored fiber placement (TFP). Both processes have great potential at the architectural scale. These techniques deploy numerically-controlled machines in digital design workflows for precision, efficiency, and mass-customization [6,7]. However, both CFW and TFP still used complicated and resource-intensive fabrication frames or molds during production (Figure 1).

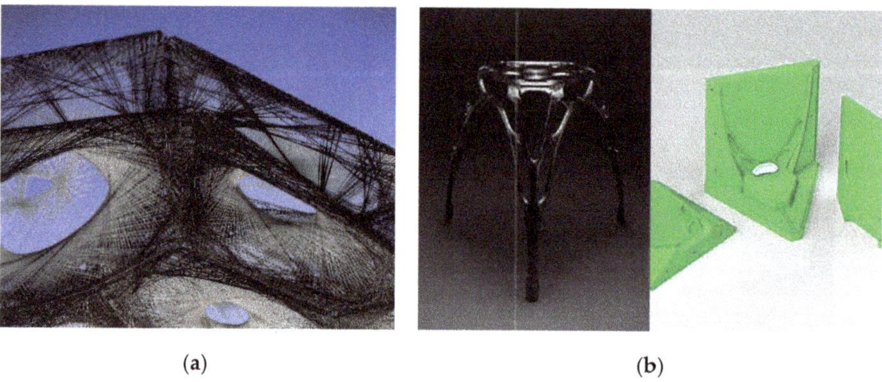

(a) (b)

Figure 1. Fabrication frames and molds: (**a**) coreless filament winding (CFW); the Institute for Computational Design/the Institute of Building Structures and Structural Design (ICD/ITKE) Research Pavilion 2013–14 and its module edges shaped by a complex reconfigurable steel filament winding frame [9]; (**b**) tailored fiber placement (TFP); the L1 Lightweight Stool alongside its elaborate rigid mold [8].

For almost 10 years, the Institute for Computational Design (ICD), the Institute of Building Structures and Structural Design (ITKE), and, more recently, the BioMat group at ITKE (the Department of Biobased Materials and Materials Cycles in Architecture) at the University of Stuttgart, have investigated composites through the lenses of computation, digital design and fabrication, structure, and sustainability. Their research has not been limited to conventional FRC based on carbon fibers (CFs) and glass fibers (GFs), but also explored possibilities of replacing them with fibers from annually renewable resources, such as flax and hemp, as well as with agricultural residues, namely straw, for the purpose of developing structural systems based on natural fiber-reinforced polymers (NFRPs) [8]. Through a multifarious series of temporary pavilions, demonstrators, seminars, and research projects, the potential of these materials at an architectural scale have been investigated [5,6]. This project dives deeper into this research context, building on the pioneering work of these institutes. The corresponding precedent projects shown below should additionally be understood through the broader goal of this project and the design and prototyping of a stool.

In their project "Tailoring Self-Formations", Aldinger and Margariti explored carbon fiber bending elements sewn onto a prestressed membrane substrate (Figure 2a). The self-formation process was driven by the release of the prestress in the membrane, which in turn acted as an actuator. The actuation sent the CFP elements into bending, determined by specific predefined 2D geometries, thereby predictably creating complex double-curved surfaces without the need for a mold. This project illustrates the possibility of creating complex forms in mold-less TFP fabrication, as well as fiber-reinforced composites (FRCs) in bent states of compound Gaussian curvature [10].

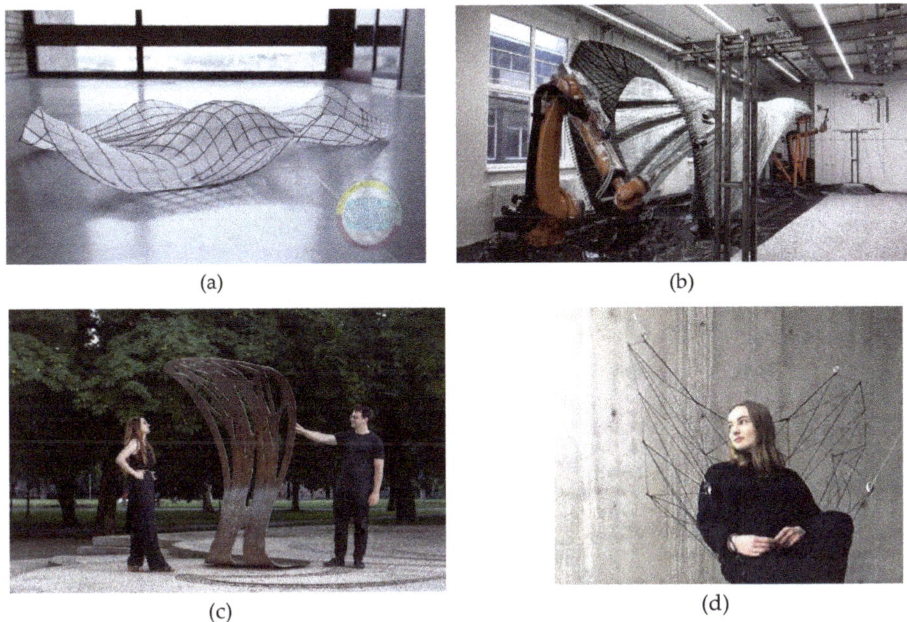

Figure 2. State-of-the-art: (**a**) "Tailoring Self-Formations" [10]; (**b**) "ICD/ITKE Research Pavilion 2016–17" [11]; (**c**) "Tailored Biocomposite Mock-up of BioMat 2019" [12]; (**d**) "Aerochair" [13].

The ICD/ITKE Research Pavilion 2016–17 is a good precedent of a biomimetic transfer for an embedded bending active filament winding frame (Figure 2b). In this case, the biological role model was the leaf miner moth (*Lyonetia clerkella*), which bends a leaf into shape with its silk and then uses the resulting geometry as the environment to create a cocoon. The abstracted fabrication process starts with a flat CFW preform, which is bent into shape, becoming the winding scaffold for the main layers of structural filaments of the demonstrator. In this case, the ratio of the winding frame to the CFW component was minimal. Additionally, a substantial portion of the fabrication setup remained embedded within the structure, taking on an additional function [11].

In 2019, BioMat successfully attempted to realize a small scale monolithic NFRP canopy using TFP preforms for precise control of fiber orientation in the structure (Figure 2c). Rhinoceros 3D plugins (Grasshopper, Galapagos and Millipede) used in the initial form-finding process of this lightweight canopy allowed us to define the most deformation-resistant geometry. Further topological optimization of the structure was conducted using Matlab and an agent-based system tool created using processing. This allowed us to calculate the optimum flax fiber paths orientation on the whole structure surface to be realized at a later TFP fabrication stage. Secondly, the structure was tessellated into several overlapping layers consisting of multiple preforms no larger than 1.0 x 1.4 m. This size limitation was imposed by the working area of the Tajima TFP embroidery machine, offered by the Institute of Aircraft Design (IFB) at the University of Stuttgart. Finally, all preforms were placed on a mold, custom-made using CNC-milled elements, for a vacuum-assisted resin transfer molding (VARTM) process [12]. A tailored biocomposite mock-up of BioMat 2019 proved the applicability of TFP fabrication in structural projects requiring topological optimization. However, at the same time, it exposed the necessity of developing alternative, more resource-efficient solutions to complex 3D molds used in the VARTM process.

Finally, Duque Estrada and Wyller investigated design methodologies for CFW of carbon fiber for ultra-lightweight furniture design (Figure 2d). Their "Aerochair" used the high tensile strength of carbon rovings as the primary design driver. "By allowing the chair to hang, instead of stand, a new design expression that properly conveys the material characteristics presents itself" [13]. This project successfully demonstrates the use of continuous resin-impregnated carbon fiber yarns, with the furniture piece weighing only 300 g. However, the fabrication process still requires a large and complex frame. Furthermore, the novel hanging typology may not be a fully representative portrayal of a load-bearing architectural system, which must typically accommodate compression and bending in addition to tension. Consequently, there still remains significant terrain for further investigation.

Therefore, the aim of this project was to create a stool with coreless filament winding and tailored fiber placement of resin-impregnated, continuous natural fiber yarn without the use of a complicated mold or frame. The design and physical production of a small furniture piece was used as a means to engage the research context beyond just the theoretical level. A small-scale demonstrator, in other words, facilitates practical access to the architectural function (ergonomics, scale etc.) and structural performance (strength, weight etc.), as well as the realities of fabrication, implementation, and the myriad of small design challenges that emerged during the prototyping process. The investigation was undertaken through a bespoke design-to-production system, comprising three conceptual stages; "Stitch", "Bend", and "Weave" (Figure 3).

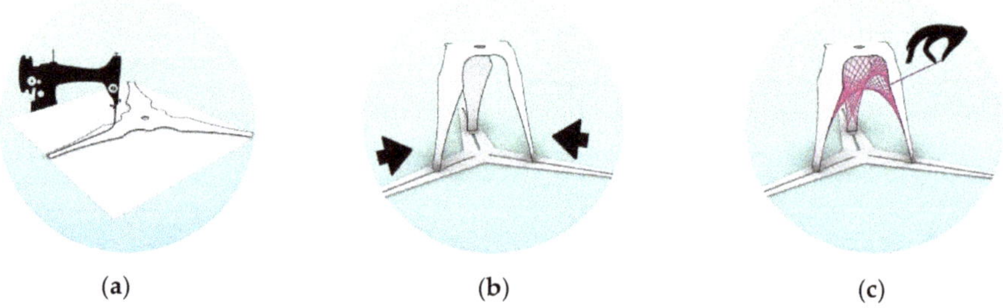

(a) (b) (c)

Figure 3. Concept: (**a**) Stitch: A tailored natural fiber textile is designed and produced in its flat form. This textile is infused with resin to create a fiber composite; (**b**) Bend: The activated polymer is formed into shape, enabled by specific fiber bending patterns; (**c**) Weave: This form becomes a permanent winding frame, upon which natural fibers are placed through coreless filament winding. Once cured, the TFP and CFW elements become a co-dependent functional and structural system in the form of a stool.

In this experimental approach, the selection of sustainable structural materials and fabrication methods took place before the step of defining the actual architectural design of a piece. In this case, applying the principles of "Materials as a Design Tool" philosophy [14] allowed for unrestricted shaping of the spatial form of the prototype, with particular focus on efficient use of fibrous composite material in an additive fabrication process, allowing for a reduction in the amount of production residues and eliminating the necessity of using disposable moulds made in subtractive fabrication processes such as CNC-milling.

2. Materials and Methods

2.1. Flax Fibre and Epoxy Resin

Research on natural fibers is a field of growing interest within the realm of the composites industry, primarily due to their production from annually renewable resources in low energy-intensive processes and their inherent potential for biodegradability and recycling at end of life; aspects which their synthetic counterparts cannot yet address as effectively at the time of writing [15–18]. This project builds upon the current trend as a means of establishing a viable structural application. As a result, non-twisted flax fiber rovings in two linear densities, 1200 Tex and 2400 Tex, have been selected for all production methods at the prototyping stage and sourced from the distributor, Group Depestele (Teillage Vandecandelaere 5, rue de l'église, 14 540 Bourguebus, France).

Although equivalent research may be found with respect to biodegradable polymer matrices, material explorations were limited to only the fiber component of the composite. Fully synthetic Epoxy Resins (3 parts EPIKOTE Resin MGS RIMR 235 + 1 part EPIKURE Curing Agent RIMH 237, Pot Life: 48 h, all produced by Hexion Inc. and provided by Hexion Stuttgart GmbH, Fritz-Müller-Straße 114, 73730 Esslingen am Neckar, Germany) were therefore selected as the base matrix, due to prior experience with the products.

2.2. Deductive Methodology

The investigations detailed in this paper build upon the previously outlined case studies with respect to production techniques of fibrous composites and product design. The project consequently utilizes a deductive research method to assess whether the addition of TFP with CFW techniques can unlock further avenues for fibrous composites via the moldless design and production of a stool. This hypothesis is further broken down into sub-investigations, pertaining to each production method respectively.

2.3. Design Process

The physical design was developed in an iterative bottom-up process that overlaps material, fabrication, and global design investigations. Initial explorations began with rapid modelling of scaled paper models to establish global geometry. This was subsequently coupled with full scale material tests, the principles of which were extracted from pre-existing literature, such as the lamination theory [19,20]. These tests investigated localized performance, such as bending or stiffness, which were then extrapolated to the overall global geometry of the stool (Figure 4). These elements were then synthesized into a consolidated design prototype and tested for performance and function. The findings from each iteration were then interrogated, explored, and developed in subsequent versions of the design. Collectively, this formed an integrative design feedback loop.

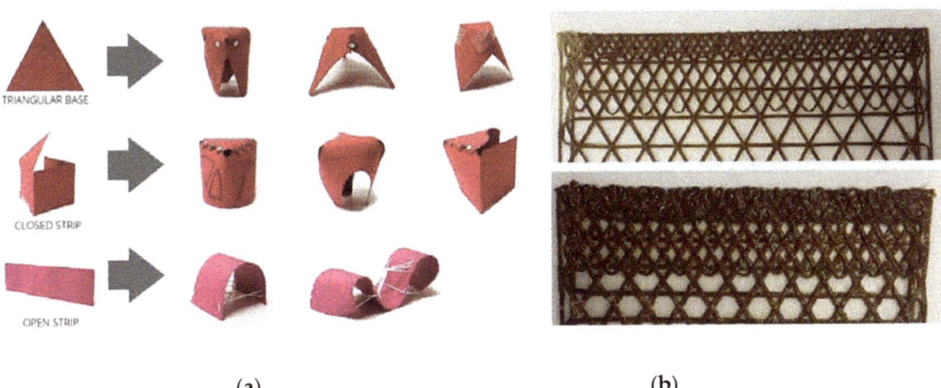

(a) (b)

Figure 4. Design Process: (**a**) global design paper models; (**b**) material test samples with vacuum-infused resin. Top to bottom: 1200 Tex Fibre and 2400 Tex non-twisted roving.

The main structural criteria that needed to be met by the design was to sustain the weight of a person inducing a load of minimum 800 N. Final testing of the prototype was conducted by the authors.

2.4. Production Process

The production process was divided into two main parts; the manufacture of a bendable scaffold, produced flat and bent into shape using a custom rig (previously conceptualized as "Stitch" and "Bend") and the reinforcement of the scaffold via fiber winding to sustain the design load (termed "Weave"). Both parts were characterized by the use of the TFP and CFW processes, respectively, and detailed further below.

2.5. Tailored Fiber Placement

The TFP process makes use of an industrial grade embroidery machine that continuously lays filament material upon a thin, stretched textile mounted on a 2D movable frame [19]. In this project, dry, non-impregnated flax roving was laid using a 4-head embroidery machine of the brand Tajima, provided through the courtesy of Institute of Aircraft Design (IFB) at the University of Stuttgart. Filament material was also prepared by being spun in small quantities over a spool attached to the head of the machine. As fibers were laid down onto the textile, a fixed sewing needle and bobbin secured the fibers in place via a second threaded spool. The automated textile frame was numerically controlled and digitally programmed to follow any given path that was fed as a continuous polyline. In this project, a file with fiber path polylines was generated using CAD software Rhinoceros 3D. Moreover, primary fabrication parameters of the TFP process included the "stitch length" and "stitch width" of the thread, which collectively influenced the resolution of the final laid pattern. Exact values for these parameters were adjusted during prototyping and tabulated accordingly in the results section. Topological optimization in solidworks was used to inform the distribution of fibers along the designed preform.

The nature of this machine allowed an infinite set of customizable patterns to be made on a 2D surface (Figure 5a). In this context, a first sub-hypothesis is formulated as to whether tailored orientation of fibers in a cured state can allow for a controlled ability to bend a surface and achieve specific geometries.

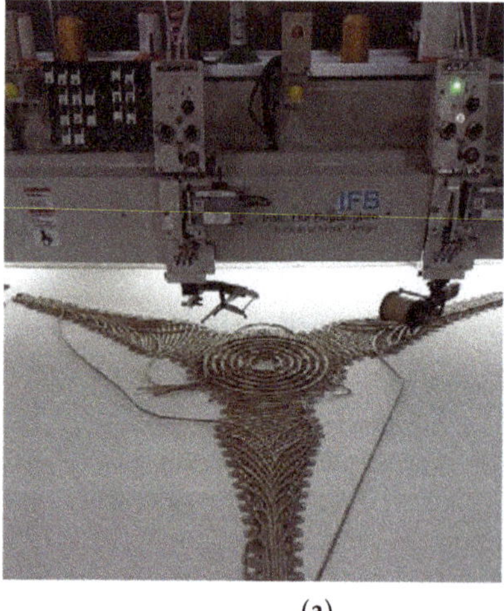

(a)

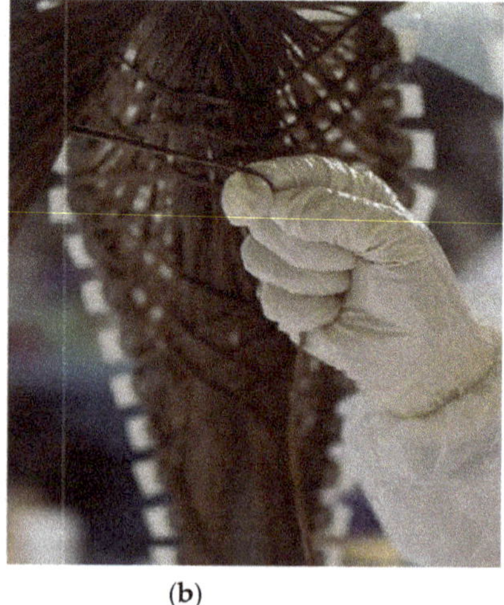

(b)

Figure 5. Production process, (a) TFP Machine Tajima, provided by the Institute of Aircraft Design (IFB), University of Stuttgart; (b) manual CFW technique.

2.6. Coreless Filament Winding

CFW entails winding a continuous resin-impregnated filament material over a scaffold, typically via robotic automation. Due to the anisotropic nature of fibrous composites, an understanding of force flow in components becomes crucial to the fiber layup design. This pattern, termed "syntax", refers specifically to the sequence in which the continuous fiber is wound. Consequently, it can allow for a localized differentiation in material count, resistance to buckling failure, and varied geometry [9,21]. Since loading conditions for a small scale product design provide a simplified testing environment, stress flows and syntax design for the stool were set up empirically. The small scale also eliminated the need for robotic automation. Instead, the winding procedure for prototyping was carried out manually (Figure 5b).

This CFW technique relies heavily on keeping fibers taut upon the scaffold for maximum fiber–fiber interaction and performance [21,22], consequently placing additional pressures on the scaffold's stiffness. A second sub-hypothesis here questions whether such a technique can be used to permanently reinforce the 2D bent geometries generated via TFP methods.

2.7. Fibre Impregnation Methods

The mechanical properties of fiber composites are heavily influenced by the matrix-to-reinforcement ratio and the corresponding interface [23]. Due to a need for repeatable prototyping and controlled bending stiffness of the TFP surface, a reliable fiber impregnation method became necessary. A vacuum infusion process was consequently adopted to minimize behavioral discrepancies stemming from production. In this process, dry TFP preforms were manually infused with resin and subsequently sealed in an airtight bag. A negative pressure was applied by means of a vacuum pump, causing the resin to disperse evenly throughout the TFP preform and drain any excess material away from the sample. Upon full infusion, the sample was removed from the bag and left to cure.

For CFW, the fibers were continuously passed through a custom-built resin bath in order to absorb sufficient resin throughout the winding process, as shown in Figure 6 Upon completion, full curing took place and the prototype was demounted from the clamping rig.

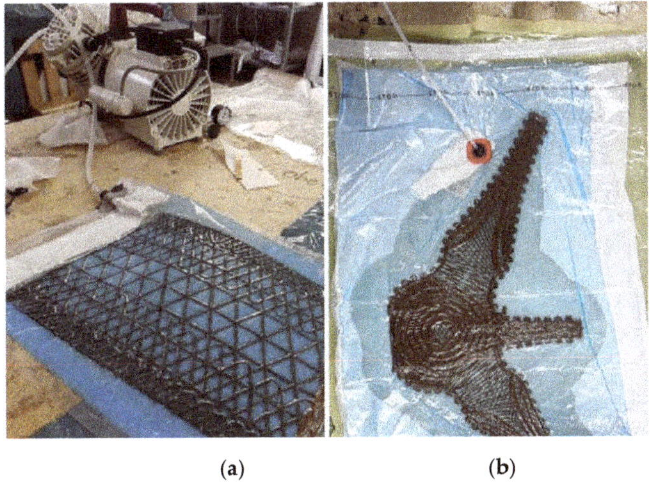

(a) (b)

Figure 6. Fiber impregnation methods: (**a**) Resin infusion of test sample using vacuum pump and airtight sealing bag. Cotton layers and blue plastic film were used to soak excess resin and facilitate infusion; (**b**) infusion of prototype in process.

2.8. Analysis Criteria

As a means to benchmark the prototyped outcomes, functional and performative criteria were restricted to three design conditions:

1. Ability to sustain the 800 N design load;
2. Minimalization of total weight of stool and corresponding material usage;
3. Qualitative comfort enabled by the geometry;

These conditions, coupled with design development, were tested, verified, and presented in further detail in the following sections as a means of validating the initial hypothesis.

3. Results

Performing the production process at the Institute of Aircraft Design (IFB) at the University of Stuttgart provided opportunities for interdisciplinary cross-pollination, where valuable insights were gained from the engineers. The TFP machine was used to produce a series of material test preforms and a first prototype to engage with the complexities of the machine and the specific data protocols required for production. After that, a final iteration was produced (Figure 7), which is the focus of this Results section. The final iteration was informed by these previous tests and addressed specific challenges of topological optimization, the TFP process, filament winding, and performance. Although somewhat independent, the integrated nature of the design and fabrication method meant that these elements were interrelated.

(a)

(b)

Figure 7. Final prototype: (**a**) view of the TFP Pattern; (**b**) view of the internal CFW syntax.

3.1. Topological Optimization

To better inform the placement of fibers in the TFP preform, the topological optimization platform in solidworks was used.

A simplified topology study tool was used to provide a rough visual estimation as to how the stresses distribute along the surface. The results allowed for a diagrammatic visualization of where the bulk of the material would be required for a given load configuration. Different amounts of material reduction were investigated for the same load case and edge condition. A selection of these studies can be seen in Figure 8.

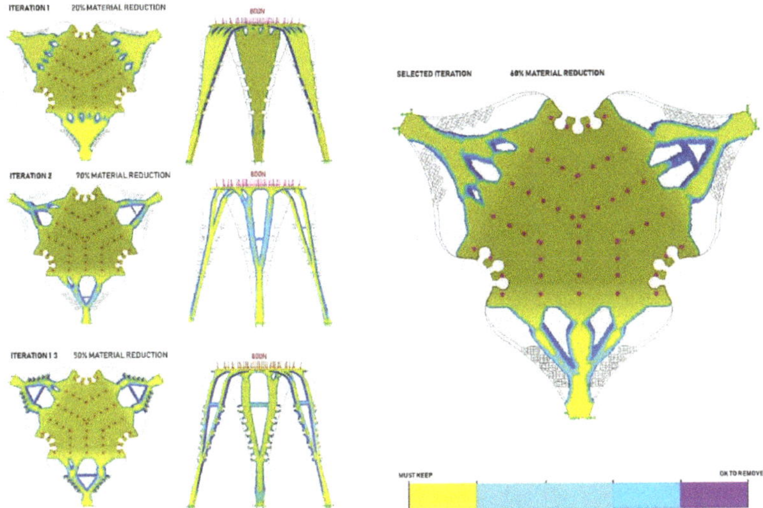

Figure 8. Topological optimization iterations.

For the topology study, an 8 mm thick single curved surface, modelled in Rhino3D, was prepared and imported accordingly. The specified surface thickness was imposed by the fabrication limitations of the Tajima embroidery machine used at the fabrication stage. The filament winding syntax was omitted from the simulation. As the exact mechanical properties of the used combination of flax roving and resin were difficult to pinpoint and were not available at the specific moment of conducting the fabrication experiment, the simulation was run using the built-in properties of glass fiber, providing that the desired stiffness criteria could be reached when lower mechanical strength of flax fiber in comparison to glass fiber was compensated by using sufficiently higher volume fractions [24].

An evenly distributed load of 800 N was applied to its horizontal seating surface. Winding pins were defined as a fabrication constraint, which forced the software to preserve the geometry during optimization.

The simulation was run using the best mass-to-stiffness goal criteria, for which various percentages, ranging from 30% to 70%, were studied. No distinct outcome was favored. Instead, the simulations, all of which delivered similar results (such as in the associated figure), were used comparatively to identify the crucial areas transferring higher forces and thus requiring more material to be deployed.

Regions in yellow indicate places where material was needed most and those in blue indicate where material was needed the least. Throughout the iterations, it was possible to notice the emergence of a Y-shape distribution of material on the legs, which suggested that these areas had the biggest fiber cross-sections, i.e., structural layers. It was also possible to conclude that the seat area needed to remain dense. Other areas of the initial input surface were excluded, generating a new edge condition for the final design. Finally, this simulation did not provide the definitive visual pattern of the surface, but rather guided the TFP surface design.

3.2. TFP Process

3.2.1. TFP Pattern

In order to determine the fiber path pattern in the final TFP preform, with particular focus on enabling the flexible behavior of the bending zones, the anisotropic nature of the fibrous composite material required combining the results of the above-mentioned simplified optimization process with an interpretation of classical lamination theory (CLT) [17,18].

The finished TFP preform, after resin infusion process and complete curing, was supposed to be formed into a single curved surface and serve as the integrated frame in a later winding process. This required planning a diversified fiber path layout, which would reinforce and stiffen the seating area and leg endings, while guaranteeing a necessary flexibility of the composite material in the areas where the TFP preform was supposed to be bend (Figure 9).

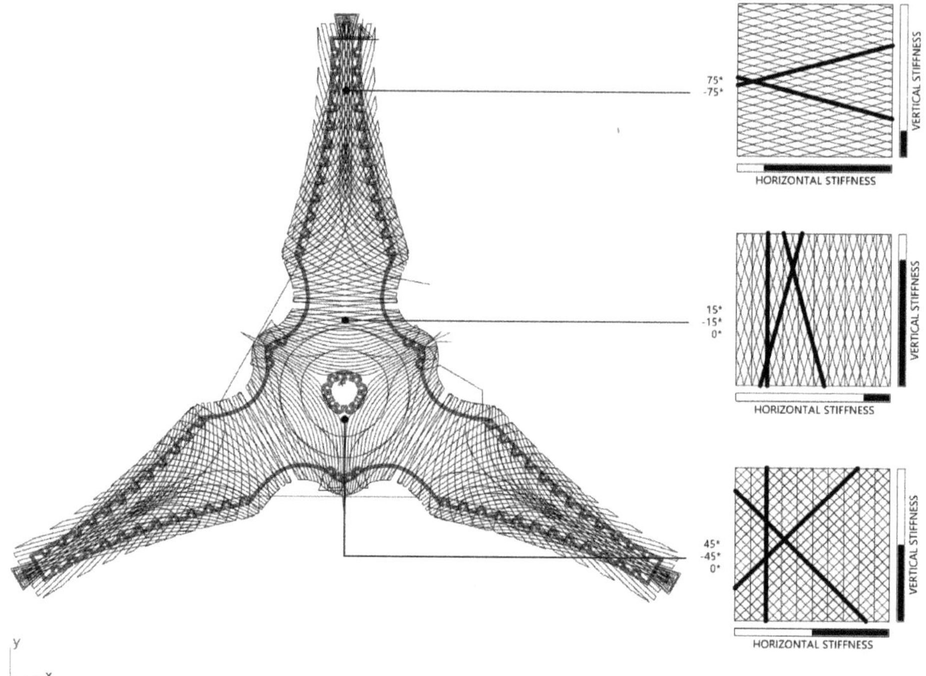

Figure 9. Final TFP pattern and interpretation of CLT.

Consequently, fiber paths in the seating area, oriented at −45°, 0°, and 45°, aim at creating a quasi-isotropic layup sequence, resulting in constant stiffness of the material regardless of the force direction, yet offering minimal elasticity of the surface for the purpose of seating comfort. Starting from the leg tips, in the leg areas, fiber paths were oriented, at −75° and 75°, thus almost parallel to the leg axis, in order to use material properties to the maximum when trying to develop a stiff structure that would not deflect under load. However, the closer to the hinge area, the more the fiber path angles smoothly translated to more perpendicular orientation, reaching −15° and 15° in the center of the hinge areas.

Secondly, this core fiber path topology, designed according to structural performance of the stool model, was complemented with additional paths, which constituted further reinforcement and design details needed at a later stage of fiber winding.

The complete design of the fiber paths in the final TFP preform structure had to be adjusted to the fabrication requirements of the machine. Thus, according to the role each fiber path fragment played in both in the fabrication process, as well as later in the structural performance of the stool, they were grouped into three layers, which were fabricated consecutively one on top of the other, using 1200 TEX and 2400 TEX flax fibers by the embroidery machine.

The final TFP pattern consisted of three layers:

1. Winding Pins Layer (Figure 10a).

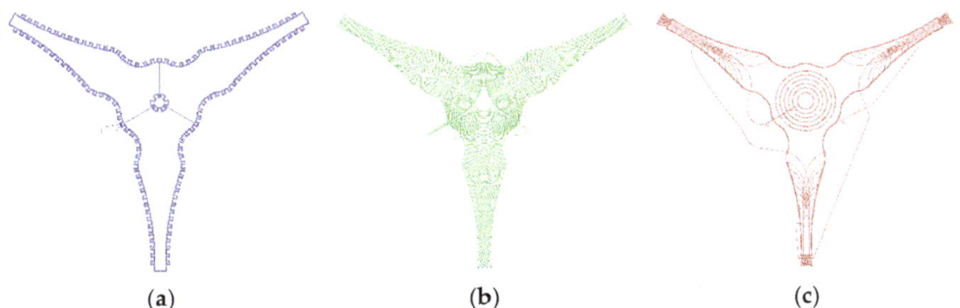

Figure 10. TFP layers: (**a**) Winding pins layer; (**b**) surface layer; (**c**) structural layer.

This layer was of special importance, since it provided the interface for the CFW process. The challenge at this step was determining the appropriate size of the pins. In order to guarantee visual integration of the pins in the final design of the FlexFlax, their dimensions should remain as modest as possible, but, at the same time, should be above the minimum fabrication size requirements of the TFP machine. The chosen geometry was a trapezoid of 2 cm (long base), 1 cm (short base), and 2 cm (height). To provide a delicate cross-section, a TFP machine preset of 1.5mm stitch width was used for 1200 Tex flax fiber rovings. To guarantee the resolution of the winding pins, a 1 mm stitch length was set. Consequently, flax roving was sewn to the substrate with a higher number of stitches. However, this compromise resulted in a significantly longer production time for this stage of the TFP process.

2. Surface Layer (Figure 10b).

The surface layer, containing the fiber layout for the bending zone, was formed from 1200 Tex Flax roving with a stitch width of 1.5 mm and a stitch length of 7 mm.

3. Structural Layer (Figure 10c).

The structural layer was fabricated using 2400 Tex roving, with a cross-sectional dimension of 3.6 mm (stitch width). In this final layer, a boundary curve was embedded in order to reinforce the thinner rovings during the fabrication process. The final TFP pattern was sewn into a cotton substrate of 1000 × 1400 mm. The used TFP machine settings, as well as the resulting fiber lengths for each layer, are presented in Table 1.

Table 1. Settings and results for the TFP process.

TFP	TEX	Stitch Count (stitches)	Stitch Width (mm)	Stitch Length (mm)	Fiber Length (m)
Layer 01	1200	31078	1.5	1	31
Layer 02	1200	8725	1.5	7	54
Layer 03	2400	6838	3.6	7	38
Total	-	46641	-	-	123

3.2.2. Vacuum Infusion

After the accomplished TFP fabrication of the preforms, the excess of the substrate fabric was trimmed, and resin was applied through the vacuum infusion process explained in Section 2.7. The used resin system was EPIKOTE Resin MGS RIMR 235 with EPIKURE Curing Agent RIMH 237, produced by

Hexion Inc. Standard polythene vacuum bags of 130 × 90 mm, commonly available in the market, were used in conjunction with a vacuum pump (model P3 produced by R&G Faserverbundwerkstoffe: max. 55 l/min. at 0.900 bar vacuum). One sheet of cotton textile was placed underneath the sample and one layer of perforated release film (model P1 produced R&G Faserverbundwerkstoffe) was placed on top. The dimensions of the sheets were the same size as the bag. Their presence ensured an even distribution of air pressure, as well as absorption of the excess of resin from the undesired areas. After the infusion process, the preforms underwent a curing process at room temperature for 48 h.

3.3. Coreless Filament Winding Process

3.3.1. Fabrication Clamp

A simple clamping rig (Figure 11) was required to stabilize the bending active TFP preform for winding. For the purposes of this project, timber was milled in a shape that corresponds to the tripod stool morphology. Each foot was held by an adjustable timber socket. These sockets can rotate vertically and slide along defined rails. This allows the bending active element to naturally find a stable geometry, as well as enable the rig to be reused and reconfigured to fabricate stools of various shapes, sizes, and heights.

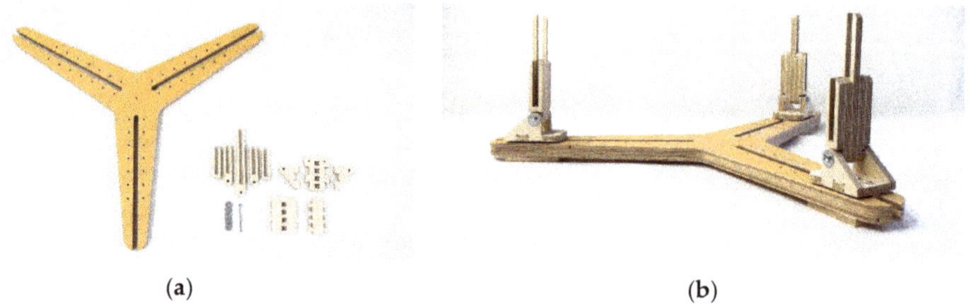

(a) (b)

Figure 11. Fabrication clamp: (**a**) Pieces prior to assembly; (**b**) assembled clamp.

3.3.2. Syntax Design

The CFW syntax used on the prototype consisted of three layers that served different strategies:

1. A first Spine Syntax transferred forces from the center of the seat to each leg, locking the bending zone in place (Figure 12a);
2. A second Bracing Syntax braced each leg individually to ensure fiber–fiber interaction on the previous syntax (Figure 12b);
3. A third and final Locking Syntax connected the legs to one another, to provide stability to the stool when it is being sat on, and the feet try to thrust apart from each other. Additionally, this layup provides geometric depth to the structure in an evocative anticlastic shape; a geometry that is extremely difficult and costly to create using traditional fabrication methods (Figure 12c).

The complete syntax layup was initially developed on 1:5 scale models and then wound using dry, non-impregnated fibers on the full-scale prototype. Once the sequences were satisfactorily defined, the prototype was wound using a resin bath impregnation system, described previously in Section 2.7. The winding process was done manually, after which the prototype was cured for 48 h at room temperature.

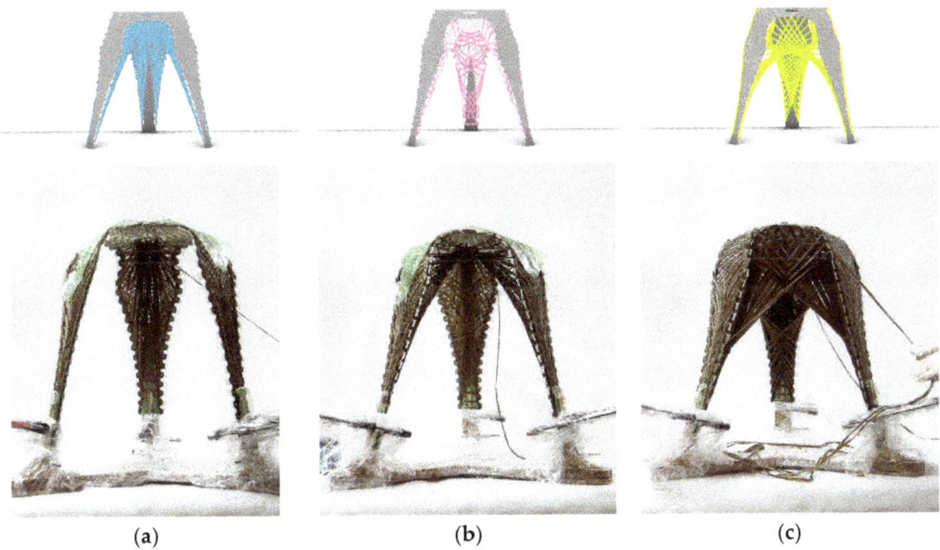

Figure 12. Syntax layers: (a) Spine layer; (b) bracing layer; (c) locking layer.

3.4. Performance

Once curing was completed, the performance of the FlexFlax could be assessed. As mentioned previously, this consisted of the analysis of the material usage and the ability of the stool to withstand loads of minimum 800 kN, i.e., an 80 kg person.

3.4.1. Material Count

The final weight of this prototype was 1080 grams. Tables 2 and 3 present, respectively, the amount of fiber and resin used. It is important to note that it was not possible to assess exact losses of the material in the fabrication process, especially in regards to the resin.

Table 2. Fiber usage of final FlexFlex prototype.

Process	Fiber Length (m)
TFP	123
Filament Winding	190
Total	313

Table 3. Resin usage of final prototype.

Type of Process	Resin Type	Ratio	Resin Usage (g)
Vacuum Infusion	EPIKOTE 235 and EPIKURE 237	3:1	780
Filament Winding	EPIKOTE 235 and EPIKURE 237	3:1	950
Total	-	-	1690

The final dimensions of the stool base were approximately 40 × 40 cm, with a height of 45 cm. The three legs formed the shape of an equilateral triangle, with a side length of approximately 42 cm (±0.5 cm).

3.4.2. Seating Test

The structural performance of the stool was done by conducting a seating test with participation of persons of different weight. Despite the light weight of the stool, volunteers whose weight ranged from 55 kg to 85 kg could successfully sit on the stool (Figure 13). Meanwhile, only slight elastic deflection of the stool feet for the highest load was observed. At the load of 85 kg, deflection at the stool feet ends measured 0.7 cm (initial distance between leg feet: 41.7 cm; during seating: 42.5 cm) (Figure 14). These tests also allowed for estimation of qualitative comfort of the stool. The stool demonstrated the same stiffness independently from the direction from which it was sat on. Both the number of legs and the span of approximately 42 cm (± 0.5 cm) between them was sufficient in providing stability of the stool during the test. As planned, the seating surface demonstrated limited elastic behavior contributing to the overall seating comfort of the user. Simultaneously, the elastic behavior was restricted to the seating area only and did not affect the performance of the legs, which remained stiff.

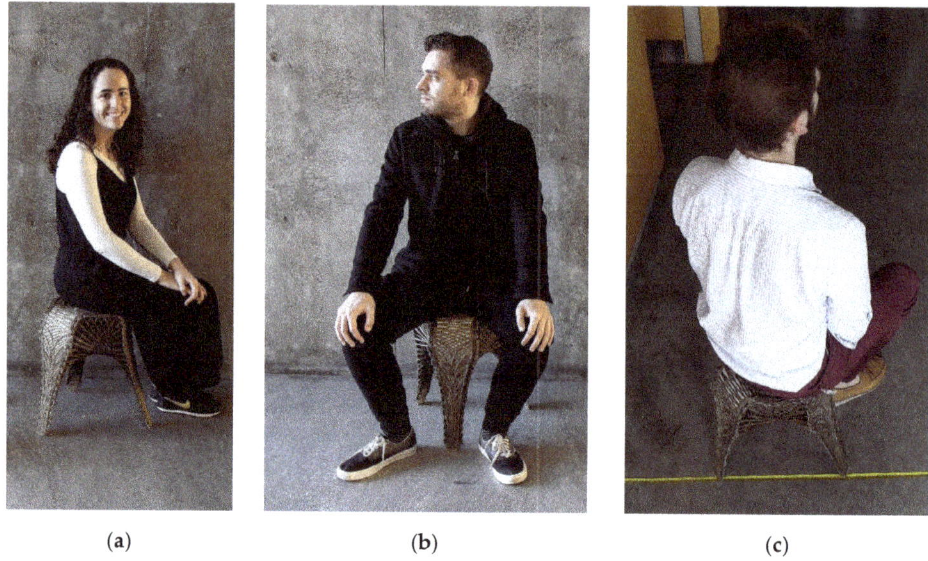

(**a**) (**b**) (**c**)

Figure 13. Preview of the seating test: (**a**) 55 kg user; (**b**) 70 kg user; (**c**) 85 kg user.

Such low deflection of the stool indicates its capabilities of withstanding even greater loads. However, the validity of this hypothesis can only be proven by submitting the stool to destructive tests conducted with loads greater than 800 N. Currently, the presented stool is the only demonstrator of the FlexFlax fabrication process realized by us and a destructive compression test has not been performed on it yet, but rather an experimental validation of the whole hypothesis through physical application of human weight, which ranged between 5300 N and 800 N, as discussed. A destructive test will be conducted in near future work once a greater number of prototypes, of the same dimensions and with the same quantity of composite material, is fabricated in order to obtain comparable results, which will be published in a separate publication.

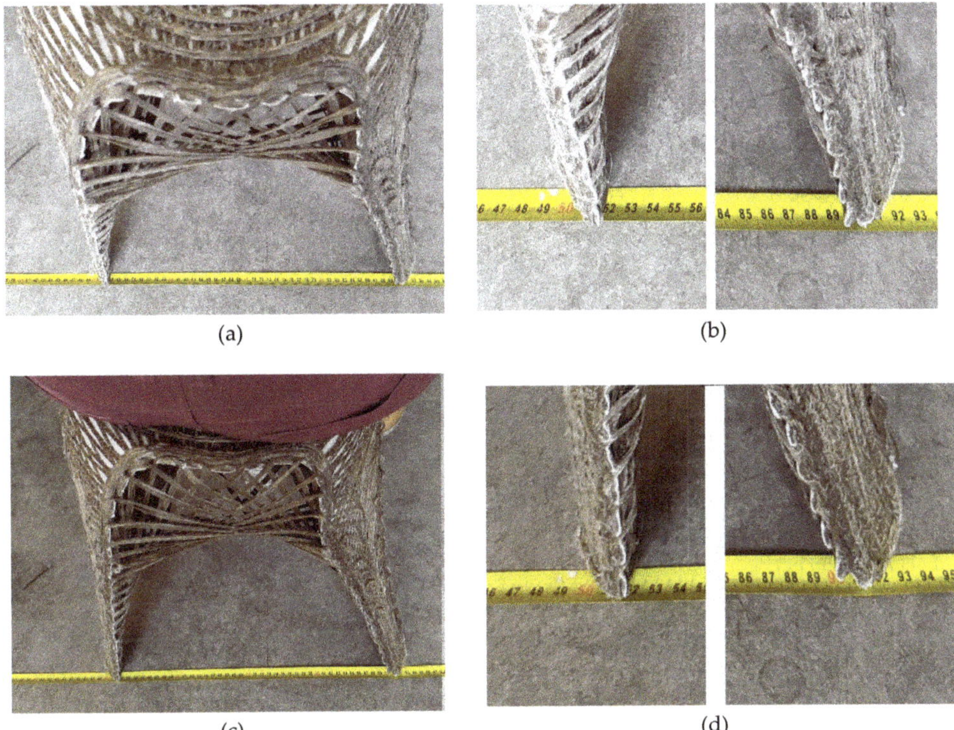

Figure 14. Preview of the seating test with 85 kg user: (a) and (b) before sitting: 41.7 cm span; (c) and (d) during sitting: 42.5 cm span.

4. Discussion

The aim of the project was to explore the potential of combining TFP and CFW techniques for the purpose of realization of spatial structures without the necessity of using complex molds or frames. The successful implementation of these techniques is demonstrated in a high-performance design object, which can support approximately 80 times its own weight. This object was realized by a novel digital design and fabrication workflow, which are expressed by a unique material tectonic.

The TFP preform, embroidered on a cotton substrate, predictably bends in the desired zones with similar bending radii for each leg. The seat and legs of the realized stool offered sufficient structural stiffness and aesthetic richness informed by the fiber pattern design. The CFW syntax in the interior of the stool offered a secondary layer of geometrical and formal complexity to complement the exterior. Collectively, the fiber patterns and ordinary tripod geometrical typology familiarized the unusual material expression. The FlexFlax visually recalls woven structures, such as ratan, but in a new material-informed design expression (Figure 15).

Although a simplified topological optimization strategy used in this experiment was sufficient for estimating NFRP material quantities and its distribution in the structure, with the aim of reaching desired structural strength, it is clear that further research using this approach and further advanced approaches are needed to investigate and deliver exact data about the weight-to-strength ratio, which can be reached using the FlexFlax fabrication approach. At this step, the prototype still remains a demonstrator of *Materials as a Design Tool* philosophy [14], rather than providing exact data about performance of NFRP structures. Undoubtedly, since the FlexFlax workflow was established in the presented paper, the following steps should focus on including more detailed structural simulation, considering exact fiber paths and optimization of the fiber infusion and winding processes. This would allow for exact control over the fiber-to-resin ratio and, consequently, mechanical properties of the generated NFRP. The above-mentioned improvement would allow for realizing several comparable prototypes and estimating their maximum strength through a compression test.

Figure 15. Final FlexFlax stool prototype: Detail of the underside.

Nonetheless, the high performance comes at the cost of a multi-method fabrication system that may become labor-intensive. Automation potentials are currently under study to overcome those potential drawbacks. There are several aspects that could still be refined, particularly with regards to the CFW step of the process. The use of prepreg rovings at this step was not explored during this project, but it is predicted that this method could decrease fabrication complexity and reduce time consumed during the winding process. The curing time could also be further reduced, both by choosing an alternative resin system characterized by a shorter pot life and by curing the components in an autoclave.

Considering all the above-mentioned conditions, it is justified to speculate about possibilities of application of this system at an architectural scale, in both discrete and continuous natural fiber application. Regarding the discretized design, a controlled prefabrication environment could help with material fabrication constraints. The light weight, in conjunction with a small-scale building component, makes it suitable for modular sustainable building constructions. Finally, as a continuous system, this process could enable partial prefabrication with on-site continuous winding realized by mobile robots [25]. Such an approach would allow for realization of larger building elements with fewer fragile connections between them. These are the avenues of research that will likely produce rich new architectural modes of production for future sustainable building systems.

Author Contributions: V.C.M., S.C., and C.v.d.H. contributed equally to this paper. Concept, design, methods, representation, prototyping, and testing of the project, as well as writing of the paper, was done by the trio together. H.D. (BioMat Department director) and P.B. (BioMat research associate) gave the students all prerequisite know-how in the field of biocomposites and sustainable architectural building elements, edited the paper, and supervised the educational course Material Matter Lab (Material and Structure, winter semester 2019/2020), in which this prototype was produced. All authors have read and agreed to the published version of the manuscript.

Funding: The project was partially supported by the Deutsche Forschungsgemeinschaft (DFG, German Research Foundation) under Germany's Excellence Strategy EXC 2120/1-390831618.) The project was also partially supported by the Agency for Renewable Resources (FNR) under the Federal Ministry of Food and Agriculture (BMEL) throughout the research project LeichtPro: Pultruded load-bearing lightweight profiles from natural fiber composites (FKZ:22027018), managed by H.D.

Acknowledgments: The project was developed in the seminar Material Matter Lab (Material and Structure, winter semester 2019/2020), offered by BioMat (The Department of Biobased Materials and Materials Cycles in Architecture) ITKE at the University of Stuttgart under supervision of Hanaa Dahy and tutoring of Piotr Baszyński and Sophie Luz. We would also like to show our gratitude to Peter Middendorf, Stefan Carosella, and Benjamin Grisin at the IFB (Institute of Aircraft Design) at the University of Stuttgart for the opportunity to access their machines and for the intensive technical support and guidance throughout the process. The flax rovings used in the project were provided at the courtesy of Group Depestele, Teillage Vandecandelaere 5, rue de l'église, 14 540 Bourguebus, France. The resin system used in the project was provided at the courtesy of Hexion Stuttgart GmbH, Fritz-Müller-Straße 114, 73730 Esslingen am Neckar, Germany

Conflicts of Interest: The authors declare no conflict of interest.

References

1. Pomponi, F.; Moncaster, A. Embodied carbon mitigation and reduction in the built environment—What does the evidence say? *J. Environ. Manag.* **2016**, *181*, 687–700. [CrossRef] [PubMed]
2. Abergel, T.; Dean, B.; Dulac, J. *Towards a Zero-Emission, Efficient, and Resilient Buildings and Construction Sector: Global Status Report 2017*; UN Environment International Energy Agency: Paris, France, 2017.
3. Lehne, J.; Preston, F. *Making Concrete Change: Innovation in Low-Carbon Cement and Concrete*; Chatham House Rep. Energy Environment Resour. Dep.: London, UK, 2018; pp. 1–66.
4. Menges, A. Computational Material Culture. *Archit. Des.* **2016**, *86*, 76–83. [CrossRef]
5. Dahy, H. Natural Fibre-Reinforced Polymer Composites (NFRP) Fabricated from Lignocellulosic Fibres for Future Sustainable Architectural Applications, Case Studies: Segmented-Shell Construction, Acoustic Panels, and Furniture. *Sensors* **2019**, *19*, 738. [CrossRef] [PubMed]
6. Menges, A.; Knippers, J. Fibrous Tectonics. *Archit. Des.* **2015**, *85*, 40–47. [CrossRef]
7. Richter, E.; Spickenheuer, A.; Bittrich, L.; Uhlig, K.; Heinrich, G. Applications of variable-axial fibre designs in lightweight fibre reinforced polymers. In Proceedings of the Materials Science Forum; Trans Tech Publications: Bäch, Switzerland, 2015; Volume 825, pp. 757–762.
8. Dahy, H. Biocomposite materials based on annual natural fibres and biopolymers—Design, fabrication and customized applications in architecture. *Constr. Build. Mater.* **2017**, *147*, 212–220. [CrossRef]
9. Prado, M.; Dorstelmann, M.; Schwinn, T.; Menges, A.; Knippers, J. Core-Less Filament Winding: Robotically Fabricated Fiber Composite Building Components. 2014. Available online: https://www.researchgate.net/publication/272784909_Core-Less_Filament_Winding (accessed on 4 May 2020).
10. Aldinger, L.; Margariti, G.; Körner, A.; Suzuki, S.; Knippers, J. Tailoring Self-Formation fabrication and simulation of membrane-actuated stiffness gradient composites. 2018. Available online: https://www.researchgate.net/publication/326552014_Tailoring_Self-Formation_fabrication_and_simulation_of_membrane-actuated_stiffness_gradient_composites (accessed on 4 May 2020).
11. Felbrich, B.; Frueh, N.; Prado, M.; Saffarian, S.; Solly, J.; Vasey, L.; Knippers, J.; Menges, A. Multi-Machine Fabrication: An Integrative Design Process Utilising an Autonomous UAV and Industrial Robots for the Fabrication of Long Span Composite Structures. 2017. Available online: https://www.researchgate.net/publication/321026463_Multi-Machine_Fabrication_An_Integrative_Design_Process_Utilising_an_Autonomous_UAV_and_Industrial_Robots_for_the_Fabrication_of_Long_Span_Composite_Structures (accessed on 4 May 2020).
12. Dahy, H.; Petrs, J.; Baszynski, P. Design and Fabrication of two 1:1 Architectural Demonstrators based on Biocomposites from Annually Renewable Resources displaying a Future Vision for Sustainable Architecture. In *FABRICATE 2020*; UCL: London, UK, 2020; ISBN 978-1-78735-812-6.
13. Duque Estrada, R.; Wyller, M.; Dahy, H. Aerochair Integrative design methodologies for lightweight carbon fiber furniture design. In Proceedings of the 37th eCAADe and 23rd SIGraDi Conference—Architecture in the Age of the 4th Industrial Revolution, Porto, Portugal, 11–13 September 2019; Volume 1, pp. 691–700.
14. Dahy, H. 'Materials as a Design Tool' Design Philosophy Applied in Three Innovative Research Pavilions Out of Sustainable Building Materials with Controlled End-Of-Life Scenarios. *Buildings* **2019**, *9*, 64. [CrossRef]

15. Van der Werf, H.M.G.; Turunen, L. The environmental impacts of the production of hemp and flax textile yarn. *Ind. Crops Prod.* **2008**, *27*, 1–10. [CrossRef]
16. González-García, S.; Hospido, A.; Feijoo, G.; Moreira, M.T. Life cycle assessment of raw materials for non-wood pulp mills: Hemp and flax. *Resour. Conserv. Recycl.* **2010**, *54*, 923–930. [CrossRef]
17. Stevens, C. *Industrial Applications of Natural Fibres: Structure, Properties and Technical Applications*; John Wiley & Sons: Hoboken, NJ, USA, 2010; Volume 10.
18. Pickering, K.L.; Efendy, M.G.A.; Le, T.M. A review of recent developments in natural fibre composites and their mechanical performance. *Compos. Part A Appl. Sci. Manuf.* **2016**, *83*, 98–112. [CrossRef]
19. Herakovich, C.T. Composite Materials: Lamination Theory. In *Wiley Encyclopedia of Composites*; American Cancer Society: Atlanta, GA, USA, 2012; pp. 1–5. ISBN 978-1-118-09729-8.
20. Fan, M.; Fu, F. *Advanced High Strength Natural Fibre Composites in Construction*; Woodhead Publishing: Sawston, UK, 2016.
21. Aschenbrenner, L.; Temmen, H.; Degenhardt, R. Tailored Fibre Placement Technology—Optimisation and Computation of CFRP Structures. 2007. Available online: https://www.researchgate.net/publication/224985701_Tailored_Fibre_Placement_optimization_tool (accessed on 4 May 2020).
22. Knippers, J.; La Magna, R.; Menges, A.; Reichert, S.; Schwinn, T.; Waimer, F. ICD/ITKE Research Pavilion 2012: Coreless Filament Winding Based on the Morphological Principles of an Arthropod Exoskeleton. *Archit. Des.* **2015**, *85*, 48–53. [CrossRef]
23. Rana, S.; Fangueiro, R. *Fibrous and Textile Materials for Composite Applications*; Springer: Singapore, 2016; ISBN 978-981-10-0232-8.
24. Duflou, J.R.; Yelin, D.; Van Acker, K.; Dewulf, W. Comparative Impact Assessment for Flax Fibre versus Conventional Glass Fibre Reinforced Composites: Are Bio-Based Reinforcement Materials the Way to Go? *CIRP Ann.* **2014**, *63*, 45–48. [CrossRef]
25. Yablonina, M.; Prado, M.; Baharlou, E.; Schwinn, T.; Menges, A. Mobile Robotic Fabrication System for Filament Structures. In *Fabricate 2017*; Menges, A., Sheil, B., Glynn, R., Skavara, M., Eds.; UCL Press: London, UK, 2017; pp. 202–209. ISBN 978-1-78735-000-7.

Article

Metamorphosis of the Architectural Space of Goetheanum

Romana Kiuntsli [1], Andriy Stepanyuk [2], Iryna Besaha [3] and Justyna Sobczak-Piąstka [4,*]

1 Department of Design of Architectural Environment, Lviv National Agrarian University, Zhovkva District, 80381 Dubliany, Lviv Oblast, Ukraine; romana.lviv@ukr.net
2 Department of Architecture and Rural Settlements' Planning, Lviv National Agrarian University, Zhovkva District, 80381 Dubliany, Lviv Oblast, Ukraine; andriy.lviv@ukr.net
3 Faculty of Training for Police Prevention Subdivisions, Lviv State University of Internal Affairs, 79007 Lviv, Lviv Oblast, Ukraina; animespirit12@ukr.net
4 Faculty of Civil and Environmental Engineering and Architecture, UTP University of Science and Technology, 85-796 Bydgoszcz, Poland
* Correspondence: justynas@utp.edu.pl

Received: 30 May 2020; Accepted: 30 June 2020; Published: 8 July 2020

Abstract: In the beginning of the XX century, political, economic, and demographic revolutions contributed to the emergence of extraordinary people. In architecture, they were Frank Lloyd Wright, Antonio Gaudí, Frank Owen Gary, Le Corbusier, Hugo Hering, Alvar Aalto, Hans Sharun, Walter Burley Griffin, and Marion Mahony Griffin. Each of them was given a lot of attention in the media resources and their creativity was researched in different fields of knowledge. However, Rudolf Steiner's work remains controversial to this day. Although many of the architects mentioned above enthusiastically commented on Steiner's architectural works, there was always ambiguity in the perception of this mystic architect. Such a careful attitude to the work of the architect is due primarily to his worldview, his extraordinary approach to art and architecture in particular, because it is in architecture that Steiner was able to implement the basic tenets of anthroposophy, which he founded. The purpose of this study is to determine the content of the spatial structure of Steiner's architecture, which makes it unique in the history of architectural heritage. The authors offer the scientific community the first article in a series of articles on the anthroposophical architecture of Rudolf Steiner and the philosophical concept that influenced the formation of this architecture.

Keywords: Rudolf Steiner; anthroposophy; architecture; Goetheanum

---

1. Introduction

The beginning of the twentieth century was marked by emotional and intellectual conflicts that have become embodied in new artistic and literary trends and philosophical views. The accumulation of such great ideas, events and their unpredictable consequences put some artists in the spotlight and forgot the others. Among the undeservedly forgotten was the work of Rudolf Steiner, whose philosophy was incomprehensible to mass perception and contained views unacceptable to Christian society because they conflicted with Christian dogma. Rudolf Steiner's architectural creativity must be seen in close conjunction with his outlook. It is through the works of Steiner that one can often learn about the motives of his architectural decisions. Despite the researchers' attention to the Goetheanum, the ancillary structures of the Rudolf Steiner Architectural Complex carry no less important information than the Goetheanum itself, yet remain unaddressed today. For the researcher of Rudolf Steiner's work, the most valuable source is without a doubt the memories of his friends and followers. Among the published memoirs today we can find the memoirs of Andrey Beliy [1], Herbert Khan [2], Asia Turgeneva [3], Margarita Voloshyna [4], and others. An analysis of Steiner's life and anthroposophical doctrine was made in the works of Bondarev G.A. [5], Rudy Lissau [6], and Karen Swassjan [7]. The analysis of Steiner's architectural works in the context of the world architectural heritage is analyzed in the works of Sokolina A. [8–11] and Elena Bogdanovich [12]. An important contribution to the study of Steiner's design work is the dissertation of Reinhold Johann Fet [13], and concerning the architecture of Goetheanum, Fiona Gray [14], Marina Agranovskaya (Emmendingen) [15], and Latief Perotti [16]. David Adams [17], Ionova O.M. [18], and others devoted their research to the Waldorf school. The aim of this study is to determine the purpose of the architectural spatial structure of Steiner's work, which makes it unique in the history of architectural heritage.

2. Main Materials

Andrei Bely wrote about the misunderstood ideas of Rudolf Steiner "who understood Paul's fiery spirit (Apostle Paul), [as] he also understood Steiner: he understood his problem—the problem of the seeds of future unexperienced flowers. He was completely sealed with the seal of fire and a real understanding of the now incomprehensible and modern problem: "thinker", "scientist ", "teacher", "yogi", "magician", "occultist", "anarchist", "crusher of the basics", "chancellor of the order", "gnostic", "cultivator"; and he did not fit into the framework of this literature! Paul was a fidget: and so was Steiner. Thundered that one; and this one rattled. When I read Paul's cry that he was everything for everyone to wake some up, I say to myself: "Oh, I understand; I saw Steiner!"" [1]. Even during his life, Steiner felt the enmity of theologians, Catholics, Protestants, occultists, Marxists, nationalists, and Nazis. There were two attempts on his life and the arson of his greatest creation, the Goetheanum (Figure 1).

(a)

(b)

Figure 1. (a) Goetheanum of 1913–1920; (b) Goetheanum of 1925–1928 [19].

Almost 90 years after Steiner's death, his anthroposophical postulates remain as keenly stimulating for ideological opponents. Despite the complexity of the anthroposophical teachings of Steiner, at the beginning of twentieth century, his ideas and doctrines, as well as those E. Blavatsky, A. Aksakov, E. Shure, P. Shenavar, Dr. Papus were popularized in art. V.S. Turchin writes in detail about it, analyzing the situation of the twenties: "Inheriting the doctrines of mysticism, esoteric and occult sciences, alchemy, dreaming of a synthesis of the religions of the West and East, both ancient and new, reading Boehme, Swedenborg, and the founders of the Masonic orders, the artists rushed to a new Theosophy, which at the beginning of the century was reborn into anthroposophy, what A. Besant and her followers spoke of... Theosophy was associated with the artists of the Nabi group and the founder of the Rose and Cross association, Sar Pelladan, members of the Free Aesthetics society in Brussels, and Italian J. Segantini, Odilon Redon, and the young K. Petrov-Vodkin, E. Munch and F. Kupka,

G. Klimt and E. Carrier. A little later V. Kandinsky, N. Kulbin, the brothers Burliuk, K. Malevich, A. Exter, P. Mondrian, and R. Delaunay... "[20].

It was Steiner who became the intellectual source from which Ukrainian cultural figures like L. Kurbas and G. Narbut borrowed their ideological and creative ideas. They organized the circle of "creators and scientists" (Kiev, 1916), where "Steiner's theories" were studied. The members of the circle were prominent representatives of Ukrainian culture (F. Ernest, M. Zerov, Ya. Stepovyi, P. Tychyna, and others), who later became known as "Executed Renaissance" [18]. The idea of the creation of a new world of Kazimir Malevich was clearly in line with Steiner's theosophical idea. "The strategy proposed by K. Malevich (for the reconstruction of urban space) was based on the fact that the artist considered urban planning, like all Suprematism, as a promising project for a rationalistic world order, the transformation of the world based on the ideas of universal harmony as Creation" [21].

Steiner's views influenced the creativity of M. Khvylov, A. Bely, K. Boguslavskaya, M. Voloshin, V. Kandinsky, B. Pilnyak, M. Sobashnikov, V. Khlebnikov, M. Chekhov, and others. The implementation of anthroposophical ideas into architectural forms was continued by Eric Asmussen and Joachim Haider, the Camphill Architects in Scotland, a Dutch architect Anton Alberts, and a Hungarian architect Imre Makovecz. Some researchers [22,23] consider Rudolf Steiner's architecture to be organic architecture. However, to date, there is no single clear and generally accepted definition of organic architecture, as well as its differences from environmental and bionic [24].

Steiner's architecture can be talked about from the works of Steiner himself and his followers (about 70 works devoted to architecture). This architecture can be called anthroposophical or Steiner's architecture, since its primary feature is a visualized idea, and all other characteristics are not minor. They follow the idea and attached to it. The architectural form of Steiner's building was to contain important information, a matrix of anthroposophical doctrine that would be readable over the centuries.

Steiner himself believed that the correct architectural forms of the artist creates a "larynx for the gods" [25], thus emphasizing the role of architecture in the spiritual development of society. Moreover, such correct forms are present in all buildings of Steiner. The building he was speaking about was Goetheanum (named after a German poet and scholar named Goethe). For the shape of the Goetheanum, Steiner chose the shape of a skull. One of the prerequisites for this choice was the idea of the metamorphosis of the image of man: "The inner unity of all nature is based precisely on the fact that, in essence, everything, even extremely dissimilar appearances, are metamorphoses of one original basic form". The Lord, or as Steiner calls him, the Wise Divine Authorities of the Cosmos, have created us all in one form, but each of us is transformed, changed according to our chosen path, and the strangers we meet. The skull is this common human form that has become zero in the direct metamorphosis: the process of human materialization, the development of the skull from the spine (common to all people), and after zero when the person becomes an individual in his own form.

In the first Goetheanum, the skull was not clearly defined. The main idea here was expressed by the intersection of two domes of different sizes. In the researcher of Steiner's architecture by D. Adams the domes of the first Goetheanum expressed the "union of spirit and matter" [17], and for Latief Perotti [16] it was a demonstration of human essence through images of the frontal and parietal lobes of the brain, the somatosensory and somatomotor cortices (Figure 2).

Analyzing the works of Steiner, we understand: the first Goetheanum is the Universe and the human projection of the world through the senses. Further, not without reason, Steiner presents the domes in different sizes: man is not able to comprehend all the greatness of the universe, which is represented by a large dome, he can comprehend only a small part of it. This knowledge is the personality (smaller dome). The intersection of the domes is also not accidental, like everything in the design and planning of the anthroposophical center in Dornach. Steiner embodies his theory of the origin of races. Man is a divine creation, he comes from cosmic energy, is a part of it, and draws strength from it. He is a product of the universe. He is its observer and participant in great actions. Joining, the contact of two hemispheres, is a direct indication of a common space, which, although invisible, is always present. The world exists because it is reflected in our senses, and our perception of the world is transmitted into space. The concept of quantum theory and its philosophical applications was perceived by Steiner at the beginning of the twentieth century: "The vitality of our building is expressed in the fact that in some sense one dome finds a consciously displayed reflection in the other, in that both domes are reflected in each other, just as existence of the outside world is reflected through the senses of man" [25]. The double dome is present in many buildings of Steiner. "If there was only one dome, then the essence of our building would be dead".

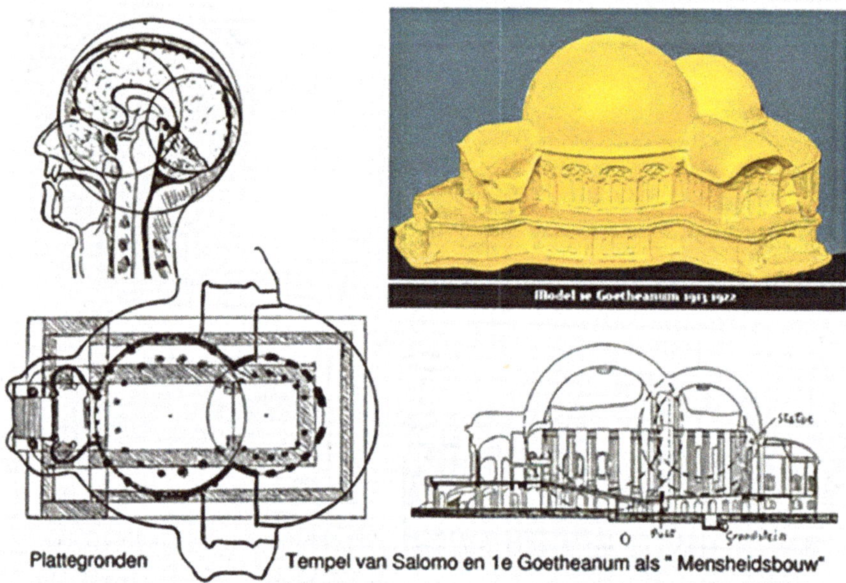

Figure 2. The philosophical structure of the first Goetheanum according to Perotti L. [16].

On the night of 31 January 1923, the Goetheanum was set on fire and burned completely [3]. Since the building was insured for 3.2 million francs, Steiner began active work on the new image of Goetheanum. In addition to the smooth shapes made possible by the use of the latest material, concrete, the new building has acquired a new symbolic meaning.

One can say that Goetheanum of Rudolf Steiner is a continuation of the use of a new material, namely reinforced concrete, in construction, the use of which was invented in Europe by Auguste Perret. In 1904 he built the first residential building in Paris with reinforced concrete, and in 1922–1923 the Church of Our Lady in Le Rennes, which became the first church in France built with reinforced concrete. In 1909 he worked at the firm of Auguste Perret and studied the basics of technical drawing by Le Corbusier.Structure of Steiner's Goetheanum, built in 1925–1928, made entirely of reinforced concrete. Impressive dimensions in differentiated organic design are unique. Steiner was convinced by the advantages of reinforced concrete in terms of fire safety, cost, and its spatial plasticity.

Hermann Ranzenberger, Otto Moser, Ernst Eisenpreis and Albert von Baravalle participated in the design and construction of the Goetheanum as architects/executioners of the project. The first three were already involved in the creation of the Goetheanum in 1913–1919. The Basel engineering office of Leuprecht & Ebbell was responsible for the load-bearing structure in reinforced concrete. Exquisite framework was created by a carpenter from the Anthroposophical Center under the management of Heinrich Ledwogel.

The basic structure with an area of 3200 square meters extends for 90 m in the east-west direction and over 85 m in the north-south direction. The building measuring 72 m by 64 m rises to 37 m in height. The internal space is 110,000 cubic meters, with 15,000 cubic meters of concrete and 990 tons of reinforcement steel. However, the massiveness of the building is only illusory. The building is built of delicate, uninsulated concrete frame structure (separate rooms are insulated from the inside). Reinforced concrete columns and beams are placed with large spacing and concrete slabs 8 cm thick. The walls of the first floor in the west side of the building have a solid thickness of 50 cm. Despite its filigree nature, the load-bearing structure, which is also the skeleton of the building, is extremely strong. The quality of the building is evidenced by the fact that during the first 50 years of operation the bare concrete facade did not require any maintenance whatsoever [26].

It should be noted that Steiner was an esoteric. The esoteric skull has always been regarded as the repository of the soul and it has been thought that the soul continues to live within the skull after a person has died. The human skull is given a special value. In the second Goetheanum, a skull shape is clearly drawn, and domes are no longer dominant in the architecture of the building. Steiner placed all visitors of Goetheanum in their primordial form, in one which, in his opinion, was the ascendant pre-metamorphose point of the people. Having composed the mystery of human creation, Steiner built a concrete skull that was reincarnated with hundreds of souls as spectators of Goetheanum. "The form, writes Rudolf Steiner, is perfect when it is filled with the proper content" [25]. He demonstrates the effect of the metamorphosis of the dead and the living, which occurs with the participation of spectators who are participants in the action. He created what he repeatedly stated in his lectures: "We really must become disciples of the creative hierarchies who have created through metamorphosis, and we must learn to imitate the creative principle of the higher hierarchies in the same way" [25].

Steiner imitates the actions of the higher powers: in the man-made space there is a life that arises through art. Both the actor and the viewer of the Goetheanum become members of the great Steiner mystery. Such a symbolic image could not have been erected just anywhere. The site for sacred structures was carefully selected and researched. Steiner was not able to choose a place for Goetheanum himself. This is why he granted a sacred meaning to the piece of land he owned. Steiner placed the Goetheanum, according to the architect Perotti [16]. The outbuildings were placed within the projection of the Plato Pentacle (ordinary pentagon), and the spiral, according to the Fibonacci mathematical series. The true proportions of this symbol are based on the sacred proportion called the golden intersection: this is the state of a point on any drawn line when it divides the line so that the smaller part is in the same proportion to the greater part as the greater part to the whole (Figure 3). So, when you look from space at the planning structure of the anthroposophical settlement, you can see that Steiner focused on the dodecahedron, which, as a geometric figure, he paid much attention in his work (the plan shows one of the planes of the dodecahedron, the tops of which fix the buildings in space).

The very foundation of the "Corner Stone" was a unification of two copper dodecahedrons, for the 1st Goetheanum of 21 September 1913, consisting of twelve pentagons, was the sacralization of the site. Two united dodecahedrons Steiner laid as a symbol of the indissoluble connection of the universe (macrocosm) and man (smaller dodecahedron) of the microcosm. When Herbert Zoifert writes about the "corner stone" in the Goetheanum foundation, he mentions a document that was placed in a copper vessel. The text corresponds to Steiner's speech: "In the name of the Seraphim, Cherubim, Thrones, Spirits of Wisdom, Spirits of Movement, Spirits of Forms, Spirits of Personality, Archangels, Angels: In the macrocosm, as a microcosm, let man live. Anthropos, represented here also as a twice-duodecimal image of the spiritual world. And inside it the expression of the Rosicrucianism expressing the meaning of our aspiration: Ex Deo Nascimur, In Christo Morimur, Per Spiritum Sanctum Reviviscimus" [27].

Perhaps coincidental is the fact that a physicist from the University of Ulm (Universität Ulm) Frank Steiner, has advanced another version [28] of the Universe structure over the centuries, according to which the latter has the form of a dodecahedron. Thus, it is likely that Rudolf Steiner's feelings, beliefs, or intuitions prompted him to locate the Goetheanum, which demonstrated the materialization of man on Earth at the center of the dodecahedron (Figure 3), symbolizing the Universe.

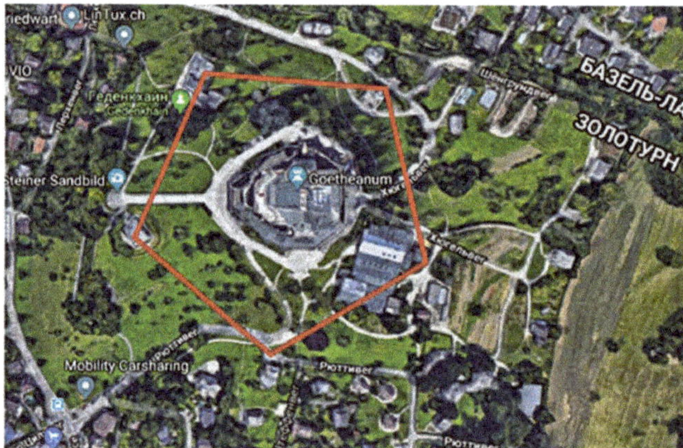

Figure 3. Goetheanum and ancillary structures.

The Universe is invisible and invaluable without the main viewer – the man. The birth of manmade the Universe visible as well. By the way, the basic idea of the four anthropic principles proposed by Carter, Hole, Pringle, and others, in the field of studies of astronomical and atomic physics, states that the Universe exists and expands only because it is in us and we observe and understand it. Steiner's understanding of the origin of man and his relationship with the divine influenced his vision of the temple. After all, if the Greek temple, according to Steiner, is the location of the god where no human presence is required, where he is a guest, then the Gothic temple is half the temple. It is incomplete without prayer and human presents. "There is no God in the Gothic temple unless there is a prayer of believers. If it is, it is filled with the divine. The Greek temple is not the house of believers, it was created as a house in which God himself lives, he can stand alone" [29].

Steiner's Temple is a message with information for those who can read, art for those who can admire, science for those who want to learn. However, Steiner with his worldview went far ahead of his time, and therefore was lost from view. The Goethanu, the anthroposophical temple, is not a reflection of the structure of the universe. It is the embodiment of creation of the universe and the evolution of man and his materialization. The main emphasis was on the mission of a materialized person in the world where the Calvary events took place. The Anthroposophical Temple is a language in itself for the disciples in the spiritual science who understand the Universe not only as a great mechanism, but as a living, Spirit-filled organism, a particle of which is present in every human being.

Steiner's creations carry important knowledge, they are filled with symbols and original details, but devoid of symbolic casuistry. Marina Agranowska (Emmendinger) in the analysis of the architecture of the Goetheanum writes that in the account of the building you can see: "the echoes of the organic architecture of the American Frank Lloyd Wright, the roll call with the architectural fantasies of the Spaniard Antonio Gaudi, we see the common with the architecture of quiet expressionism, however the originality and uniqueness of the Goetheanum are undeniable" [13].

However, it should not be forgotten that all these representatives belong to the beginning of the twentieth century, characterized by motifs of organic and bionic architecture, ornaments in the style of modernity. Their contacts, communication could not but affect each other's architectural works. However, an important fact is that each of the architects, and in particular Steiner, contributed something unique to architectural science.

Despite the original dynamic forms, the author's interpretation and the pedantry of the interior first denied the well-known thesis of Louis Sellivan (or Sullivan) that the form follows function.

He did not go along with Antonio Gaudi, who combined the idea, function, and constructive elements. Steiner put the idea above function, and this idea was present not only in the Goetheanum, but also in the auxiliary structures. In order to harmonize the aura of the occult space, Steiner constructed the building of the boiler room in the form of a human spine which symbolized the further evolution of man, or rather his materialization (Figure 4).

Figure 4. The boiler room (1915) [19,30].

If we follow the theory of transformation (metamorphosis) of Steiner, the skull developed in a person from the spine. It is the eve of conscious perception of reality. The skull, which arose from the expansion of one of the vertebrae, became a housing for human consciousness to store information about the Universe. "If you imagine that such a vertebra is expanding, expanding in the way, so that the hole through which the spinal cord passes, because the vertebrae are located one above the other, becomes larger, and the bones accordingly become thinner, and also expanding like something elastic, not only in the horizontal direction, but in other directions, then from these spinal bones a shape arises, which is nothing more than the shape of the bones that make up the shell of our skull. The bones of our skull are, therefore, transformed bones of the spine" [29].

The spine, according to Steiner, is an anatomical element that appeared in humans as a result of his materialization. So, initially, man was a spirit. The effect of human materialization is holistic when operating a boiler room, namely, when smoke rises above a building, giving it life. Important elements of the boiler room are domes that symbolize Luciferic and Ahrimanic in humans: Luciferic, i.e., emotional, provoking killings and wars, aggression, and Ahrimanic, i.e., practicality and calculation. The domes are symmetrical and thus testify to the balance of these two demonic principles, which are alongside of materialized person. Steiner divided spirit of Mephistopheles from Renaissance and Satan from Christianity into two characters who represent two opposite poles of evil in man. Arild Rosenkranz in "The Fruits of Anthroposophy—An Introduction to The Work of Dr. Rudolf Steiner" [31] writes that with the penetration of a true visionary, Rudolf Steiner gave each power an individual shape. In the Goetheanum, these two negatives are shared by a pure soul, very reminiscent of the figure of Christ. Describing the sculptural group of the Gethanum, Arild Rosenkranz concludes that Christ's own power establishes the right balance in man.

During construction of the boiler room, Rudolf Steiner makes the image presented in the Goetheanum more complicated. Between the two domes that represent sin, he places man, not formed or materialized, but who must already fight for the right to seek spiritual progress. The boiler room resembles female forms in the shape of the sphinx. At the level of the spatial solution of the boiler room, the author presented a large canvas of the spiritual world, according to which the backward, imperfect on the spiritual or mental level cannot find eternity, and always goes into decline. According to Steiner on the Sphinx has not reached perfection on the spiritual level. It has been reborn, and its astral body comes to people in other forms. In his 11th lecture "Egyptian Myths and Mysteries" Steiner writes that, during a sunstroke, "The etheric and astral bodies are freed from one part of the physical body, and these people are transferred to the astral plane and see the degenerate last descendant of the sphinx. It is called by various names. In some places, it is called the midday woman—*Mittags-Frau*" [29]. Midday woman is a well-known character from the folklore of Western Slavic tribes. This is a woman in white clothes who comes to those who do not stop working at noon on the field. As a result, the person receives a sunstroke and, as a rule, can see a reborn sphinx in the form of a woman asking many questions. If the person cannot answer them, the midday woman punishes such a person with severe headache.

The building of the Steiner boiler room is information embodied in concrete forms. It is a ready-made surreal picture of the transformation of a person from a spiritual being into a material being, as well as an image of a soul seeking perfection.

3. Conclusions

The peculiarity of Steiner's architecture lies in the reproduction of the anthroposophical worldview in the spatial and planning decisions of the Goetheanum building and additional structures. The creation of an architectural complex dominated by a philosophical idea of function.

Rudolf Steiner's genius is in his unique ability to convey in architectural forms a segment in time and space without having points A and B. Rudolf Steiner captures the image not as other painters do, not as photographers do, he did not convey emotions like music does. Rather, with his art of architecture, he presents to generations a long process and the size of eternity. Steiner looked to the distant future, and he created an architecture that was able to unite humanity through spiritual science.

Steiner's teachings went beyond classical thinking about architecture. Despite the ambiguous perception of his doctrine and his eccentricity, he forces one to look at architecture as a model for the construction of the universe and the place of man in it.

Author Contributions: Conceptualization, R.K. (25%) and A.S. (25%); investigation, I.B. (25%) and J.S.-P. (25%). All authors have read and agreed to the published version of the manuscript.

Funding: The authors declare that there is no any funding source for this paper.

Conflicts of Interest: The authors declare that they have no conflict of interest.

References

1. Bely, A. *Memoirs of Steiner*; Editions La Presse Libre: Paris, France, 1982; Chapter 1; Available online: http://bdn-steiner.ru/modules.php?name=Books&go=page&pid=220 (accessed on 23 June 2019).
2. Khan, H. Rudolf Steiner. How I saw and knew him, M.: Starklight, translation from German by Victor Medvedev, Russia. 2006, p. 96. Available online: http://bdn-steiner.ru/modules.php?name=About_Steiner&fbclid=IwAR3wP6rMai8tKrnl-DgoO32y9OF4DzppqCdEcSihm0QjDE8H4jQJs_eODQE (accessed on 23 June 2019).
3. Turgenyeva, A.A. *Memories of Rudolf Steiner and the Construction of the First Goetheanum*. Library of Spiritual Life. 2002. Available online: http://bdn-steiner.ru/modules.php?name=Books&go=page&pid=7501 (accessed on 4 April 2019).
4. Voloshyna, M.; Sabashnykova, M.V. Green Snake. The Story of One Life. Translation from German by M. N. Zhemchuzhnikova. Enigma, Moscow. 1993, p. 412. Available online: http://bdn-steiner.ru/modules.php?name=Books&go=page&pid=5702 (accessed on 3 April 2019).

5. Bondarev, G.A. The experience of understanding the methodology of the science of spirit. In *Triune Man Body, Soul and Spirit*; University Book: Moscow, Russia, 1999; p. 688.
6. Lissau, R. *Rudolf Steiner: Life, Work, Inner Pathand Social Initiatives*; Hawthorn Press: Stroud, UK, 1987; p. 178.
7. Svasyan, K. Rudolf Steiner. The Philosophy of Freedom. Preface to the New Edition of 1918. Available online: http://lib.ru/URIKOVA/STEINER/freedom.txt (accessed on 15 June 2019).
8. Sokolina, A.P. (Ed.) *Arhitektura i Antroposofiya*; KMK: Moscow, Russia, 2010.
9. Sokolina, A.P. *The Goetheanum Culture in Modern Architecture*; Science, Education and Experimental Design; Shvidkovsky, D.O., Yesaulov, G.V., Ivanovskaya, V.I., Aurov, V.V., Bazhenova, E.S., Bayer, V.E., Bgashev, W.N., Volchok, Y.P., Efimov, A.V., Ivanova-Veen, L.I., et al., Eds.; MARKHI: Moscow, Russia, 2014; pp. 157–159.
10. Sokolina, A.P. Biology in Architecture: The Goetheanum Case Study. In *The Routlegde Companion to Biology in Art and Architecture*; Terranova, S., Tromble, M., Eds.; Routledge: New York, NY, USA; London, UK, 2017; pp. 52–70.
11. Kugler, W.; Baus, S. (Eds.) *Rudolf Steiner in Kunst und Architektur*; Dumont: Cologne, MN, USA, 2007.
12. Bogdanović, J. Evocations of Byzantium in Zenitist Avant-Garde Architecture. *J. Soc. Archit. Hist.* **2016**, *75*, 299–317. Available online: https://online.ucpress.edu/jsah/article/75/3/299/60956/Evocations-of-Byzantium-in-Zenitist-Avant-Garde (accessed on 14 June 2019). [CrossRef]
13. Fäth, R.J. Rudolf Steiner Design: Spiritueller Funktionalismus-Kunst. Ph.D. Thesis, University of Konstanz, Konstanz, Germany, 2004. Available online: http://kops.uni-konstanz.de/handle/123456789/3624 (accessed on 7 May 2019).
14. Gray, F. Rudolf steiner: Occult crankor architectural mastermind? *J. Archit. Theory Rev.* **2010**, *15*, 43–60. Available online: https://www.tandfonline.com/doi/abs/10.1080/13264821003629246 (accessed on 7 July 2019). [CrossRef]
15. Agranovskay, M. Anthroposophy and Rudolf Steiner Goetheanum: Architectural Image of the Universe. 2017. Available online: https://www.partner-inform.de/partner/detail/2017/9/240/8678/gjoteanum-arhitekturnyj-obraz-vselennoj?lang=ru (accessed on 25 July 2019).
16. Perotti, L. Gewijde Bouw. Available online: https://architect-latief-perotti.jouwweb.nl/ (accessed on 4 July 2019).
17. Adams, D. Organic Functionalism: An Important Principleof the Visual Artsin Waldorf School Craft sand Architecture. *Res. Inst. Waldorf Educ. Res. Bull.* **2005**, *X*, 28.
18. Ionova, O.M. Rudolf Steiner in educational and cultural space. In *Pedagogika, Psychology and Medical and Biological Problems of Physical Education and Sports*; Kharkiv: H.S. Skovoroda Kharkiv National Pedagogical University: Kharkiv Oblast, Ukraine, 2011; pp. 60–63.
19. Foto C Holtermann Goetheanum Tour. Available online: https://www.obretenie.info/txt/sht/geteanum.htm (accessed on 11 March 2019).
20. Turchyn, V.S. *The Image of the Twentieth... In the Past and Present*; Progress-Tradycyya: Moscow, Russia, 2003; p. 681.
21. Gryber, Y.A. Urban Planning and Kazimir Malevich, Agreement Moscow. 2014, p. 180. Available online: http://color-lab.org/files/283/malevich-oznakomitelnyi-f.pdf (accessed on 3 January 2020).
22. Lokotko, A.I. *Architecture: Avant-Garde, Absurdity, Fantasy*; Minsk: Belarusian Science: Navuka, Belarus, 2012; Volume 206, p. 68.
23. Zymina, S.B. Stylistic features of the organic and environmental. *Mod. Probl. Archit. Urban Plan.* **2018**, *47*, 71–88.
24. Bistrova, T.Y. The Evolution of Organic Architecture: Interpretations and Modifications. Chast 2. KyberLenynka. 2013. Available online: https://cyberleninka.ru/article/n/evolyutsiya-organicheskoy-arhitektury-traktovki-i-modifikatsii-chast-2 (accessed on 28 May 2019).
25. Shtejner, R. Art in the Light of the Mystery Wisdom. Eight lectures delivered in Dornach from December 28, 1914 to January 4, 1915. Translation from German by A. Demidov. Available online: https://www.rulit.me/books/ga-275-iskusstvo-v-svete-mudrosti-misterij-read-434991-1.html (accessed on 8 May 2019).
26. Die Narben der Baugeschichte Sichtbeton, Instandgesetzt. Available online: https://www.espazium.ch/de/aktuelles/die-narben-der-baugeschichteBAUKULTURSEIT1874 (accessed on 20 April 2019).
27. Zojfert, G. Michael and Iron, Anthroposophy in the Modern World. Translation from German by O. P. 2002. Available online: http://anthroposophy.ru/index.php?go=Pages&in=view&id=219 (accessed on 16 January 2019).

28. Steiner, F. Can One Hear the Shape of the Universe? Universität Ulm. 14 January 2005. Available online: http://www.matfys.lth.se/seminars/old/abstracts/steiner.html (accessed on 14 March 2019).
29. Shtajner, R. Lecture: Egyptian Myths and Mysteries. Lecture 11. Available online: http://www.lib.ru/URIKOVA/STEINER/lec_egypt.txt_with-big-pictures.html (accessed on 3 April 2019).
30. Heizhaus. Available online: https://upload.wikimedia.org/wikipedia/commons/6/62/Heizhaus_2013_002.JPG (accessed on 19 March 2019).
31. Rosenkrantz, A. *The Fruits of Anthroposophy—An Introduction to the Work of Dr. Rudolf Steiner. The New Impulse in Art*; George Kaufmann Threef Old Commonwealth; The Threefold Commonwealth: London, UK, 1922; p. 151.

MDPI
St. Alban-Anlage 66
4052 Basel
Switzerland
Tel. +41 61 683 77 34
Fax +41 61 302 89 18
www.mdpi.com

Applied Sciences Editorial Office
E-mail: applsci@mdpi.com
www.mdpi.com/journal/applsci

www.ingramcontent.com/pod-product-compliance
Lightning Source LLC
LaVergne TN
LVHW071032170726
843515LV00006B/1264

* 9 7 8 3 0 3 9 3 6 9 9 4 2 *